COLLECTED WORKS OF

BERNARD LONERGAN

VOLUME 18

PHENOMENOLOGY AND LOGIC

COLLECTED WORKS OF BERNARD

LONERGAN

PHENOMENOLOGY AND LOGIC

The Boston College Lectures on
Mathematical Logic and Existentialism

edited by

Philip J. McShane

Published for Lonergan Research Institute
of Regis College, Toronto
by University of Toronto Press
Toronto Buffalo London

© Bernard Lonergan Estate 2001
Printed in the U.S.A

ISBN 0-8020-3578-7 (cloth)
ISBN 0-8020-8448-6 (paper)

Reprinted 2005, 2013

Printed on acid-free paper

Requests for permission to quote from the Collected Works of Bernard
Lonergan should be addressed to University of Toronto Press.

Canadian Cataloguing in Publication Data

Lonergan, Bernard J.F. (Bernard Joseph Francis), 1904–1984
   Collected works of Bernard Lonergan

   Partial contents: v. 18. Phenomenology and logic / edited by Philip J.
   McShane. Includes bibliographical references and index.
   ISBN 0-8020-3578-7 (v. 18: bound)   ISBN 0-8020-8448-6 (v. 18: pbk.)

   1. Theology – 20th century.   2. Catholic Church.   I. Crowe, Frederick E.
   II. Doran, Robert M., 1939–   .   III. Lonergan Research Institute.   IV. Title.

   BX891.L595 1988   230   C88-093328-3

The Lonergan Research Institute gratefully acknowledges the generous
contribution of THE MALLINER CHARITABLE FOUNDATION, which has made
possible the production of this entire series.

The Lonergan Research Institute gratefully acknowledges the financial
assistance of the family of VERONICA CECELIA LIDDY toward the publication
of this volume of the Collected Works.

University of Toronto Press acknowledges the financial support for its
publishing activities of the Government of Canada through the Book
Publishing Industry Development Program (BPIDP).

Contents

General Editors' Preface

The general editors of the Collected Works of Bernard Lonergan wish to
extend their deepest thanks to Philip McShane for his extensive and erudite work in bringing to publication this volume of Lonergan's 1957 lectures on Mathematical Logic and Existentialism.

The volume originally scheduled to be number 18 of the Collected
Works of Bernard Lonergan was listed in the prospectus as *Lectures from the
Roman Years*. That did not mean only those lectures that were given in
Rome; most of the lectures, in fact, were given in the United States or
Canada, as Lonergan took the opportunity to lecture in summer institutes
that would fund his transatlantic journey and enable him to summer on this
side of the ocean. Thus he conducted institutes in Boston (1957), Halifax
(1958), Moraga (1961), Toronto (1962), Spokane (1963) – all of two weeks
or more – and one of a week in Washington (1964); further, he gave single
lectures or shorter series at various institutions, notably Thomas More
Institute in Montreal.

The title *Lectures from the Roman Years* could therefore mean only a
selection from materials that altogether would fill several volumes. The
passage of time has enabled us to reassess that original plan: there were sets
of lectures that formed a unit too valuable to be simply chopped up, and
that is the case with the present volume 18, which contains the Boston
College lectures of 1957, a week of ten (or five two-part) lectures on
Mathematical Logic and another week of ten (or five two-part) lectures on
Existentialism. Both sets of lectures were to prove important in Lonergan's
development, and study of them is essential to understanding what was

going forward in his thinking during this period. To these lectures there is added, in two short appendices (A and B), material taken from or related to a 1961 institute in Dublin on 'Critical Realism and the Integration of the Sciences.' This material is added because it is intimately related to the 1957 material, and helps the reader come to grips with certain elements in the lectures themselves.

This is the earliest material in the Collected Works to depend almost entirely on the tape-recording of the lectures as delivered, and it presented special problems. The recording was not as acoustically clear or as complete as it was a year later for the Halifax lectures on *Insight* (CWL 5: *Understanding and Being*); further, there were no known notes taken by participants, nor has any list of participants who might be consulted been found. The recorded history of negotiations with Boston College is minimal in the extreme, and the press reports that are available for other sets of Lonergan's lectures are nonexistent for this set. On the other hand, Lonergan provided his own schematic notes for parts of the course, and this compensates at least in part for what is lacking in other ways.

The style of these summer lectures and institutes, which derived from their purpose and was very informal, presented further difficulties. This was especially the case at Boston in 1957, for it is evident from the content that Lonergan was lecturing to fellow Jesuits, and so felt very much *en famille*. The topics were not chosen by him but were requested by Boston College, and so led Lonergan into areas that he had not previously treated at length or in depth.

For mathematical logic, indeed, we have remarks scattered through *Insight*, the fruit of a life-long interest in mathematics, which might be collected and organized. For existentialism we are both better off and less well off; less well off, for there is little that is thematic on that question in the early Lonergan; on the other hand better, for there is a short chapter *De existentia* in the volume Lonergan had published a year earlier *De constitutione Christi ontologica et psychologica* (Rome: Gregorian University Press, 1956, pp. 14–19) which treats the theme more explicitly.

For the negotiations with the College and the actual history of arrangements for the two weeks of lectures we have only a few indications in letters to Frederick Crowe. Thus, in a letter of 28 March from Rome Lonergan reveals that he was asked to speak on these two topics and accepted the invitation because the institute would take care of his transatlantic fare. It came out also that he was operating on a tight schedule: he hoped to finish exams at the Gregorian University by 6 July, had an airline reservation for

7 July that would bring him to Boston on 8 July, with his two weeks of lectures to start 9 July.[1] No allowance for jet lag! or did they have jet liners in 1957? they certainly had time lag. On the completion of the lectures he planned to go to Halifax, where his Jesuit brother Gregory was stationed at Saint Mary's University High School; and this had its fateful importance in Lonergan's history, for while in Halifax he was persuaded by Fr John Belair to return in 1958 for two weeks of lectures on *Insight* (CWL 5).

A letter from Boston written 17 July when the lectures were into their second week gives the lecture schedule: 'from 9.30 to between 10.15 and 10.30 and then again from 11.00 to between 11.45 and 12.00 five days a week, with an evening session thrown in once for each week.' He indicated his satisfaction with the results: 'From all reports the summer school is going extremely well,' and specifies his hidden agenda: 'Quietly selling the ideas in *Insight*, and some results.' He mentions in particular the interest of the late Timothy Fallon; this is the only name we have from among the participants, though remarks here and there (pp. 106, 227, 239, 293) show that he is speaking to fellow Jesuits.

A small further piece of information in this letter: 'I sent you two copies of the mimeographed notes that were given out here. One of the copies is intended for Rev. R. Eric O'Connor ... Lecture notes were mimeographed for the first week on Math Logic and for half of one lecture on Existentialism; then the secretarial pool broke down, and though all avenues were tried no results were obtained.'

So much for the history of the institute. The history of the book, our own project, is presented by Professor McShane at the beginning of section 1 of his Editor's Introduction, and to that account we add only that this history begins with the then Lonergan Center of Regis College obtaining a set of cassettes, copied in 1973 from the reels Fr Joseph Flanagan had at Boston College.

Certain major decisions were naturally the province of the volume editor, Philip McShane, in consultation with us: for example, the title of the volume, the location of Lonergan's notes between the lectures on Mathematical Logic and those on Existentialism, the introduction to the whole volume, and the contents of the appendices. But over and above these major determinations, along with his attention to the minutiae of editing, and his provision of an index and English translations of Lonergan's Latin

---

1 It would seem that Lonergan was off a day in this report. The lectures began on Monday, 8 July.

and Greek, there is the very extensive work of the footnotes, all of which are editorial, tracing Lonergan's sources, and supplying links with the wider academic and scholarly world.

A word is in order regarding the title of the volume and the first two appendices. The title was suggested by Professor McShane, and the General Editors readily agreed that it was a good choice. While the second week of lectures was devoted to existentialism (as is reflected in the subtitle of the volume), there are very good reasons for the title *Phenomenology and Logic.* The chapters on Husserl and on phenomenology in general are arguably the most important chapters in the third part of the volume. But the main justification for the title is that Lonergan's reading of Husserl in preparation for these lectures was an important step in his own development, more important, it would seem, than his engagement with the thinkers usually referred to as existentialists. This volume provides perhaps the earliest source of data on the influence that the reading of Husserl had on Lonergan. There is one mention of Husserl in *Insight* (p. 440 of the Collected Works edition), but clearly Lonergan came to a new evaluation and appreciation of phenomenology in general and of Husserl in particular as he prepared to deliver these lectures. The influence of Husserl was to be reflected in a number of lectures over the next eight years or so. Lonergan comes back again and again to certain themes from *The Crisis of European Sciences and Transcendental Phenomenology,* and the present volume provides the earliest account of his reading of that book.

The material presented here in appendix A and appendix B is related to the 1961 set of lectures in Dublin, 'Critical Realism and the Integration of the Sciences.' The first diagram in appendix A is taken from the typescript of these lectures. The second diagram is based on the first. Both are reproduced here by photocopy. They are diagrams of (1) Lonergan's cognitional theory and (2) its relation to existential decisions. The first diagram is similar to one he used in the Halifax lectures that became *Understanding and Being* (CWL 5), in the education lectures that became *Topics in Education* (CWL 10), and in the two sets of lectures that appear in the present volume. Professor McShane's introduction to appendix A (below, p. 319) indicates the pages in the text of the lectures to which the first diagram is particularly relevant (and see also below, p. 226, note 25, and p. 245, note 10). The second diagram is included here because it calls for precisely the developments in Lonergan's thinking on the fourth, existential level of consciousness that are inchoately present in part 3 of the present volume but that are also foreshadowed in the mention of 'the decision problem' in the lectures on logic.

The short selection that is appendix B shows Lonergan attempting to develop his notion of science. It was filed by Lonergan with other notes, also apparently related to the Dublin lectures, in the same folder with the autograph of his distributed notes on mathematical logic and existentialism. Lonergan was very careful about the way in which he filed his papers, and so it is probable that there is a significance other than convenience in his filing these notes together with the material on the Boston College lectures. In particular, in the last lecture on logic Lonergan states the relation of his cognitional theory and of his views on logic to the issue of science and to the changing nature of all the sciences. Appendix B shows him dealing with the same issue, as he did for a number of years during this period of his development.

Our decision to publish here the material in these two appendices is prompted also by the consideration that, unless further data become available on the Dublin lectures of 1961, there will not be enough material to warrant separate publication of what we have on those lectures. Thus, this may well be the only place in the Collected Works in which the items that are here appendix A and appendix B will find their way into publication. These items (and especially the diagrams) should be made available, and it is our judgment that they fit nicely into the present volume and complement the material of the lectures recorded here.

As always the General Editors assume responsibility for the minutiae of style, continuing to take as their authorities the *Oxford American Dictionary* and the *Chicago Manual of Style.*

It remains now to acknowledge our editing debts. Richard Walker of Audio-To-Go in Toronto has helped us on more than one occasion with improved tapes. Deborah Agnew provided a good deal of assistance with typing and photocopying. Besides the particular debt we have acknowledged to Philip McShane, we must thank those already mentioned: Joseph Flanagan, from whom we got our first set of tapes; Nicholas Graham; John Dadosky; Des O'Grady; Thomas More Institute. And there is a debt internal to the General Editors, of Fred Crowe to Robert Doran for revising and integrating the input of the various workers and checking many of the footnote references. In turn, Robert Doran wishes to acknowledge the help of Daniel Monsour in tracking down some references, the hospitality of the Jesuit Community at Marquette University, where he did a great deal of his work on the volume, and the assistance of the library staff at Marquette for their constant and extremely friendly assistance.

FREDERICK E. CROWE

ROBERT M. DORAN

Editor's Introduction

Volume 18 of the Collected Works of Bernard Lonergan makes available the bulk of his two weeks of lecturing on mathematical logic and existentialism, in July 1957. The lectures were given in Boston College at the end of his fourth year of teaching theology in Rome, and the audience did not differ greatly, except perhaps in age, from that of his Gregorian University courses. The 'men in black' referred to in the first week[1] may well indicate Lonergan's view from the podium.

While the lectures represent a solid piece of Lonergan's drive of the Roman period, 1953–1965, towards his refined view of method, these lectures were deliberately tailored to his audience, relatively innocent of the previous one hundred years of logic and perhaps not vastly familiar with the parallel movements of the phenomenological and existentialist traditions. So, the lectures are a peculiar mixture of introductory description and profound intimations.

The editorial comments are conveniently separated into two sections. The first section, 'Text,' deals with the structuring of the material. The second, 'Context,' gives some indication of a fuller perspective on the material.

1   Text

The editing of the volume has aimed at providing a faithful, useful, and readable presentation both of the spoken words of those weeks and of the

---

1 Below, p. 106.

outlines that Lonergan prepared beforehand and made available at the beginning of each week. A major difficulty here was the quality of the recordings of more than forty years ago. We are fortunate, of course, that the lectures were recorded at all: one must remember that Lonergan was not recognized by many as a major figure. *Insight* had just appeared, and various articles in *Theological Studies* showed the range of his interests, but only to a limited audience. At all events, the old tapes are part of the Toronto Lonergan Archives, and the spoken words have been recaptured in large part, both in upgraded recordings and in typescripts.

The first typescript version of the lectures is that of Nicholas Graham, done in Toronto in the early 1970s. It was a magnificent achievement considering the state of the recordings, the difficulty of the topics, the obscurity of the references, the absence of clues regarding blackboard work. In the early 1990s I began work on the lectures of the first week, those on mathematical logic, and two versions have been available since then: a verbatim version and a preliminary edition that included the diagrams that are reproduced in the present volume. Both these versions, and further listening, brought forth the version that appears here, in chapters 1 to 5.

In the summer of 1997, John D. Dadosky successfully tackled the task of refining Graham's work on the second week of lectures, with the help of improved tapes, and it is from the text of Graham and Dadosky and further listening by the general editors and myself that we moved towards the edited version presented here in chapters 9 to 14.

The reader has already noticed, perhaps, a peculiar asymmetry in the ordering of the material of this volume. Following the initial chapters on the lectures on logic come three chapters of lecture notes. Chapter 6 presents Lonergan's outline for the logic lectures of chapters 1 to 5. Chapters 7 and 8 give the outline of the lectures on existentialism, and I note immediately that, apart from the division into two chapters with editor's titles, the notes are presented in a format as close to Lonergan's as possible. The problem of the order of the presentation of the notes will be touched on below. For convenience, the main section numbers that appear in the table of contents of chapters 7 and 8, corresponding to the final ordering, have been added to these notes. Chapters 9 to 14 present the lectures of the second week.

Various considerations led me to the order and presentation in three parts. First, the recordings of the treatment of existentialism do not include all the lectures: in fact a good deal of the third day's lectures are absent from the tapes, as well as occasional beginnings and transition sections.

What is missing, indeed, seems to be a continuation of the presentation of Heidegger's views and the introductory considerations of 'Horizons' that were to follow that presentation.[2] Second, there is a lack of symmetry in the relation between the notes and the lectures of the two weeks. The notes on mathematical logic are sketchier than those on existentialism, while the lectures are fuller, more coherent, more introductory. By contrast, the notes on phenomenology and existentialism have the fullness of a summary text, although a text in evolution – I will return to this point shortly. The recorded lectures give an additional fullness to sections of that summary. So, the logic lectures lead to a more comprehending reading of the logic notes, while the notes on existentialism and phenomenology provide a full readable context for the incomplete treatment of the lectures as recorded. Third, there is the obvious benefit of keeping Lonergan's notes in a central block, symbolizing his existential orientation before the event of the two weeks of lectures. For these reasons, the order of the material took the present shape.

The evolutionary state of the notes on phenomenology and existentialism is most immediately seen from the use of capital Roman numerals in the notes to designate each day's outline. The notes of the five days on logic are clearly divided, I to V, with Arabic page numbers in each section. The section headings are those given in chapter 6 of this volume, which in turn are the headings of chapters 1 to 5. In the notes for the second week, only VI and VII are used, and the rest of the notes lack typed (I will return to this) numerals, Roman or Arabic. Further, VI has three Arabic numbered pages followed by five unnumbered pages, 'On Being Oneself,' these being followed by VII, pp. 1–4, with title 'The Late Husserl.' The remainder of the notes lack either numeration, but they could easily be ordered by internal orderings and comparison with the lectures. However, this was not a necessary editorial task: the various sets of notes available to me were already in almost identical order, and I follow in the volume the pagination of Fr Crowe's autographed copy (paginated by him on receipt of the notes from Lonergan at the end of the lectures), with one slight adjustment.[3] Lonergan's own autograph copy of the notes was discovered by Robert Doran at a later stage of the editing: it was among archival material that Lonergan left

2 The part of Lonergan's treatment of Heidegger that was recorded is partly bibliographic. It is to be found in appendix D.

3 A simple matter of relocating the single-page critique of Husserl so as to bring this set into correspondence with other sets of the notes, and in particular with Lonergan's own set.

behind at his death. This copy did not lead to any modification of the order already established and presented in these two chapters. But there are modifications of the text in Lonergan's own hand, presumably from the time of lecturing. These have been included in the text, and such inclusions have been noted.[4]

Lonergan's copy of the lecture notes gives further grounds for the suspicion that the notes were not complete, or completely ordered, when he set out for Boston in the summer of 1957. First, there are the three pages on 'Phenomenology: Nature, Significance, Limitations,' which are at the beginning of Lonergan's own copy, but located later in other sets of the notes. It seems likely that these pages were typed in Boston as a supplementary handout for the participants. Neither the typewriter (nor the typing?) are Lonergan's: the archives contain neither a handwritten version from him, nor a typed version from his typewriter. Second, while the order of Lonergan's notes for the last two days follows that given in this volume, he has four interesting numberings for those days that give the impression of a late change of mind. The pages beginning 'Subject and Horizon,' 'Horizon and Dread,' 'Horizon and History,' and 'Horizon as the Problem of Philosophy' have, respectively, the handwritten numbering 'IX A,' 'X A,' 'IX B,' and 'X B' at the top, which is different from that of the final delivery. Third, there are the comments added by hand, already mentioned.

One is left with the impression of a divided interest, a shift from the original intention of the lectures towards an enrichment of his own view, a creative energy with regard to the topics of this second week. It is in this zone that one would find details of the contribution of the study of these Europeans to the development of Lonergan's own perspective. Lonergan had been invited, by Fr Adelman of the Boston College Philosophy Department, to give lectures on the two topics. In the letter to Frederick Crowe of spring 1957, Lonergan indicated that this was the reason he was tackling the two topics, and that his talks would not presuppose that his listeners were acquainted with the subjects but would lead them into the areas. But it is evident that the dynamics of the preparation and the lectures led Lonergan towards a development of his own existential horizon analysis.

A very clear stroke of fortune regarding Lonergan's preparation for the two weeks is recorded in jottings on Lonergan's biography by Fr Crowe, which he kindly shared with me. The lectures began on Monday, 8 July. On

---

4 Lonergan's copy of the notes includes, at the conclusion, two extra pages that do not belong to this period: these are reproduced in this volume as appendix B.

Wednesday, 3 July, Lonergan received Ladrière's freshly published 705–page book.[5] Thursday was the usual class holiday; Friday was also a holiday, being the Feast of the Sacred Heart. Lonergan reported working hard on the book for three days and revising the second lecture on logic (I suspect that he meant the second day's lectures) in the light of that work. Unless he carried his typewriter with him to Boston that weekend, the notes presented in chapters 6 to 8 here were completed before Lonergan flew to Boston that weekend to begin his jet-lagged lecturing.

Like the lectures, the question sessions that occurred at the end of some of the lectures suffered from poor recording. An edited version of them is included in appendix C, prefaced by an editorial comment. Appendix E presents bibliographies that Lonergan distributed to participants. Appendices B and D have already been mentioned in notes. Appendix A gives diagrams that relate to Lonergan's presentation and blackboard work at various stages of the lectures. The degree of convergence of these with Lonergan's own diagramming of his views is discussed there. A lexicon of Greek and Latin words is given at the end, to preserve the flow of the text: it saves bracketed or footnote translations. Finally, a single index covering the entire volume was considered appropriate: an introductory comment there explains peculiarities regarding symbol entries, etc.

2   Context

Indication of a context for these two sets of lectures of over forty years ago is a difficult matter that I must leave to future functional-specialist scholarship.[6] How the lectures contributed to Lonergan's own developing perspective is certainly a topic for such later work. How the two fields of inquiry that he dealt with have fared in the years since, and whether his views are still significant for them, these too are large topics beyond the scope of this introduction. But a paragraph on each of the two topics, logic and phenomenology, noting broad developments and helpful recent works, seems advisable.

Mathematical logic has moved forward in the past fifty years in sophistica-

---

5 Jean Ladrière, *Les limitations internes des formalismes* (see below, p. 49, note 22).

6 There is, of course, the relevance of functional specialization (see *Method in Theology*, chapter 5), discovered by Lonergan in 1965, to both fields of inquiry. I will deal with this, and with further issues of context, in *Lonergan's Phenomenology and Logic: Contexts* (Halifax: Axial Press, 2002).

tion both of incompleteness theorems and of category determinations.[7] Around that development there is a penumbra of what we may call philosophic issues, which, however, are generally not allowed to interfere with a steady technical achievement. The question of truth has become more precise.[8] The issue of foundations has become more explicit, even inclusive of problems of the logician or mathematician as ground, as foundation.[9] *Questioning* has emerged as an explicit topic.[10]

While there is a certain unity of progress in the field of logic, existentialist and phenomenological studies at century's end are scattered both topically and geographically. Existentialism, as a particular philosophical perspective of the mid-century, is still a zone of scholarly inquiry, but insofar as any

7   References will be given at appropriate places in the text. However, some initial pointers are immediately useful. For those unfamiliar with the present issues in mathematical logic I recommend Roger Penrose, *The Emperor's New Mind: Concerning Computers, Minds, and the Laws of Physics* (Oxford and New York: Oxford University Press, 1989). For those who wish to push forward for precise understanding without heavy mathematics there is Richard Jeffrey, *Formal Logic: Its Scope and Limits* (New York: McGraw-Hill, 1981). This work leads the reader all the way from elementary considerations to Gödel and Skolem. For pointers to advanced category problems, see note 9 below.

8   One might take as focus the continued discussion of Paul Benacerraf, 'Mathematical Truth,' presented at a symposium on Mathematical Truth, sponsored jointly by the American Philosophical Association (Eastern Division) and the Association for Symbolic Logic, 27 December 1973, originally published in *The Journal of Philosophy* 70 (1973) 661-79, and regularly reprinted, most recently in Paul Benacerraf and Hilary Putman, *Philosophy of Mathematics* (Cambridge and New York: Cambridge University Press, 1993) 403-20. This volume also provides a useful context for Lonergan's discussion of enumerability and for related problems of the continuum hypothesis. The Benacerraf perspective is central to Stewart Shapiro's recent effort: 'Much of contemporary philosophy of mathematics has its roots in Benacerraf' (Stewart Shapiro, *Philosophy of Mathematics: Structure and Ontology* [New York: Oxford University Press, 1997] 3).

9   For example, Jean Bénabou attempts foundations for category theory in 'Fibered Categories and the Foundations of Naive Category Theory,' *Journal of Symbolic Logic* 50 (1985), but he is wise enough to admit that while 'we would claim that this is enough to recapture all the richness and complexity of naive category theory, such a claim *cannot* be justified on the basis of the analysis itself because it "speaks about" the analysis and "says" that it has been thorough enough' (12). A more recent presentation of the problem of categories is Elaine Landry, 'Category Theory: The Language of Mathematics,' *Philosophy of Science* 66 (1999) S14-S27. (The peculiar pagination indicates that the paper was delivered at the Sixteenth Biennial Meeting [March 1998] of the Philosophy of Science Association.)

10 See below, p. 103, note 26.

such tradition lives on in a manner relating to the present volume, it lives on in its phenomenological methodology.[11] The spread of the general orientation of phenomenology, evident from journals and monographs, is symbolized well by the volume *Phenomenology East and West*.[12] The progress it represents may be characterized briefly as a continued heightening of attention to the complexity and richness and global variability of the psychically given. But limitations to such attention have also become a precise topic,[13] and on the other hand the patterns of phenomenological attention have acknowledged fresh mediations.[14] The issue of truth and objectivity, so central to Lonergan's reflections in both areas, is still a discomforting presence. The most forward-looking reference of Lonergan in the lectures on this topic is to the anticipated work of Merleau-Ponty on the origin of truth,[15] a work that failed to escape the tradition of obscurity, nor are there serious breaks with that problematic tradition.[16]

I must conclude my comments on context here. A concern of the publisher's readers for clarity and accessibility to the average reader dictated the reduction of earlier elaborate contextualizing efforts to these few paragraphs. Still, it seems to me that a final editorial comment on clarity and on readings would not be amiss.

---

11 A single reference would not do justice to the complexity, but a recent English reviewing of the European scene is Robert D'Amico, *Contemporary Continental Philosophy* (Boulder, CO: Westview Press, 1999). The volumes referred to in the following note give a perspective on other geographic zones.

12 *Phenomenology, East and West: Essays in Honor of J.N. Mohanty*, ed. Frank M. Kirkland and D.P. Chattopadhyaya (Dordrecht [The Netherlands] and Boston: Kluwer Academic Publishers, 1993). See also *Phenomenology and Indian Philosophy*, ed. D.P. Chattopadhyaya, Lester Embree, and Jitendranath Mohanty (Albany: State University of New York Press, 1992). The volume *Phenomenology and the Formal Sciences*, ed. Thomas M. Seebohm, Dagfinn Føllesdal, and Jitendra Nath Mohanty (Dordrecht: Kluwer Academic Publishers, 1991), is particularly useful in providing a background to both sections of the present volume. Michael H. McCarthy, *The Crisis of Philosophy* (Albany: State University of New York Press, 1990) gives a background for the entire volume that is sympathetic to Lonergan's perspective.

13 A context is provided by John C. Baird, *Sensation and Judgment: Complementarity Theory of Psychophysics* (Mahwah, NJ: Lawrence Erlbaum Associates, 1997), where the focus is on experimental limitations.

14 See below, p. 14, note 16.

15 See below, p. 278, text and note 23.

16 An up-to-date context is provided by the articles in *Phenomenology and Skepticism: Essays in Honor of James M. Edie*, ed. Brice R. Wachterhauser (Evanston: Northwestern University Press, 1996).

Certainly, Lonergan's presentation in these lectures is somewhat accessible to the ordinary cultured reader. But my concern is with the inaccessibility of the content and perspective that mediated his words. It is with the further obscurity of a functional-specialist perspective that crowned his life's achievement and that would replace Husserl's search for a rigorous science with a collaborative empirical humility. The seriously cultured reader should not miss the challenge to grapple with the existential gap,[17] the existential distinction, between Lonergan's comfortable presentation and his discomforting pointing to horizons quite unfamiliar to the cultures of the new millennium. Those unfamiliar horizons are needed to meet the desperations of our modern and postmodern times.

PHILIP J. McSHANE

17 See below, pp. 281–84.

Phenomenology

<u>Phenomenology</u>

    deals w concrete data

    moves towards invulnerable insight

    <u>hence</u>, presupposes a philosophy | data
                                      | insight
                                      | judgment

    & rejects a counter philosophy
                          | that neglects concrete data
                          | does not understand
                          | does not require invulnerable

    <u>hence</u>, <u>nonsense</u> to dispute phenomenological results
        because they suppose a philosophic position

        <u>nonsense</u> to attempt phenomenological study
        under pretence that it is voraussetzungslos

PART ONE

Lectures on Mathematical Logic

1

The General Character of Mathematical Logic[1]

This week we will consider mathematical logic from a philosophic viewpoint. I will not expound to you in detail the mathematical-logical systems with their considerable number and complexity. Rather, the question before my mind is, What about it? What is there in mathematical logic for us? What are we to think about it? What help can we get from it? What are we to be on our guard against with regard to it?[2] The main point I want to get across in these preliminary lectures on mathematical logic is the attitude of the people who are doing it, the aim that they have. Our point is to understand what they are up to. It is not so much a matter of asking whether we agree with them or not as it is of discovering what they are trying to do, understanding what is going on. So I will do more of the talking these earlier days, and the point to the discussions and questions will be, more or less, Well, what precisely did they mean by this? Would you or can you go into this, make this a little clearer?[3]

1 Opening lecture, Monday, 8 July 1957.
2 Lonergan indicated here that he would be providing notes containing his major points, so that his listeners would not have to take their own notes on his lectures. For Lonergan's notes see chapter 6 below. It will be clear later in this chapter that the notes had been distributed in time for the second half of the lecture. See below, note 23.
3 Lonergan fielded questions after some of the lectures. Most of his comments are recorded below, in appendix c. There is also evidence of at least one evening discussion session (see below, p. 350), but there is no record of this session.

There will be five topics through the days. Today, we are treating the general character of mathematical logic. Tomorrow, we will speak of the development of mathematical logic. On Wednesday, our topic will be the truth of mathematical-logical systems, the question of truth. On Thursday, we will deal with the foundations of logic: given that there are all sorts of systems, how do you tell which is right? Why do you take them all on? And on Friday, we will speak about the relationship between mathematical logic and Scholasticism.[4]

The title of my lecture this morning is 'The General Character of Mathematical Logic.' Mathematical logic is also known as symbolic logic. I divide the presentation into two parts: 'A Descriptive Approach' and 'An Analytic Approach.' In the descriptive approach we will be doing comparisons, obviating difficulties. In the second part, the analytic approach, we will be trying to get hold of *the* fundamental idea, the idea that is as much a key to the whole business as anything.

1   A Descriptive Approach

1.1 Traditional Logic and the Questions of Mathematical Logic

First, then, traditionally logic has been concerned with terms, propositions, and inferences. You will find that any book on logic has those three fundamental chapters. It may expand them or divide them, but it is talking about terms, propositions, and inferences. In traditional logic, the investigation of whole deductive systems, of axiomatic systems, has been considered inevitably too complex to be attempted. The original and valuable contribution of mathematical logic is that it makes the study of deductive systems possible.

A deductive system consists of axioms, rules of derivation, and conclusions. Mathematical logic considers *that* as a single whole which possesses certain properties. What are the properties of a deductive system? What can you do starting out from axioms, having rules of deduction or derivation, and moving from initial propositions to later propositions? Consider such a totality of propositions. What are the properties of such a totality? This sort of question cannot be asked on the basis of traditional logic. The main

---

4 This paragraph has been moved from a point much later in the lecture. See below, note 22.

value of mathematical logic lies in the fact that it makes such a discussion possible.

Moreover, that discussion is enlightening. We can see what can be done when one has a rigorously deductive system, and what cannot be done; and we can revise our own notions accordingly. Thus a question that will come up at the end of the week – it will take some time to get to it – is, In just what sense is our Scholastic philosophy a deductive philosophy?

That, perhaps, will serve as a general notion of what this is all about. It is a discussion of deductive systems, of LF, Logical Formalizations.[5]

The types of questions that are asked with regard to deductive systems, Logical Formalizations, are mainly three.

The first type is the question of coherence. Is such a system, such a totality of propositions deduced from initial axioms, coherent? If some proposition, say P, can be constructed in the system, will it be possible to deduce both P and its contradictory? Can you show that it is not possible to deduce both P and its contradictory, where P is any proposition that can be constructed out of the basic terms of the system? This is quite an elaborate question. The question of coherence is one of the fundamental questions asked about logical systems.

Next there is the problem of completeness. If P is any proposition that can be constructed within the system, will it always be possible to prove either P or its contradictory? In other words, the logical system is conceived as something in which you can prove all the propositions that belong to the system and never prove both a proposition and its contradictory. Coherence and completeness are two fundamental ideas.

The third problem that arises is called the decision problem. Practically, it is the question, Can you construct a mechanical computer that will enable you to solve any problem that arises in the system?

These are the fundamental questions asked about deductive systems. First, can you prove all the propositions in the system [completeness]? Second, if you can prove all of them, are you sure that you will not prove too many, namely, two contradictories [coherence]? And third, can you solve mechanically all the problems that arise in the system? This is called the decision problem: is there some standard procedure that will solve any problem that arises on this level of logic?

5 Lonergan's expression at this point was 'logical formulations.' At most times in the lectures he spoke of 'logical formalizations,' and this is the expression that appears in the notes that he distributed. The text will use the latter expression throughout.

1.2 Logic and Mathematics

The investigation of deductive systems, in which those questions are raised, has been very closely linked with mathematics. The two have not been confused, but there has been a very close link. Three reasons can be given for this connection.[6]

First, the principal advantage of such a link is that it is much more concrete, much more open to exact control, to ask how much *mathematics* can be deduced from a given set of premises than to suppose some set of premises, to pick some set out of the air, and start deducing from them. In the latter case, you do not know whether or not you are getting everything out of the premises. You have no control. But mathematics is a deductive system or a set of deductive systems, and if you put your questions – What premises do you need to be able to deduce the whole theory of geometry? What premises do you need to deduce the whole theory of arithmetic (namely, the theory of numbers)? – then you know what the whole theory is, and you can examine its premises and set up premises from which the whole can be deduced or cannot be deduced. You can work towards finding the axioms from which it might be deduced. There is quite a history to such procedures, but the questions are asked about deductive systems in concrete fashion: how much mathematics can be deduced from what basis? In other words, this is not a discussion of deductive systems in the air. A concrete, known deductive system is used as a check, as a basis.

In the second place, the question is linked with mathematics because in general mathematics is deductive or putatively deductive, and so the general theory of possible deductive systems includes as a particular case the totality of mathematics. The theory of deductive systems is a more general theory that includes among its branches the totality of mathematics, so that mathematical logic is mathematics on its most general level.

The third reason for the link between the study of deductive systems and the study of mathematics lies in the fact that at the beginning of this century, mathematicians were involved in difficulties in what is called set theory.[7] An example of the sort of paradox they encountered regards the notion of infinity. One may ask, 'Are there more points in line *AB* than in line *CD*?' [see figure 1]. Imagine a dotted line swinging from *AC* to *BD*. It will intersect all the points on *AB* and will intersect all the points on *CD*.

---

6 This sentence is an editorial addition.

7 Alejandro R. Garciadiego, *Bertrand Russell and the Origins of the Set-theoretic 'Paradoxes'* (Basel, Boston, Berlin: Birkhäuser Verlag, 1992) is a readable context.

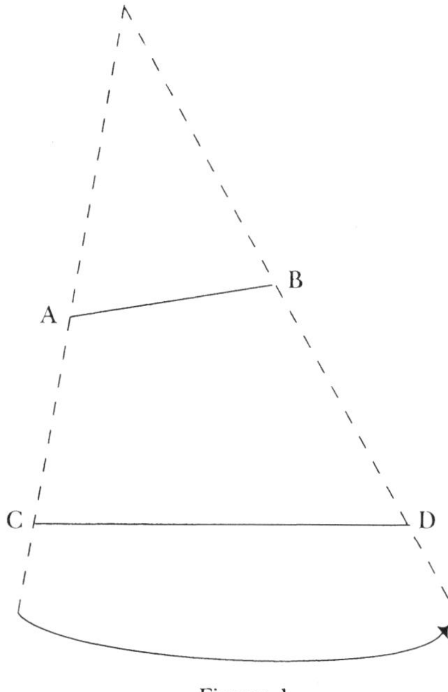

Figure 1

The two will intersect in each case only at one point: two straight lines intersect in only one point. For every point on *AB*, there is a point on *CD*, and for every point on *CD*, there is a point on *AB*. So there is the same multitude of points in both cases.

There is a more interesting construction when the lines *LM* and *LN* are at right angles and you take point *P* [see figure 2]. If you swing the extended line *PL* through an arc to *M*, you will have a one-to-one correspondence between the points on line *LM* and the points on the infinite line. Such a one-to-one correspondence did not cause any difficulties, but you can see how you will start getting into difficulties if you think about what are called infinite sets. So in Zermelo's set theory there is an axiom of choice,[8] and there is also a paradox involving the set of all sets.

8 In popular language the axiom of choice claims that, given a collection of sets, one may form a set consisting of precisely one element from each set of the given collection. Like the parallel postulate in geometry, it seems not just plausible but irreplaceable. A readable account of the matter, linking the problem with the continuum hypothesis, is 'Non-Cantorian Set Theory,' in *The Mathematical Experience*, by Philip J. Davis and Reuben Hersh (Boston: Houghton-Mifflin, 1981) 223–36. The same book presents the axioms of set theory on p. 138.

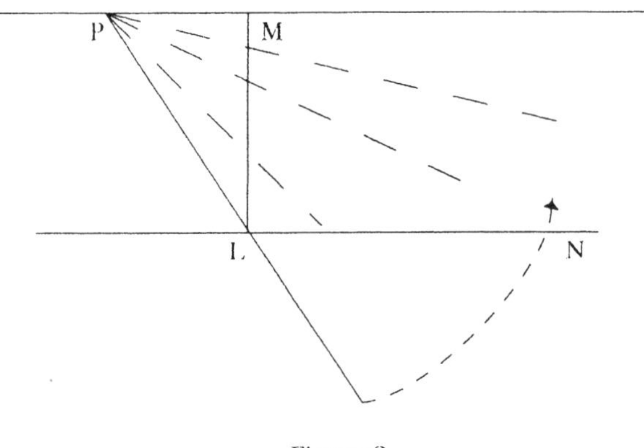

Figure 2

At the beginning of the century, mathematicians were pretty well stumped by problems that arose in the theory of sets, and logic seemed a solution. Since all the points in the line can be considered as a class, mathematicians saw a connection between mathematics and logic, and they thought of going to logic to find a solution.

That type of difficulty, as a question, has waned, at least among some mathematicians. There is what is known as the Bourbaki group. Notable among them are Dieudonné and Cartan in Paris. Their attitude is, 'Let's forget all this stuff about mathematical logic and do straight mathematics.'9 However, that is just one school, and it does not eliminate the other links between mathematics and logic.

So much, then, for the kind of question that asks about deductive systems. Are they coherent? What are the criteria of coherence? What are the criteria and the possibility of completeness? What is the possibility of mechanical solution? What is the link between those questions about deductive systems and, on the other hand, the use of mathematics as a control?

9 See L. Fang, *Bourbaki. Towards a Philosophy of Modern Mathematics*, vol. 1 (New York: Paideia Press, 1970), on the movement, its methods, and its founding members, among whom were Henri Cartan and Jean Dieudonné. The nonexistent founding father, Nicholas Bourbaki, is given the last section of William Ewald's two-volume work *From Kant to Hilbert: A Source Book in the Foundations of Mathematics* (Oxford: Clarendon Press, 1996). This book would provide the reader with the more mathematical context for Lonergan's reflections. For a contemporary context of logic, see the work by Shapiro mentioned below in note 6 of chapter 2, p. 41.

1.3 Symbolic and Technical Investigation

The next point is that this investigation has been symbolic and technical.

First, it has been symbolic. Mathematical logicians do not talk about 'All men are mortal.' They talk about any proposition P or any other proposition Q or any other proposition R, and that is as near as they get to particular propositions. And so it is with predicates, classes, and relations: the work is done in symbols. Why is it necessary to do the work in symbols? When the expression is symbolic, it is possible to consider the totality of possible chains of deduction from a given basis. If sentences were the basic axioms, it would be a bit difficult to say, 'There is this set of possible chains of deduction from this basis.' But if you arrange the symbolism in a suitable fashion, it is possible to pin down just what the possible deductions from that basis are. The deductive system is under control, and you can see, more or less, the thing as a whole – at least there is something of that possibility. The advantage of symbolism is that it is brief and that you can control what will be possible within the symbolism, and so be able to envisage, in some fashion, a totality of results.

The notion of symbolism is coupled with another notion, 'technical.' That means that there are certain permissible symbols, operations, changes of order, omissions, substitutions, additions, and none but those specified. One must state in advance everything that is allowed and nothing but what is allowed, and all one's operations have to be in accord with that statement.

What does this 'technical' mean? If you were subjected, as I was in elementary school, to the process of taking square roots, you will get the point from this example of what is meant by 'technical.' The point is that you do the thing without understanding why you are doing it or why it works.

Take the number 85849. We want the square root. We are taught first to draw a line down the side and another across the top. Divide off by 2's from the end: 8 58 49. 3 times 3 will be too big, so you write 2 and square it and write 4. Subtract 4, and pull down the 5 and the 8, and double the 2 at the side so you get 4, and that becomes 49 because it will go into 458 nine times, so you add the 9 at the top and multiply 49 by 9. Nine 9's are 81, nine 4's are 36 and when you add the 8 you have 44: 441. Then subtract, so that you get 7, 1, 4, 9. Double 29 and write 58, and that becomes 583 because it will go three times. The square root is 293.

$$\begin{array}{r} \overline{2\ 9\ 3} \\ 2\,\sqrt{85849} \\ \underline{4} \\ 49\quad 458 \\ \underline{441} \\ 583\quad 1749 \\ \underline{1749} \end{array}$$

Why does it work? Well, you do not have to know why it works. All you have to do is obey the rules. That is technique. Anyone can be taught the rules and do it. You can get a machine to do it. Various types of machines are being developed at the present time that can do precisely this. As long as you observe the rules, you get the right result. If you needed to know why it works, you could not teach square roots in elementary school.

The reason it works is the following. 293 can be expressed as $a\,b\,c$, with $a = 200$, $b = 90$, and $c = 3$. Now $(a + b + c)^2 = a^2 + (2a + b)\,b + (2a + 2b + c)\,c$. 49 is your a, 200, multiplied by 2, to get 400 and adding on the b: 490, and you multiply by b, by 90, and the two zeros are taken care of, and so on down. Again, your $2a + 2b$ is 580: add on the c: 583, and multiply by 3.[10]

But if you had to teach this in elementary school, you could not teach square roots. The students would never get on to this algebraic formula. But you can teach them to do it, and that is the advantage. One aspect of technique is that you can do it whether you understand it or not.

Doing things rightly whether you understand them or not is quite different from the way in which Euclid worked out his *Elements*. Euclid wrote down a set of axioms and definitions and postulates and then proceeded to solve problems and set up theorems. But it was discovered sometime in the course of the eighteenth century or at the beginning of the nineteenth century that Euclid was not a rigorous logician. Euclid's first proposition is to construct an equilateral triangle on a given base [see figure 3]. He takes the center A and the radius AB and draws a circle, and the center B and radius BA and draws a circle. The circles intersect at point C. You join CA and CB. AC and AB are equal because they are radii of the same circle; BC

---

10 As he is explaining, Lonergan is also writing on the board and saying 'this' and 'that' and 'here' and 'there.' The transcription is not complete. For a more complete explanation of the same matter, see Bernard Lonergan, *Understanding and Being*, vol. 5 in Collected Works of Bernard Lonergan, ed. Elizabeth A. Morelli and Mark D. Morelli (Toronto: University of Toronto Press, 1990) 55–56.

and *BA* are equal because they are radii of the same circle. So you have an equilateral triangle. But what Euclid does not have among his axioms is the fact that, if the centers of two circles are apart by a distance less than the length of the sum of the radii and greater than the difference between the radii, then the two circles will intersect. He has no such axiom. He has no proof that those two circles will intersect at the point *C.* You can see that it will happen any time you draw the diagram, but it is not logically rigorous. He should have another axiom in his system if he is going to be logically rigorous.[11]

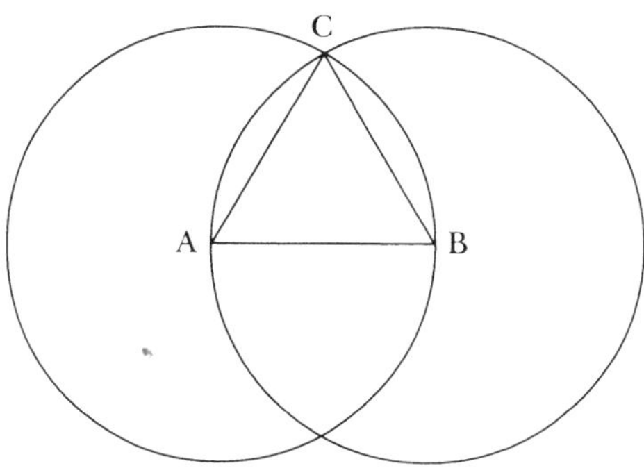

Figure 3

Again, Euclid proves that the external angle of a triangle is greater than the internal opposite angle: angle *BCE* is greater than angle *ABC* [see figure 4]. Bisect *BC* at *O.* Join *AO* and produce it to *D* so that *AO* = *OD.* Join *DC.* Opposite angles are equal, so angle *BOA* = angle *DOC. BC* has been bisected: *BO* = *OC. AO* = *OD.* So the triangles BOA and COD are similar in all respects. Therefore, the angle ABO is equal to the angle DCO. But the angle *ECO* is greater than the angle *DCO.* Therefore, the angle *ECO* is greater than the angle *ABC.*

11 For a fuller discussion, including comments on axiomatization and computerization, see *Understanding and Being* 23–28. Lonergan's audiences regularly found this baffling. Imagine the possibility of 'holes' in the lines. What is needed is some such axiom as 'suppose a line belongs entirely to a figure which is divided into two parts; then, if the line has at least one point common with each part, it must also meet the boundary between the parts.' *The Thirteen Books of Euclid's Elements,* trans. and commentary by Sir Thomas L. Heath (New York: Dover Publications, 1956) vol. 1, p. 235.

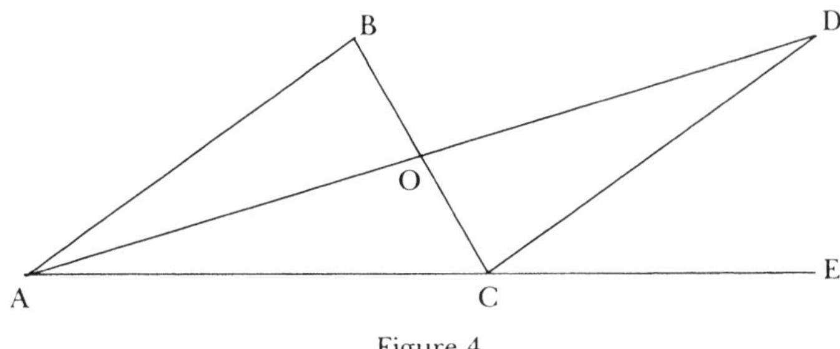

Figure 4

How did he know that the line *DC* would fall within the angle *BCE*? Well, you can see how it would work, but he has no postulate to the effect that this line must fall within. He is not logically rigorous.[12]

However, if you are doing mathematics on the basis of a technique which works whether you understand it or not, which does not depend on what I feel about the business or what my opinion is or what I am inclined to think, then the technique works objectively the way a machine works. Then you do not introduce surreptitiously axioms that you have not acknowledged at the beginning of the proof or at the beginning of the deductive system. In other words, technique eliminates casual insights. You have a casual insight when you see that the two circles are going to intersect. You have another insight that you have not made explicit when you see that the line will fall within the angle. You have a casual insight every time Euclid says there are two and only two cases, or three and only three. You can see that there are three and only three, but you have no way of proving that there are three and only three. His list of axioms is not complete.

Moreover, there are more fundamental difficulties. The whole of Euclidean geometry is metrical, and that metrical character of Euclidean

12 It might help the reader to consider the problem when the triangle *ABC* is on a basketball. The diagram has *BC* as equal to a quarter of a great circle, so *F* lands on *D*. With *BC* longer, *F* gets out from between *CA* and *CD*.

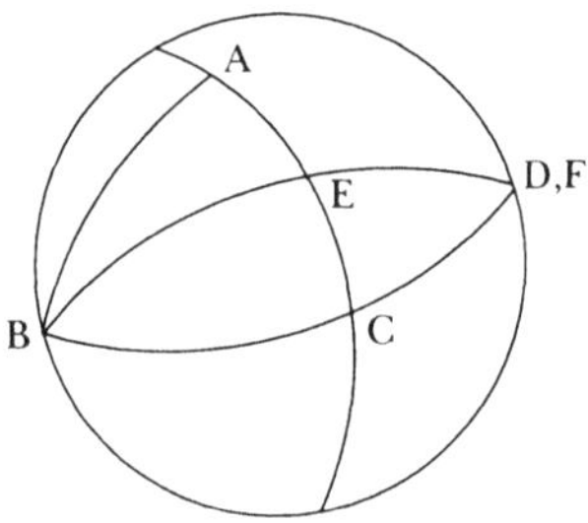

geometry is not recognized too explicitly.[13] Not only does Euclid lack logical rigor in following out the deduction strictly from the set of axioms laid down – Euclid presumably proposed to settle everything from his axioms, his definitions and postulates, but he brought in new ideas as he went along. But also, as you know, his parallel axiom is not necessary. An insight into an image is infallible, *if* you have the image.[14] But no one has an image of infinite space. You have an image of as much space as you care to imagine, but that is not infinite. You can stand on a mountaintop looking up at the sky and all round the horizon, and you can imagine more space

13   As the text to follow points out, the 'obviously objective' character of the Euclidean metric, within its geometry, clouded the physics of space (and indeed the metaphysics of grace: see Bernard Lonergan, *Grace and Freedom: Operative Grace in the Thought of St Thomas Aquinas*, vol. 1 in Collected Works of Bernard Lonergan, ed. Frederick E. Crowe and Robert M. Doran [Toronto: University of Toronto Press, 2000] chapter 1-4). One must couple the oversight with Aristotle's understandable failure to conceive momentum as act (not imperfect act) to glimpse the ground of the obscurity that prevailed, and still prevails, in physicists' search for a valid geometry and statistics of space-time. The context of that ground is given in Lonergan, *Insight*, vol. 3 in Collected Works of Bernard Lonergan, ed. Frederick E. Crowe and Robert M. Doran (Toronto: University of Toronto Press, 1992) chapters 4, 5, 15, and 16. While the point lifts the discussion into the 'more fundamental difficulties' just mentioned by Lonergan, it is worth adding his fundamental perspective on actual geometry. The following was found in the Lonergan archives, Batch II, file 48, on the last page of the five-page item A296: 'The error is that local motion as a constant velocity is not conceived as essentially distinct from local motion as an acceleration. For on that point, though this is not the place to argue it out, I believe there is no alternative to the Aristotelian theory of the universe and the Newtonian or similar theories: for if local motion as such, that is, change of place without other change, for example, of energy or momentum, is a transition from imperfection to perfection, from potency to act, then there must be imperfect places and perfect places, and the whole Aristotelian theory follows; but if places are indifferent, then change of place as such is not a transition from potency to act but simply being in a peculiar kind of act; local motion becomes a state, and the Newtonian type of theory follows. This much is said to show that St Thomas's position on the influence of the heavenly bodies is far more reasonable than that of the cosmology manuals that hold all places to be indifferent and yet define local motion as a transition from potency to act.'
14   See Bernard Lonergan, *Collection*, vol. 4 of Collected Works of Bernard Lonergan, ed. Frederick E. Crowe and Robert M. Doran (Toronto: University of Toronto Press, 1988) index, under Infallibility; also *Verbum: Word and Idea in Aquinas*, vol. 2 of Collected Works of Bernard Lonergan, ed. Frederick E. Crowe and Robert M. Doran (Toronto: University of Toronto Press, 1996) 75, 185–86, 188.

than that. But if you want to go on to infinite space, you have to construct it, add it on. If you add it on in the form of Euclid's axioms, Euclid's parallel axiom will be evident, but if you add it on in the form of a different axiom, the other axiom will be evident. In other words, the parallel axiom is a *possibility*, a possible way of constructing space, not a necessary way of constructing space.[15]

Those examples illustrate the danger of using casual insights, as illustrated by the parallel axiom, and the lack of rigor in using casual insights, because you are using concealed axioms. The insights that come along the way are concealed axioms, unknown presuppositions.

What happens when you use casual insights that are unknown presuppositions? First of all, because you do not make an explicit axiom of every fundamental insight that you introduce, you will not exploit it to the full; you will not get all the good out of it that there is in it. It just comes in casually and may recur later on casually in a more complex form, but you are not working systematically to your goal; you are not getting everything out of it.

In the second place, there is the danger of unknown presuppositions, concealed axioms. Euclid was not aware of an axiom with regard to the intersection of circles, or of the axiom regarding lines falling within a certain angle. In general, he failed to define the idea of betweenness,[16] as in

---

15 See Bernard Lonergan, 'A Note on Geometrical Possibility,' in *Collection* 99–101. The phenomenological perspective of this short article is extremely relevant to the large contemporary project of 'constructing space,' the reach for a geometry/topology of the entire universe. Without that perspective, and the positional context of *Insight* that qualifies its content and that of *Insight*, chapter 5, tinges of fantasy intrude, even in serious work. For context, see Marc Lachièze-Rey and Jean-Pierre Luminet, 'Cosmic Topology,' *Physics Reports* 254:3 (1995) 135–214. See also volume 2 of the reference below at note 13, p. 356.

16 Something missed for over two millennia obviously is not easily brought to personal evidence. If you bisect the angle *ABC*, is it not evident from the diagram that the divider cuts the join of *A* and *C*? Yet this is a missing postulate in Euclid. The Postulate of Pasch (1882) is that if a line intersects a triangle not at a vertex, then the line intersects two sides of this triangle. Modern books on geometry treat postulates of betweenness, starting, for example, with the symbol (*a b c*) as indicating that '*b* is *between a* and *c*.' For a good introduction see Walter Preurwitz and Meyer Jordon, *Basic Concepts of Geometry* (New York: Blaisdell Publishing Company, 1965). Chapter 1 focuses on deficiencies in Euclid regarding betweenness; chapter 10 develops axioms of betweenness. For problems of the early history see Heinrich Guggenheimer, 'The Axiom of Betweenness in Euclid,' *Dialectica* 31 (1977) 187–92.

more modern formulations. He was not aware of his real presuppositions, and that is a logical defect.

In the third place, when you use concealed axioms, presuppositions of which you are unaware, not only are you not getting the full good out of what your real premises are, but also you may be engaged in a contradictory enterprise. If you define the barber as the man who shaves other people, then does he shave himself or not? You are in a contradictory enterprise just by the way in which you define the barber. While that may seem very simple with regard to the barber, still, in very complex mathematics this is the type of thing that may lead one to work for years until one discovers one made a stupid mistake at the beginning.

Consequently, symbolic logic, mathematical logic, is technical: you observe certain rules that you lay down at the beginning, and you do not use any casual insights that you get along the way. After you have set down your axioms and your rules of derivation, you do not use your head; that is what is excluded. You just observe the rules. Why? Because if you start using your head on the way, you will really be operating on a different basis from the one you set out from. If you are operating on a different basis from the one you set out from, you will think your conclusions follow from the latter basis when in fact they do not. You will not be getting all the good you have out of these casual insights that you have introduced, and you may be engaged in a contradictory enterprise.

These are some of the reasons that led mathematicians to concern with a logic that is symbolic and technical. The logic proceeds with symbols; it uses symbols to represent fundamental terms and operations. It proceeds according to rules, and it proceeds blindly according to those rules. 'Do not use your head as you go along.' You have to use your head a good deal as you go along, of course, but just to observe the rules, as you would to take the square root properly.

Now, with regard to those two characteristics, 'symbolic' and 'technical,' we can make two notes for later reference, for our own purposes. First, modern logic is at times called 'formal,' but in the sense that it is symbolic and technical, not in the old sense of 'formal logic.' This is a new idea of logic, as symbolic and technical, and you should be aware of that change in meaning. It may be simply an evolution of the old idea, but it is not exactly the same idea. A.N. Prior put out the book *Formal Logic* in 1955, and it is mathematical logic, and 'formal' means 'symbolic and technical.'[17] Again,

17 A.N. Prior, *Formal Logic* (New York: Oxford University Press, 1955; 2nd ed., 1962). The 1955 edition was, in fact, Lonergan's first introduction to mathematical logic.

Fr Bochenski, a Dominican, put out a 600-page book on logic called *Formale Logik,* and he is saying that he really arrives at formal logic when he arrives at mathematical logic.[18] So the word 'formal' has a new meaning.

A second note is that there is quite a harmony between the symbolic, technical approach to logic and the empiricist, pragmatic viewpoint in philosophy. This business of not using your head as you go along eliminates or makes irrelevant the human subject. And the fundamental issue, as we will see, with regard to mathematical logic is the opposition between mind, on the one hand, and technique, on the other. In Aristotle, logic is an assertion of rationality against sophistry or against mere rhetoric. But the modern opposition is not 'rationality versus sophistry,' 'rationality versus rhetoric.' It is rather the fundamental opposition that arises between mind and its own creation. Will the creation take the place of mind, eliminate it, shoulder it out of the way? There is more than a little danger along those lines.

1.4   Contrast with Scholastic Procedures

We have said, first, that mathematical logic is concerned with deductive systems; secondly, that it is closely linked with mathematics; and thirdly, that it is symbolic and technical. Now, in the fourth place, this symbolic and technical character of mathematical logic gives rise to problems unknown in the Scholastic or classical tradition of logic. It is a new, different viewpoint, with consequences that are unknown to the Scholastic or the classical logical thinker. All that I want you to grasp at present is that there is an entire and radical difference in mentality between the two approaches. The *relative* advantages of both we can consider later. What I want to get across now are the advantages of mathematical logic.

When a Scholastic comes to a difficulty, he draws a distinction, and that is the right thing to do within the Scholastic tradition; that is the way they do things. But a symbolic technique cannot draw distinctions. All the distinctions have to be drawn from the start. You are not to use your head after you have started; just obey the rules. If the results are bad, then you start again. If you feel the need of a distinction after you have gone so far, you start all over again. This is one concrete difference between the Scholastic mentality

18 I.M. Bochenski, *Formale Logik* (Freiburg, München: Verlag Karl Alber, 1956); English translation, *A History of Formal Logic,* trans. and ed. Ivo Thomas (Notre Dame: University of Notre Dame Press, 1961).

and mathematical logic. Distinctions have to be included in the initial definitions. Why? Because they are equivalent to the introduction of new insights. You are shifting your ground when you make a distinction. It can be, of course, that you have already made the distinction at the start of your thesis, and you are merely recalling what you previously said. But it often happens that a distinction involves something new, something that had not been thought of before but that is needed to get around some difficulty; and when that occurs, you are changing the ground on which you are operating. You are moving from Logical Formalization 1 to Logical Formalization 2, and when you draw your next distinction, you are moving to Logical Formalization 3, and so on. You are operating on a series of systems, and no one knows what system you are in. Unless distinctions are excluded in the deductive process, you are not dealing with one deductive system. You are shifting your ground, moving from one to the next, and on to a third, and so on. Mathematical logic eliminates that. A lot of Scholastic controversies are connected with this procedure of introducing distinctions. (The high point occurs, of course, when someone says, 'Well, what I am saying is what St Thomas said.' His mind is St Thomas's!)

Again, the Scholastics accept the principle of excluded middle: not only can you not say, 'Both *A* and not-*A*,' but you must say, 'Either *A* or not-*A*.' That is the principle of excluded middle: one side of a contradiction must be right. However, while Scholastics accept the principle of excluded middle, their acceptance does not mean that they waive the right to draw distinctions. The principle of excluded middle is quite all right, but if you ask a man, 'Have you stopped beating your wife yet?' he demands the right to reformulate the question. He does not feel bound to say either yes or no. Similarly in Scholastic questions generally, the principle of excluded middle is accepted, but they reserve the right to make distinctions. But in a symbolic logic either you accept excluded middle or you do not accept it. If you accept it, you have one system, and if you do not accept it, you have another. It is built in right from the start. If it is so built in, every time it becomes applicable it will be applied more or less automatically.

Consequently, there is quite a debate about the principle of excluded middle among mathematical logicians. There are systems that have it and systems that do not have it. There are discussions about just how far you can take it in a weakened form. And so on. But the point to observe is that the principle of excluded middle has a different meaning according as you are in the context of Scholastic thinking, which draws distinctions when need arises, or within the type of thinking of symbolic mathematical logic, in which all distinctions are put in at the start and after that one goes along systematically according to one's premises.

Again, while the Scholastics have distinctions of content, that is, while they define the precise meaning of a term, in this sense and in that sense, besides these distinctions of content there are also structural distinctions that the Scholastics use all the time: *re et ratione, signate et exercite,* first and second intention, formally and materially, *virtualiter et formaliter,* and so on. Since symbolic mathematical logic does not draw any distinctions that are not there from the start, its structural distinctions are built in. First, there is an object language or a set of symbols that refer to the objects that one is talking about: mathematical entities or physical entities or whatever it is one is talking about. There is a first language to talk about the things. Then there is a metalanguage to talk about this language. It will be syntactical insofar as one is talking about the internal relations of the object language, and it will be semantic insofar as one is talking about the relation between this language and its object. And if one wants to talk about the syntactical or semantic language, then one goes on to a meta-metalanguage. In other words, the structural distinctions that Scholastics put in, namely, distinctions that apply to everything – *re et ratione* applies to every concept, and so on – appear within mathematical logic as a series of strata. Logical Formalization 1 deals with the object. Logical Formalization 2 deals with the internal relations in that logical formalization. Logical Formalization 3 deals with the relations between this formalization and the object. And further Formalizations constitute further strata. The reason they put in these stratified distinctions is that they would be in constant confusion unless they formalized such separations. In other words, distinctions that Scholastics find necessary and constantly recurring appear in mathematical logic as strata, as language and metalanguage.

Finally, Scholastics employ analogous terms, terms that have a different meaning according to different contexts. For example, existence is what is proportionate to essence, and as you change essences there are different types of existence. There are many analogies in Scholasticism. Now, the discovery of something that seems to be analogy is rather late in mathematical logic, but it is extremely important. It shows that they are getting hold of things that we consider fundamental. You find in Church's mimeographed lectures at Princeton about 1935 a series of meanings of implication and quantification. What is meant by implication and quantification on this level and on that level and when you go up a level? The levels spread.[19] Again, in Curry's fundamental notion of canonicity there is an infinite, perhaps a transfinite, series of meanings of canonicity.[20] Hao Wang has a similar series of meanings for enumerable sets, and also different meanings of construction.[21] Analogy appears again as stratification; it appears as indefinite stratification, an indefinite series. In other words, each one of the analogous meanings gives you a different Logical Formalization.

1.5 Paradoxes

A final point on the general descriptive approach to mathematical logic is that the development of mathematical logic has been closely connected with paradoxes, and in particular with two paradoxes. First, there is the paradox of the Cretan who said he was a liar. It can be put in this form:

The proposition

in this square

is false.

If it is false, it is true, and if it is true, it is false. It is a paradox.

Second, there are the paradoxes connected with reflexive relations. The

19 The mimeographed notes to which Lonergan refers were not available in editing this volume. The bibliography in Jean Ladrière, *Les limitations internes des formalismes* (see below, p. 49, note 22) gives the following information (p. 509): Alonzo Church, *Mathematical Logic. Lectures delivered at Princeton University, October 1935 – January 1936* (mimeographed, Princeton, University Mathematics Department, 1936). Evert W. Beth, *The Foundations of Mathematics: A Study in the Philosophy of Science* (Amsterdam: North-Holland Publishing Co., 1969) 688, remarks that there was a mimeographed version in 1944 of Alonzo Church's 1956 book *Introduction to Mathematical Logic* (Princeton: Princeton University Press); this same version is mentioned in Ladrière's bibliography, ibid. It is clearly an item distinct from the 1936 notes. Volume 1 of Church's book (p. v) describes this edition as 'a revised and much enlarged edition' of the first part of a work of the same title 'published in 1944 as one of the Annals of Mathematics Studies.'
20 Haskell B. Curry, 'The Combinatory Foundations of Mathematical Logic,' *The Journal of Symbolic Logic* 7 (1942) 49–64. He deals with the notion of canonicity or 'canonicalness' in section 8, pp. 62–64.
21 Hao Wang, 'The Formalization of Mathematics,' *The Journal of Symbolic Logic* 19 (1954) 241–66, esp. sections 6 and 7, pp. 247–51.

class of all classes does not contain itself. The barber shaves everyone but himself. Who shaves the barber if he is the man who shaves other people? There are series of paradoxes based on a reflexive relation.

Paradoxes that seem just logical puzzles have played a fundamental role in the development of mathematical logic. One can take a wrong view of those paradoxes. At first sight, to a Scholastic, they seem to be just a little catch that a few proper distinctions would settle; all one has to do is make a few distinctions about any of these paradoxes, and one has the solution. For example, Scholastic philosophy says of the reflexive relation, *Relatio alicuius ad seipsum est ens rationis.* This is a summary way of dealing with reflexive relations. But a mathematical logic does not use distinctions; it is concerned to bring out what follows from a given set of axioms. You can put your distinctions in at the start, as many as you want, and you can start over again as often as you like, but the distinctions do not come in just because you think it is convenient to bring them in to get around a difficulty. What mathematical logicians want to know is what follows from these premises, or those, or any set. They do not want to know how you might go so far in your system and say, 'Oh, yes, you add this on here.' No, you add it on at the beginning. Consequently, while a distinction with regard to those paradoxes may be the proper approach from a Scholastic viewpoint, mathematical logic has effected enormous developments and refined thought tremendously.

The fundamental difference is that the Scholastic follows the natural spontaneous dynamism, expansiveness, tendency to develop, of the human mind. If you are following that, you will introduce the distinctions as need arises, and you will gradually understand the thing better and better. As far as personal development goes, this business of drawing distinctions is fine; after all, what counts is whether the other fellow understands the view with you or not. On the other hand, from the viewpoint of communication, from the viewpoint of a large number of people who do not give two hoots about one another personally, but who are working together and getting the same results, this other method has tremendous advantages. In this other method, where you use the rigorously deductive system, you do not operate in accord with the spontaneous development of your intelligence. You operate in terms of the objective rules and premises that have been laid down.

Finally, to repeat a point, if you were to introduce distinctions to solve the paradoxes that arise when deducing the consequences of your system, you would not be investigating *a* system. You would just be shifting your ground from one system to another, never finding out the properties of any of them.

1.6 Summary

That will do for the descriptive approach to mathematical logic.[22] The aim of the descriptive approach has been to give you a general idea of what mathematical logic is all about. Let me conclude this section with a brief summary.

First, we could discuss at length the subject of terms, propositions, and inferences in mathematical logic, but what is really significant about it is its ability to discuss systems. That is where it starts moving and where it is interesting. If you just pick up introductory texts on the subject, you will find that they spend most of the time on terms, propositions, and inferences, and you will not see what on earth it is heading to, what is the good of it, and so on. Its interest and its real significance lies in its ability to enable you to discuss whole systems, deductive systems, such as was thought for a long time to have been achieved by Euclid in his *Elements.*

Second, this logical investigation of systems is closely linked with mathematics as a control. It is easier to ask how much mathematics you can deduce from a given basis than just to ask what can be deduced from some arbitrary basis. Because mathematics is a particular case of deductive system, the possibility of deductive systems in general determines mathematics at a most general level. Another connection between mathematics and logic comes from problems of set theory that existed at the beginning of this century.

Third, the procedure is symbolic. Because of its tightness you can put so much down in so little space that you can see all of it at once. And if the symbolism is controlled, if you know just what can be done with that symbolism, you can see the possible chains of deductions from a given set of axioms, and so you can discuss the system as a whole. The use of symbolism is essential to discussing a whole system. If the system is spread out over a whole book, you never see it all at once.

Fourth, the procedure is technical: you obey the rules laid down at the start; you do not use your head. The results are that you do not introduce distinctions the way the Scholastics do when relying on natural spontaneous intelligence. What is wanted here is to investigate one system. It may be

22 It was at this point that Lonergan presented the prospectus of the week's lectures that appears above, p. 4. He then took a break in the lecture. He began the next part of the lecture with what is here presented in the next paragraph. The concluding words in the present paragraph have been adapted from the words that he spoke at the very end of this summary.

as general as you please; you may start over again as often as you please; but you are examining just one system. If you are doing that, then you set up different levels in the system, different strata as it were: languages and metalanguages and meta-metalanguages, ad infinitum. But you do not start off on one basis and then shift gradually to another and on to a third, and so on, because then you will never know where you are.

Finally, while paradoxes may seem to a Scholastic the sort of thing that is set aside with a few neat distinctions, de facto the significance of a paradox is that one has started a machine running and it has come to a jam. Consequently, one must redesign the machine. And the amount of redesigning that has gone on has been the development of mathematical logic.

2   An Analytic Approach

We will turn next to a more analytic approach. While the aim of the descriptive approach was to present a general idea of what mathematical logic is all about, what we have to do now is get hold of a fundamental idea and do so with a certain degree of accuracy. If we can get hold of that idea we will see the *possibility* of mathematical logic. When what is vital is getting hold of one basic idea, we are about something that is somewhat more difficult than the sort of thing we were doing in the descriptive talk.

2.1 The General Character of Technique

We start out from the general character of technique. It is worth noting that this provides a link with what we will see next week on existentialism. The existentialists are in revolt against technical society. Each individual wants to be himself, but in technical society, the social mechanism pretty well settles everything. If you want to see the world, well, you are a tourist, just one of a group, and the best way to see the world is to sign on with some agency; they will do it much better for you than you can do it yourself. Every aspect of life is just like that. We live in a technical society. Our food is all in cans; we do not have to know how to cook anymore; and so on. The existentialists are in revolt against that. So our consideration now of the general character of technique provides a link between what we are doing this week and what we will be reacting against next week. This week, we are technicians.

First of all, we will consider pretechnical activities. You see how to do something and you do it yourself, and your doing it yourself is a nonexpressed sequence of movements. You just go ahead and do it without thinking about it. You get the idea, and you puzzle a bit about how you might do it; you get interested and have great satisfaction in seeing how it can be done, and so on all along the line. But basically it gets done without any great advertence to it on your part. People are always solving little problems all day long and executing their solutions in this pretechnical fashion.

The movement towards the technical arises insofar as there is an increasing accumulation of practical insights, a whole complex of practical insights in science, applied science, engineering, and so on, involving technicians and skilled workers. The complex of practical insights comes out as a detailed plan of operation. The detailed plan of operation will have two aspects: the organization of people and the structuring of equipment. The organization of people will include workers, foremen, supervisors, superintendents, managers, directors. The organization of colleges and universities tends that way, even though different names will be given to the people involved. This is so because of the enormous number of pupils. The population of the world has increased by a thousand five hundred million in the last 150 years. That terrific increase of population gives you the masses, and the existence of the masses makes technical society, the technical approach to everything, from education to food to pleasure, inevitable. If we were to eliminate technical society, most people in the world would simply starve.

So, on the one hand, there is an organizational side, and on the other hand, the mechanical. Simple tools are implements that do not contain within themselves the design of what is to be done, but a machine is a tool that has within itself the design of what is to be done. The motorcar contains the design of how petrol is transformed into locomotion. The design is in the tool; it is the tool that contains the design. The power-driven machine that executes the design without any use of human power, that is, automation, more or less eliminates ordinary human intelligence.

There is, then, the movement towards the technical: an accumulation of practical insights that are expressed in an organization of men and an organization of machines.

The fundamental character of that technique is that, while you have a highly complex process controlled throughout by intelligence, only a fragmentary understanding of the total process is found in any individual. The manager knows how to do his job, but he will not be able to tell the supervisor what has to be done. The supervisor will know that. The superintendent will have a general notion of the way the different departments come together, but he will not have the detailed knowledge of the supervisor in each particular case. The supervisor will not have the intimate knowledge that is present in the foremen, the skilled workers, the ordinary workers, and so on. There is an intellectual division of labor. No single individual, either at the top or at the bottom, understands the whole working of the big plan, but together they understand how it works. Similarly, in a technical civilization no one understands how the whole thing works, but all of them together understand how it works. It is like taking square roots, in that you do the right things; but here it is spread out over the whole of society. Therefore, the general characteristic of technique is that the process is throughout intelligible, controlled by intelligence, and yet there is no individual that does the understanding. The understanding is done by a whole lot of individuals, and they are tied together by rules. And it is very hard to change it all.

2.2 Symbolism as Technique

Symbolism is a particular case of technique. The pretechnical stage is illustrated by mental arithmetic. We reach a technical stage when a series of operations is guided simply by the rules. I can understand how to take the square root, but it makes no difference whether I understand it or not. The thing to do is to observe the rules. Understanding the algebraic formula may give me a certain amount of satisfaction, but people can take the square root just as well if they know nothing about it. That is an example of the technique represented in symbolism.

The advantage of such symbolism is threefold. First, there is an economy of intelligence. One can do it equally well though one does not understand. And even if one does understand, still one can do it without any effort of intelligence. One just goes ahead and follows the rules automatically, as a matter of routine. One does not have to take any trouble to try to understand. You do not bother thinking about this algebraic formula every time you take the square root. You save the effort, so there is an economy of intelligence.

In the second place, there is an economy of reason. You do not have to worry whether it is right or wrong. You do not bother about any theoretical questions. You go ahead and take the square root, and you check afterward, and you see, if you square your result, whether you come back to the number with which you began. It works: what more do you want? That is the economy of reason.

In the third place, there is an objectification of mind. There is a liberation from individual variations and individual aberrations. You can say, 'I see it this way,' but no one asks you how you see it, and no one cares how you see it. You can say, 'I'm inclined to think,' and 'It would seem,' and all the rest of it, but no one pays any attention to that; it just does not count. There is a pure objectification of mind. There is an objective process of taking square roots, and you go ahead and follow the rules and get the results, or you do not follow the rules and do not get the results, and that is all there is to it. That has terrific advantages from the viewpoint of communication.

As you can see, I am just summing up things that I said before on the general idea of technique; but they are very important for understanding our topic. Traditional logic claimed to be engaged in investigating the immutable laws of the mind, the laws of thought. Symbolic logic is doing something concerned with what a machine can do. The two may or may not be equivalent, but we have to know the difference. We have to know exactly where the difference lies if we are to discuss the advantages and disadvantages of mathematical logic.

2.3 Symbolism as Model

A third point in this analytic approach is to consider symbolism as a model. First of all, we considered symbolism as a technique; it runs along like a machine. But there is another aspect to it, and that aspect is quite different.

I note[23] that I am not using the word 'model' in the sense used by mathematicians and mathematical logicians. As a matter of fact, the sense in which they use the word is not too clear, because they do not have too clear a distinction between an image or a sensible sign and a concept. When Hilbert, for example, is doing mathematics, he is dealing with concepts, but he says that his mathematics is a study of the sign. So you can see how a philosophic point will influence the way these people think. There are different meanings given to the word 'model.' The simplest is the one given by Suppes.[24] But we are using 'model' in our own sense.

---

23 Lonergan is referring to the notes that he distributed. See below, chapter 6, p. 145.

24 Patrick Suppes, *Introduction to Logic* (Toronto: Van Nostrand Reinhold Co., 1957) 253. Briefly, 'When a theory is axiomatized by defining a set-theoretical predicate, by a *model* for the theory we mean simply an entity which satisfies the predicate.' However, Suppes continues with a useful discussion of the meaning of 'model' in various fields.

By 'model' I mean something that we see or imagine, something sensible. The mathematician not only produces the symbols according to the rules, but he also perceives the symbols that he produces. The machine can produce them, but the machine cannot see what it is producing. The mathematical logician not merely produces the symbols, he also sees them; and he not merely sees them – anyone can look at them – he also understands them. So the symbolism is a model insofar as it is a potential object of intelligence, a potential object of insight. You have the symbols in front of you, and you see what can be done with them. The tutor whom I had in mathematics in England, Fr O'Hara, said that to do mathematics you need an X-ray mind. You want to see the skeleton. You want to see a quadratic in a long algebraic expression, wherein de facto there have to be squares at one point, first powers at another, constants at yet another. Unless you have what he called an X-ray mind, you will not see that it was a quadratic. 'Get the form of the thing, and you will be able to see what you can do with it from there on.' That idea of seeing an intelligible form in the sensible is, of course, right down our alley. Plato in the *Meno* has Socrates summon a slave; Socrates draws a square on the sand, and he asks the slave to draw a square that is twice the area of the first one. The slave immediately goes to work, doubles this side and doubles that side and draws a square.[25] But Socrates produces the lines so that the square is seen to be four times as big, not twice as big, and the slave says, 'Yes.' Then Socrates fools around a bit and finally gets the slave to draw the lines that form the triangles. You can see that the new square is just half the size of the one that was four times as big as the original one. The square on the diagonal is twice the area of the square that he started with [see figure 5].

Because the slave was able to do that by being given not too obvious hints by Socrates, Plato introduced the doctrine of anamnesis. Plato scholars offer different interpretations of what it means, whether recalling from an earlier state or the spontaneity of intelligence. But Aristotle did not worry about any anamnesis or recollection. What he attended to was the diagram. He says in his *De anima*, 'Mind grasps forms in images.'[26] Mind grasping forms in images is the psychological, epistemological counterpart to hylomorphism, form in matter. Because our minds grasp forms in images,

25 Plato, *Meno* 80d–86c.
26 Aristotle, *De anima*, III, 7, 431b 2. Lonergan was to say a bit later in the lecture, in an offhand remark, 'The exact reference to Aristotle's *De anima* is on the title page of *Insight*.'

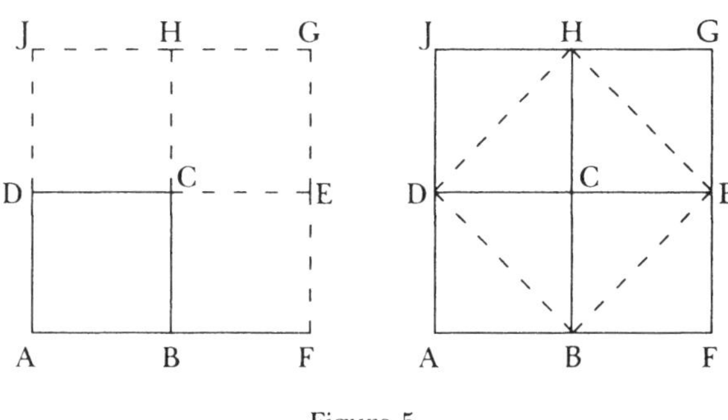

Figure 5

the object proportionate to our knowledge consists of form in matter. With this we are right at the root of Aristotle's system, psychological and metaphysical at the same time.

When we speak of symbolism as a model, we mean that the sensible signs are sensible data in which a form can be grasped. So the mathematical logician not only produces the symbols in a rather mechanical fashion, but at the same time sees them, and not merely sees them but grasps form in them, understands, gets acts of understanding.

The next point is to distinguish model and object. The model is the sensible marks, the series of sensible marks. But the object is what is conceived, thought about, considered, as a result of insight into the model. You can have a Euclidean diagram: that square will do.[27] But you can *think* of the square such that the square does not have lines with thickness as its boundaries. It does not have dots at its corners, but points, and a point cannot be imagined or represented. If there is any mark there at all, it will have some size; it will be a dot, not a point; a point has no size. If you have a line, it will have some thickness; the image always has some thickness, but what you are thinking about has no thickness at all. The object is point and line as defined; the image is something in a different order. This is what I mean by the distinction between model and object. In the first example we have the Euclidean diagram, on the one hand, and, on the other hand, what you are thinking about. What you are thinking about is not the diagram. The diagram is that into which you have insight, to think about things that are universal and nonimaginable. There is no image of a point as defined, and there is no image of a line as defined. If the line you are

27 Lonergan is probably gesturing toward his own *Meno* diagram.

imagining has no thickness, then you have no image. It just vanishes when you get it right down to nothing in the way of thickness.

Again, there are double models. For example, there is an image of two cones meeting at a point. You can cut a plane through the two cones, and the cut will be two straight lines [see figure 6].

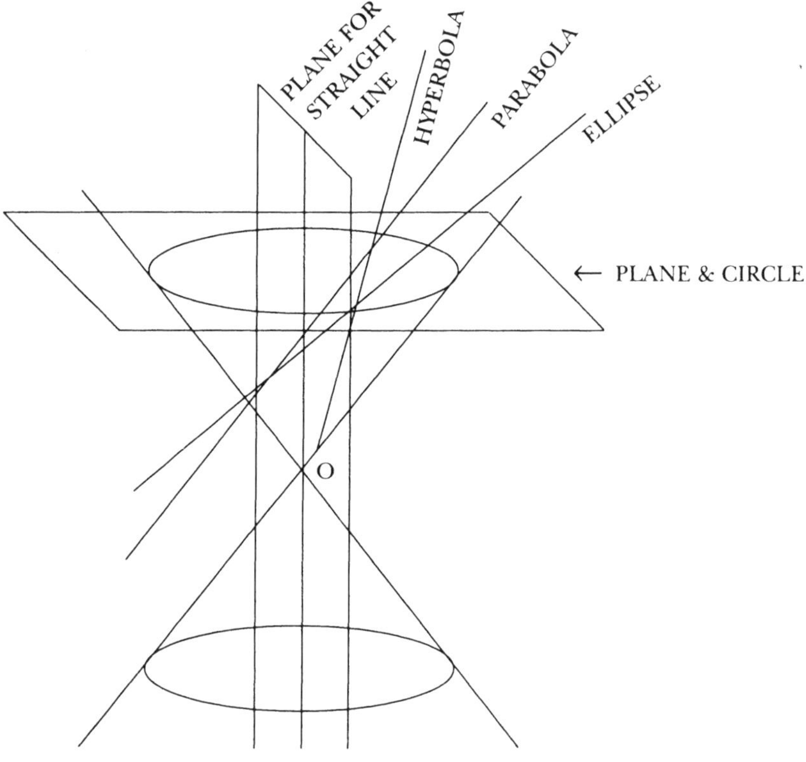

Figure 6

The plane cutting the cone will cut down the straight edge, the outside of the cone, and you will get one straight line on one side and one straight line on the other. If you cut across absolutely flatly, then the cut will be circular; if you cut another way, you get an ellipse; if you cut parallel to the generating line, you get a parabola; and if you cut through the two cones, you get a hyperbola. Those are known as conic sections, and here I am giving a two-dimensional representation of what you have to imagine in three dimensions. That is one model, the diagram.

But there is another model, the general equation in analytic geometry for all conic sections. By giving different values to the constants A, H, B, G, F, C, we can derive any of the conic sections: straight lines, hyperbolas, parabolas, ellipses, and circles.[28] That is another set of symbols, another set of things to be imagined. So there are two models for what you think of when you talk about conic sections. One is in symbols, and the other is in an image. Therefore, there are two models but one object. In our first case, we just had one model and one object, the image of this geometrical figure and the concepts of line, point, and so on. But the second case gave us two models.

There is also a third case. Consider the physics of the state of a free fall. If you are standing on the roof and toss something off and there is a vacuum for it to fall in – no air there to give you air resistance – you will have a free fall. There is an image of the free fall in a diagram. But there is also a symbolic model of the free fall. The movement out is represented by $x = ct$ and the movement down is represented by $y = -\frac{1}{2} gt^2$. Those algebraic expressions are another model of the free fall. The free fall is what you think about. It is a concept, a universal. But there is a concrete image of the free fall, and there is an algebraic representation of the free fall. The latter too, precisely as a concrete image, as something you write down and see, is a model. What we are talking about, then, is what Plato was talking about with his anamnesis in the *Meno* and what Aristotle was talking about when he said mind grasps forms in images. We are considering different cases of double models, in this case a diagram of the free fall and symbols that represent the fall.

Next, we have to distinguish between explicit and virtual elements in the model. Explicit elements in the model are $x = ct$, $y = -\frac{1}{2} gt^2$. They are all explicit. But give those two equations to a mathematician, and he will start getting results out of them. He will eliminate the t from between the two equations, and he will get the equation for a parabola and conclude that this free fall is parabolic in shape. Those are virtual elements in the model. If you give those two equations to a mind that knows how to handle them, you will get many more symbols. They also belong to the model. They

---

28 A standard form of the general equation for a plane conic is

$$Ax^2 + 2Hxy + By^2 + 2Gx + 2Fy + C = 0.$$

The type of conic involved can be determined by the properties of certain combinations ('discriminants') of the capital letters.

belong to the model as its deductive consequences within a given mathematical set of assumptions. So we have not only double models (diagrams and symbols). There is also a distinction between what is explicitly presented in the symbols and what is added on to the symbols by someone who knows how to handle them. What is added on is the virtual part. Again, you can eliminate the t by saying that $t = x/c$ and so that $y = -\frac{1}{2} g (x^2/c^2)$. When you have y and x^2 in an equation just like that, you have a parabola. The mathematician will be able to tell you that. Those are virtual elements in the symbolism, what you can do with the symbols.

2.4 Symbolism as Both Technique and Model

Now we have to put the two elements together, technique and model.

On the one hand, there is the technique: start out from specified rules, and work without using your head. Working without using your head means that you do not introduce anything new. But what you do is still done by somebody intelligent, who sees what can be done at the next step; merely to gape at the symbols does not help. Not all problems are routine problems that you solve the way you do square roots, and then the symbolism is both model and technique. In other words, when you are proceeding in technical fashion there are many possibilities of transforming the symbols from which you start out, and the role of insight, of grasping form in matter, is to select the possible symbolic transformations that will prove significant. If you give two men a problem, they both may translate the problem into the same set of written symbols, but one may just gape at it and wonder what he will do next, while the other fellow may immediately write it all down. The fellow who writes it all down grasps the virtual elements in the symbols. He does not write down anything that is not in strict accord with the rules, but what he writes down leads to a solution. The other fellow can write a lot of things down too, but it just goes on and on without ever arriving at the solution. Both people may be observing all the rules, but one fellow gets something out of observing all the rules, and the other fellow does not. So our analysis of this symbolic process shows that, on the one hand, there is a technical aspect – you only make moves in accord with predefined rules – but, on the other hand, you have to know the game to know what moves to make. When you play chess, you can make only certain types of moves. But the fellow who wins is the fellow who makes the right ones in the situation created by the moves that the other person made.

2.5 Isomorphism

Now we come to the big point, which is called 'isomorphism': *isos* and *morphê*, equal form. The simplest way of introducing the notion of isomorphism to Scholastics is to talk about the analogy of proportion. The essence of A is to the existence of A as the essence of B is to the existence of B as the essence of C is to the existence of C: analogy of proportion. You have an isomorphism when you prolong this analogy of proportion. Instead of saying A is to B as C is to D, you can say A is to B is to C is to D is to E ... as P is to Q is to R is to S is to T ... You have an indefinitely prolonged analogy of proportion. The proportions may not be the same all along the line. It would not be strictly mathematical if they were not, of course, but the general idea is that of a network of relations. The thing to get hold of is the relations in between: R_1, R_2, R_3, R_4, R_5; you can call those[29] R_{11}, R_{12}, R_{13}, R_{14}, R_{15}. Then there are R_{21}, R_{22}, R_{23}, and so on. There is isomorphism insofar as these *relations* are similar. It does not make any difference what the matter is.

Here we reach the idea of the artificial form. What is an artificial form? Well, if you put bricks and stones and wood together in a certain way you get a house. But a house is not bricks or wood or stone. It is the order imposed upon bricks and wood and stone. That order is a sequence of relations, and in a housing project in which the company builds the same type of house, the same plan all along the street, you have an isomorphism in the structures. There are different materials all along; you do not use the same wood in two houses; you use different pieces of wood. But the structure is the same. And that idea of similarity of relational structure is what is meant by isomorphism.

That is a general statement of the idea. It is worked out differently in different contexts, and it should not be used simply as I gave it to you now. All I want to convey is the general idea. Whenever you are dealing with texts in mathematical logic, you must attend to just the precise way in which the different authors understand isomorphism. They distinguish between isomorphism and representation and interpretation, and so on, and the vocabulary will differ from one to the other.

Again, you may be considering relations of valid conclusions. If B represents a proposition, and C represents a proposition, and D represents a proposition, and so on, you will have isomorphism when, if A is true and as

---

29 'Those' would be the relations between A, B, C ..., and the next group would be between P, Q, R ...

a consequence *B, C, D, E*, ... are true, then if *P* is true, all of *Q, R, S, T*, ... will be true. This is an isomorphism in terms of valid inference. So 'isomorphism' may have all sorts of different meanings, but the general idea is an identity or similarity of relational structures. With that idea of similarity of relational structures, we can see that symbolic or mathematical logic exists.

For example, geometry can be done in the way Euclid did it, by putting down diagrams and working in terms of axioms to build up his *Elements*. The same geometry can be done all over again in Cartesian fashion: put down axes; call the horizontal *x* and the vertical *y*; measure off; and so on. All of geometry can be done over again on this basis. Since it is the same geometry in the two cases, these two types of reasoning out geometry are isomorphic. Cartesian geometry deals with equations, mathematical expressions. Euclidean geometry deals with images. But the results are the same in each case. It is the same relational structure between represented lines and points *and* algebraic equations. Moreover, that isomorphism can be shown. For example, David Hilbert, a major figure in mathematical logic, defines point and line by this expression: 'A straight line is determined by 2 and only 2 points.'[30] What is a point? A point is what two of will determine a straight line. What is a straight line? A straight line is what will be determined by two points. That is called implicit definition. The total statement defines two terms by their relations to one another, without committing you to any particular notion of point or line. Consequently, when you do Hilbert's geometry, you do not mean Euclid's point and Euclid's straight line, and you do not necessarily mean Descartes's point (*a, b*), an ordered pair, or Descartes's straight line. You mean what is common to both. You are operating in virtue of the isomorphism of intuitive geometry and algebraic geometry. The two have the same structure because they give the same results all along the line. Certain types of theorem are easier to prove in Euclid, other types are easier to prove in Descartes's geometry, but there is the same geometry in both cases. And if you can move back to a more general situation in which you use implicit definition, your straight line is

30 More fully, there are, for Euclidean plane geometry, three axioms of connection: (1) any two different points *A* and *B* determine a straight line *a*; (2) any straight line is determined by any two of its points; (3) on any straight line there are at least two different points; there are at least three points which are not on the same straight line. Hilbert's drive is nicely captured for the lay person in his remark of 1891, 'It must be possible to replace in all geometric statements the words *point, line, plane*, by *table, chair, mug*' (quoted in Constance Reid, *Hilbert* [New York: Springer-Verlag, 1970]) 264. See also note 14 of chapter 2, below, p. 45.

something more general than either Euclid's or Descartes's, your point is something more general than either, and it can be either one. You are exploiting the isomorphism by these implicit definitions.

Again, to take perhaps a simpler and more familiar example, if someone is doing physics and you open his book, what do you find? You find just mathematical equations. He is solving problems, and what is it? It is more mathematics. Why do you say he is doing physics? He seems to be doing mathematics all the time. It is because there are regions of mathematics that are isomorphic with physical reality. There is the same relational structure between a given mathematical theory or system as there is between events that can be observed.[31] This is another case, a big case, of isomorphism: on the one hand, mathematical expressions, and on the other hand, physical events. There is the same relational structure. But in the mathematical case, the relational structure links symbolic expressions, or mathematical concepts, with one another, while in the physical case what are related are concrete physical events, wave-lengths that you observe through a machine, and so on.

So there is an isomorphism of geometry, algebra, physics; the same relational structure can be found in all three. Consequently, one's symbolism can be given a geometrical interpretation, or an algebraic interpretation, or a physical interpretation. You may be studying along with an equation like this: $x^2 + y^2 = c^2$. You can say that you are studying this second-degree equation in two variables, and so you are studying algebra. But you can be doing the same operations with the equation and be studying the circle, because that is what that expression means in analytic geometry. Or, with the same symbols, you may be dealing with a body moving around in a central field of force in a perfect circle. You are studying the movement, where the initial velocity is just right, so that the movement will not be an ellipse. In each case, you are dealing with the same concrete symbols, which can be interpreted algebraically, geometrically, and physically. All you have to do is add one step.

Geometry is deductive, and so your symbols will represent the deductions in geometry. Mathematics is deductive, and so your symbols will represent

31 *Insight*, chapter 5, brings out Lonergan's classic isomorphism regarding the geometry of space-time. The past forty years of particle physics enrich that perspective. See Lochlainn O'Raifeartaigh's two works, *The Dawning of Gauge Theory* (Princeton: Princeton University Press, 1997) and *Group Structure of Gauge Theories* (Cambridge and New York: Cambridge University Press, 1986).

the deductions in algebra. Physics, once you know the laws, is deductive – finding the laws is inductive, but once you know them physics is deductive – and so manipulating the symbols will represent the deductive element in physics. Therefore, de facto these symbols are representing logical relations: representing the logical relations of geometry, representing the logical relations of mathematics, representing the logical relations of physics. So why not use them to represent logic as such? There is no reason why the symbols should be used merely to represent geometrical deductions, algebraic deductions, physical deductions; they can represent deductions *überhaupt.* When that happens, one is in symbolic logic. When one says, 'Let the symbols stand for logical relations, logical operations,' then one has introduced a new term in the isomorphism. The same symbol has not only a geometrical, an algebraic, and a physical interpretation; the same images that one looks at, the same images to which one can add on in terms of technical rules, and add on intelligently the way one gets solutions – those images can correspond to geometrical concepts, algebraic concepts, physical concepts, but they can correspond also to logical concepts. When symbols are taken according to this fourth interpretation, one is doing mathematical logic, symbolic logic. So the notion of isomorphism, of structural relational similarities, similarity of relational structures, enables us to see that manipulating symbols intelligently in accord with fixed rules is a way of studying logical relations, deductions, implications, and all the rest. Just as a mathematician will invent a symbolism that meets his own purposes, so the mathematical logician will invent symbolism that will satisfy him and help him to solve his problems in logic.

2.6 Elements of Mathematical Logic

That is the idea of mathematical logic. Mathematical logic is the investigation of the field of logical relations through the development of suitable symbolic techniques.[32] The symbolism can be interpreted in terms of logical relations, just as well as in terms of geometry, algebra, or physics.

Now mathematical logic has developed along different lines. The three main types of symbolism (fundamental symbols) are the Russell-Whitehead type, the Hilbert type, and the Polish type, started by Łukasiewicz.[33] In

---

32 Lonergan did not actually give this definition in his lecture. It is reproduced here from his distributed notes. See below, p. 146.

33 On varieties of symbolisms, see note 11 on p. 166.

general, if you are going to get into the subject, you have to be able to move from one to the other, and not be put off. The symbols are fairly simple in any case. Prior, in his *Formal Logic,* has an appendix in which he outlines thirteen systems.[34] Some of them are equivalent, and some are not. By arranging the symbols in different ways, we obtain different logics. Prior enumerates thirteen different systems, and to those thirteen we can add another type, combinatory logic – Henry Curry is the main figure here – in which the fundamental properties of symbolism are explored. Here the concern is not immediately with logic but with the fundamental possibilities and properties of symbolism as such. There is an outline of combinatory logic in Feys, 'La technique de la logique combinatoire.'[35] There is also a much briefer outline in an appendix to Fitch's *Symbolic Logic.*[36] However, we will become concerned with that considerably later.

Consider, too, the plurality of systems. There is what is called the Classical Propositional Calculus. It starts off with two fundamental notions: *not* and *or,* where *or* means *not neither.* In other words, just two simple notions like that, the negation of a sentence and a connective between sentences, give a system. There is also a system based on *and,* and there is a system based on *if-then* in a peculiar sense. So the Classical Propositional Calculus is worked out from different sets of axioms in different ways. But in general, they are all equivalent. On the other hand, Lewis's modal system adds strict implication, implication in the ordinary sense of *if-then.*[37] There is an intuitionistic logic connected with Brouwer in mathematics, which omits the principle of excluded middle: the principle of excluded middle is not built into the

34 See the reference in note 17, above. The revised appendix of later editions (pp. 301–17, 'Postulate Sets for Logical Calculi') contains twelve systems.

35 R. Feys, 'La technique de la logique combinatoire,' *Revue philosophique de Louvain* 44 (1946) 74–103, 237–70.

36 Frederic Brenton Fitch, *Symbolic Logic: An Introduction* (New York: Ronald Press, 1952).

37 See C.I. Lewis, *A Survey of Symbolic Logic* (Berkeley: University of California Press, 1918) chapter 5, pp. 291–339, 'The System of Strict Implication.' Lewis had worked on it during the decade but this is a 'complete system ... not previously printed' (291, note). See also note 22 of chapter 3 below, p. 83. A 1960 republication (New York: Dover) drops chapter 5 (and chapter 6, 'Symbolic Logic, Logistic, and Mathematical Method'). The material of chapter 5 was reworked in Clarence Irving Lewis and Cooper Harold Langford, *Symbolic Logic* (New York: Dover, 1959) chapter 5 and appendices 2 and 3. See Lewis's preface to the 1960 Dover edition.

The 'peculiar sense' of 'if *p* then *q*' is best captured for the nonprofessional in the translation, '*p* not without *q.*' It ceases to be peculiar, however, when located within a consideration of formal mathematical structures.

system.[38] There are three-valued logics. They do not consider propositions merely as true or false; they add that there are some propositions that are neither, such as those related to the contingent future. As Aristotle says, it is neither true nor false that there will be a naval battle tomorrow.[39] And on that basis, Thomas discusses the problem of God's foreknowledge.[40] The proposition is not only true or false but also neither – at least at the present moment.

In general, the masters in the field are not tied down to any particular system. If they have a problem, they go ahead and invent the symbolism that will be apt for handling this problem, and they will explain what their symbolism is before they start out and then start using it. However, one must be familiar with systems in general to be able to see what they are doing.

So there is a multiplicity of systems, and there are all sorts of possibilities of further development. And the Scholastic will ask, 'Well, which one is

38 Chapter 18 of Constance Reid's *Hilbert* gives a historical context. Hilbert recognized the Intuitionist program as the rising of the ghost of Kronecker (see Ewald, *From Kant to Hilbert*, vol. 2, pp. 942–55). For an introductory perspective on the movement, see *Philosophy of Mathematics: Selected Readings*, ed. Paul Benacerraf and Hilary Putnam, 2nd ed. (New York: Cambridge University Press, 1983; this part of the volume was not changed from the first, 1964, edition), articles by Heyting and Brouwer. J.J. de Jongh remarks, 'The most important advantage of intuitionist mathematics is that it distinguishes in every instance between directly and indirectly proved propositions and analyses the mathematical concepts into sequences of concepts with different degrees of indirectness.' J.J. de Jongh, 'Restricted Forms of Intuitionistic Mathematics,' *Proceedings of the Tenth International Congress of Philosophy* (Amsterdam: North-Holland Publishing Co., 1949) 746. According to its editor, chapter 4 of the work cited in note 34 of chapter 2 below (p. 57) 'contains one of the most lucid accounts of the Brouwer viewpoint' (ibid. 18). See also the following note, and notes 16 and 17 of chapter 2 below, pp. 45 and 46.

39 Aristotle, *Peri hermeneias*, 19a 27. Stephen Cole Kleene, *Introduction to Metamathematics* (New York: D. Van Nostrand Co., 1967) 317–55, places considerations of three-valued logic in a context relevant to the development of Lonergan's perspective, the context of 'Partial Recursive Functions' (the title of chapter 12), Church's thesis (see quotation below in this note), and Gödel numbers. Later in the work Kleene arrives at a result regarding 'recursive realizability': 'the result provides a connection between Brouwer's logic as formalized by Heyting and Church's thesis that only general recursive functions are effectively calculable. Both developments arise from a constructivist standpoint, but were previously unrelated in their details' (516).

40 Thomas Aquinas, *In X Peri hermeneias*, lect. 14, §§ 20–24.

right?'[41] This multiplicity does not come out clearly in most introductory books of symbolic or mathematical logic. But it is a point of extreme significance to us because it brings to light for us things that we would not otherwise suspect, matters that the symbolic logicians themselves never suspected when they set out on their enterprise. It also reveals limitations to their undertaking that provide us with an opening. In other words, it provides a means of leading them on to something that, from the philosophic viewpoint, most of them would not think of acknowledging. The matter is in itself extremely complex and difficult, and all I am going to attempt to give you is a general outline, an indication, some insight into the sort of thing that happens.[42]

41 The final words in this chapter were actually the opening words in the next lecture.
42 See the quotation from L.G. Kattsoff with which Lonergan concluded his lecture notes for this first lecture. See below, p. 147.

2

The Development and Limits of Mathematical Logic[1]

1 The Pursuit of an Ideal

The development of mathematical logic has been the pursuit of an ideal. Mathematical logicians have been aiming at rigorously hypothetico-deductive systems from minimal suppositions, and their key test has been the possibility of deducing the whole of mathematics. In other words, they considered the problem of studying deductive systems not in the abstract but as a problem whose solution all along the line has been tested by its capacity to deduce mathematics in the form of a single grand deduction. The deductive ideal is put to the test.

2 Logical Formalization

This ideal has been formulated as an axiomatic system, a logical formalization. You will find under '2' in the notes a few indications of the general idea of such an ideal.[2] It becomes more precise, takes on different forms, and so on, as we get more closely into the matter. It is important to remember that I am not trying to give you a basis for arguing with these people or even for discussing exactly what they mean and what they say. I am giving you only an indication that I hope will be sufficient for you to understand the general idea.

1 The second lecture, Tuesday, 9 July 1957. The opening words of the lecture have been moved to the end of the first chapter.
2 See below, pp. 147–48.

What determined the hypothetico-deductive approach in mathematics was the discovery that Euclid was neither rigorous nor necessary. Those who advocate this approach do not start from self-evident truths, because that is what Euclid did, and he came a cropper. Rather, they distinguish terms and propositions into two classes: derived and non-derived. Terms are either derived from other terms or they are not. Propositions are either derived from other propositions or they are not. The non-derived are named primitive.

Derived terms are defined by primitive terms, either directly or indirectly: directly, if they can be defined solely by primitive terms; indirectly, if they can be defined by primitive terms and other derived terms that already have been defined. Similarly, derived propositions are deduced from primitive propositions, either directly or indirectly: directly, if the deduction is immediately from primitive propositions; indirectly, if the deduction is from propositions that already have been deduced. Finally – and this is where the symbolic, technical aspect on which we spent so much time yesterday comes in – rules of derivation must be stated explicitly. No derivations are admitted except in accord with the stated rules, and all derivations that would occur in accord with the stated rules have to be admitted. In other words, you can use your rules just as far as you please, but someone else is perfectly entitled to go on and use them right to the limit, even though that makes nonsense of your position. You must commit yourself to everything that your primitive terms, propositions, and rules of derivation imply.

So much for the general idea.

Certain technical terms are employed, and I note here that the definitions that follow will be given different twists in different presentations and in different attempts. But let us use the letters P and Q to denote two Logical Formalizations, two LFs, where a logical formalization means a set of primitive terms and propositions, a set of rules of derivation, and everything that follows from them. We have two formalizations, and we name them P and Q. Then P and Q are said to be equivalent systems, equivalent LFs, if what is primitive in P can be derived in Q and what is primitive in Q can be derived in P. Each one has everything that the other has, the two are equivalent, if from either you can obtain the whole of the other. If you can obtain the primitives of P in Q, you can obtain the whole of P in Q. Similarly, if you can obtain the primitives of Q in P, you can obtain the whole of Q in P.

Now, consider the statement that p denotes any proposition that can be constructed in the system. That may seem a very vague expression in terms of ordinary language, but when you set up a logical formalization, you begin by listing all the terms (including the commas) that you are going to use in the system. You have a complete list of all the terms that occur in the system, and you are supposed to have full rules for derivation. Thus, what can be constructed in a system is settled by the starting point, and the starting point has to be complete enough to settle all questions that arise. So 'p is any proposition that can be constructed in the system' has a definite meaning. That definition of completeness and coherence is weakened in various attempts, but this will do for our purposes. Again, I am here suggesting a general indication of what they are doing; I am not giving you a basis from which you can step into this field and pronounce conclusions and draw deductions and say what must be – nothing like that!

If p is any proposition that can be constructed in the system, then the system is consistent or coherent if you cannot prove both p and its contradictory.[3] Again, the system is complete if for any proposition p you can prove either p or not-p. The whole system has to come out: you have to be able to handle all propositions that pertain to the system on the basis from which you started out.

The last two characteristics listed under '2' are less important:[4] they express more artistic than strictly logical concerns. Thus, the primitive propositions of P are independent if no one can be derived from the others. The primitive propositions of P are elegant if they provide the shortest way of reaching all the propositions of the system.[5]

3   Principal Lines of Endeavor

We have expressed an ideal, and the question arises, Can it work? Does it work? The mathematical logicians found that it works for certain fields, and they can demonstrate that it will work for those fields. And they have found that it does not work for other cases. The development of mathematical logic, the question we are trying to get something across on today, regards just where it works and why, and where it does not work and why. Their control has been in deducing mathematics, and there have been a variety of

---

3 With a reference to the notes he had distributed, Lonergan added, 'Np is the Polish notation for the contradictory of p.' See below, p. 166, note 11.
4 See below, p. 148.
5 Or as it is put in the notes below, ibid.: 'The primitive propositions of P are elegant if they offer the simplest basis for deriving in the simplest manner all the propositions of P.'

lines of endeavor. It is worth while knowing something about these different lines of endeavor, both because they have led to different logical systems or different logical attempts and because we are apt to think that, after all, all mathematicians agree about everything. It is well to see that they do not, that there are differences among them.

3.1 Axiomatic Set Theory

A first line of development is called axiomatic set theory. It is pretty well outside the logical field. It is an attempt to deduce the whole of mathematics from what is most advanced in mathematics, and it works out quite smoothly. It is associated with the names Zermelo, Fraenkel, and von Neumann, and it is not of too much interest to logicians.[6] Of course, there are people who object because of difficulties in set theory. But this stream of thought has existed, and it has been of less interest from the logical point of view because it has tended to be purely mathematics.

3.2 Whitehead-Russell

The next line of development is associated with Whitehead and Russell's *Principia mathematica*, which first came out around 1911. There was a second edition around 1927. I think there has been a reprint of the second edition in this decade, in the 1950s.[7] It aims to base the whole of mathematics on logical axioms: start out from logical axioms and do the whole of mathematics on that basis. Mathematics is viewed as a department of logic. The work provided a magnificent unitary view of the whole of mathematics, and it remains one of the principal directions. The work of Quine, for example, who is a major figure in mathematical logic, has generally been along the lines of Whitehead-Russell.

However, not everyone was satisfied with this work. In the first place, not all the axioms were logical. There was in both editions an axiom of infinity. If you are going to deal with numbers, you have to deal with infinities, and it

6 In the more than forty years since Lonergan's lectures, interest has increased considerably. For a context, see Stewart Shapiro, *The Limits of Logic: Higher-order Logic and the Löwenheim-Skolem Theorem* (Aldershot, Eng.: Dartmouth Publishing Co., 1996).
7 Alfred North Whitehead and Bertrand Russell, *Principia mathematica*, 3 vols. (Cambridge: Cambridge University Press, first ed., 1910–13; 2nd ed., 1925–27; regularly reprinted, including at least twice in the 1950s [1950, 1957]).

is not a strictly logical axiom that deals with properties that are not found in any finite set, any finite group. For that consideration, an axiom of infinity is needed. Whitehead and Russell needed a not-strictly-logical axiom of infinity.[8]

There was a more fundamental difficulty connected with the class of classes that do not contain themselves. 'The class of classes that does not contain itself' leads immediately to a contradiction because if you say that it belongs to that class, then it should not, and if it does not, it should. It steps immediately into a contradiction. It is a case of a paradox. It is what I spoke of yesterday in discussing the paradoxes. It is a primary instance of the paradoxes, and it is one that was discovered as soon as people started on the way of logical formalization.

Russell's first attempt to handle that paradox was to introduce a theory of types. The theory of types was a stratification of attributes.[9] However, by introducing the theory of types, he ran into a difficulty with the mathematics. There is a famous basic definition of real numbers provided by Dedekind, and introducing the theory of types precluded the use of Dedekind's definition of real numbers.[10] So Russell had to introduce another axiom to be able to reinstate Dedekind's definition while he was also introducing the

---

8   Such an axiom would be, 'There exists a set containing o and containing the successor of each of its elements.'

9   The lowest type consists of all individuals. The next level of logical types consists of all attributes of individuals, the next of all attributes of attributes of individuals, and so on. Type i can be predicated of type j only if $i = j + 1$. Ramsey, who was the first (1926) to distinguish logical paradoxes and semantic paradoxes explicitly, split, ramified, each *type* into different *orders* depending on how quantifiers occur in the propositional functions. See the reference below at note 12.

10 Briefly, Dedekind defines a set of real numbers as a set of sets of rational numbers, which set is infinite and cannot be specified except by mentioning a propositional function that all the members satisfy, and the relevant functional expression (it involves specifying the least upper bound of an infinite set of real numbers) involves reference to the totality of propositional functions that specify real numbers, *including that function which it expresses.* I am compacting here the discussion of William Kneale and Martha Kneale, *The Development of Logic* (Oxford: Clarendon Press, 1978) 662. The section there, pp. 657–72, gives a good account of the fate of the theory of types. However, to get beyond this or Lonergan's compact expression, a problem that haunts this volume, requires detailed mathematical and logical work. One might start with Lonergan's favorite irrational, root 2. One can begin to define it with the sequence of rationals 1, 14/10, 141/100, 1414/1000, 14142/10000 ..., and ask about the regularity and the convergence of the sequence, and so on.

theory of types. The other axiom was called the axiom of reducibility: namely, attributes on any level can be re-expressed in terms of attributes on a lower level.[11] This was a way of getting around the difficulty, but the axiom of reducibility seemed neither very evident to anyone nor very satisfactory, and the result was that, in the second edition, Russell tried a different approach.[12] A man by the name of Ramsey had worked out two approaches, breaking down the theory of types into a more simple theory of types and a more complex theory. The simple theory of types solves difficulties that are named syntactical, that is, internal difficulties of the system. It eliminates the internal paradoxes, but it does not eliminate what they call the semantical paradoxes, that is, the difficulties that arise when you start considering what this system means.[13] In the second edition, Whitehead and Russell took the step of using the simplified theory of types, dropping the axiom of reducibility, and not worrying too much about the semantical difficulties. People continued to work along this 'Russell line' after the second edition.

So this was a major attempt to work out the whole of mathematics on a purely logical basis, and it ran into some snags.

11 Briefly and descriptively, the axiom claims that to any function of any type there corresponds a formally equivalent first-level function. Two functions are formally equivalent when, for any admissible argument, they are either both true or both false. A popular notion may help. Consider the equivalence of 'Mary had all the qualities of a leader' and 'Mary was intelligent, inspiring, compassionate, quick to respond adequately ...'

12 The second edition adds an 'Introduction to the Second Edition,' pp. xiii-xlvi, but states (xiii), 'In preparing this new edition of *Principia mathematica* the authors have thought it best to leave the text unchanged ... even where they were aware of possible improvements ... It seemed preferable ... to state in an introduction the main improvements which appear desirable.' In a note on the same page Mr F.P. Ramsey is acknowledged as contributing 'valuable criticisms and suggestions.' *12 (pp. 161–67) deals with the hierarchy of types and the axiom of reducibility, but revisions are matter for the two appendices. 'The new chapter *8 (given in Appendix A) should replace *9' (p. xxv). Appendix B (pp. 650–58) begins with the remark, 'The difficulties which arise in connection with mathematical induction when the axiom of reducibility is rejected have been explained in the Introduction to the present edition.' That explanation appears on p. xiv, where it is also acknowledged that 'it cannot be said that a satisfactory solution is as yet obtainable.'

13 A quite readable account of the matter is appendix C, 'The Ramified Theory of Types,' in Irving M. Copi, *Symbolic Logic* (New York: Macmillan, 1979) 344–53. Semantic paradoxes are exemplified by the Grelling Paradox: the adjective 'short' is short; the adjective 'long' is not long. If *heterological* designates the attribute a word has when it designates an attribute that it does not have itself, is the adjective 'heterological' heterological?

3.3 *Hilbert's Two-level Approach*

David Hilbert tried another approach. While he did not disagree with the Whitehead-Russell proposal of deducing the whole of mathematics from a logical foundation, still he went about the matter in a different way. He distinguished two levels: on one level you do mathematics, and on another level you work out the logical properties of mathematics. The logical property of mathematics in which he was particularly interested was the consistency of all propositions in the mathematical system. Consistency, noncontradiction, was taken by Hilbert as the one thing necessary to establish satisfactorily a mathematics.

Others would object to that criterion, notably at the present time the intuitionistic school. But that was Hilbert's idea: if we can get the whole thing consistent, we will have all we want. But, he insisted, we must not mix up logic and mathematics. We will do the mathematics separately on a strictly mathematical basis, and then after we have it all done, we will consider the logical properties of the system. So on one level there is mathematics, and on another level there is meta-mathematics, in which the logical properties of the mathematical performance are considered.

The reason for Hilbert's distinction was that, while the mathematics would use infinities of objects and infinities of operations, the logic would be on a strictly finite basis. Hilbert felt that it is hard to make things so evident that no one can possibly quibble over anything. But the best way is to put logic on a strictly finite basis. If in the logical part all consideration of infinities of objects and infinities of operations is excluded, then there is something that can be studied in order to see if it really works, if it really holds, or if it does not. If the logical part is established in a finite way, there is also established the consistency of logical properties in the mathematics, even though the mathematics itself has to involve infinities of objects and infinities of operations.

The results of Hilbert's proposal have been most interesting. As a matter of fact, they form the basis of the remainder of our discussion today.

We will distinguish the results into short-term and long-term results. The short-term results have to do with both geometry and arithmetic. Hilbert found, first of all, that his proposals worked splendidly for geometry. He was able to propose a set of axioms for geometry and to verify his axioms intuitively in a model. The model, however, does suppose the validity of the counting numbers, so if there is a difficulty about the counting numbers, there will be a bit of a scruple about the geometry. But in general it was felt that Hilbert had put geometry on its feet.[14]

Then he turned to arithmetic. By 'arithmetic,' of course, we mean not simply the performance of arithmetical operations, but the theory that justifies them, not merely taking square roots but also the theory that explains why it works. Here it was especially the theory of prime numbers, the fundamental things in arithmetic – the resolution of any number into its prime factors, the unicity of such resolution, the theory of division – that led to difficulties. When providing axioms for arithmetic, Hilbert found that he could establish the consistency on a finite basis if he omitted some of the axioms or if he weakened all of the axioms. But it did not seem possible to do this for the axioms that were needed to base an arithmetic.

Such were the short-term results. The long-term results are what we come to in section 4.[15] They show the fruitfulness of Hilbert's proposal to separate mathematics, done with an infinity of objects and an infinity of operations, from the meta-mathematics, the logic of the mathematics, done on a strictly finite basis, that is, involving a finite number of objects and a finite number of operations.

3.4   The Intuitionist School

As far as the mathematics went, both Russell-Whitehead and Hilbert held for the same sort of mathematics. However, there existed mainly in Holland a group known as the intuitionistic school, led mainly by Brouwer. In the logical field the chief exponent is another mathematician, a member of that school, Heyting.[16] They insist that mathematical logic is only a tool; it is

14 See note 30 on p. 32 above, and chapter 8 in the work of Constance Reid cited there. This was the achievement of Hilbert's lectures of 1898–99, *Grundlagen der Geometrie*; there is a 'sole authorized' English translation (*The Foundations of Geometry*) from the tenth edition by Leo Unger (La Salle, IL: Open Court, 1971, 3rd printing, 1987). See Arnold Schmidt, 'Zu Hilberts Grundlegung der Geometrie,' in *Hilbert. Gesammelte Abhandlungen*, vol. 2 (Berlin: Springer-Verlag, 1933; reprinted, Bronx, NY: Chelsea Publishing Co., 1965) 404–14.
15 See pp. 49 and 149, below.
16 See A. Heyting, *Intuitionism: An Introduction* (Amsterdam: North-Holland, 1956, with revisions in 1966 and further revisions in 1971). The book contains a substantial bibliography, updated in the later editions. Useful also is A. Heyting, ed., *Constructivity in Mathematics: Proceedings of the Colloquium Held at Amsterdam, 1957* (Amsterdam: North-Holland, 1959).

not the whole show. For Hilbert, for example, what is mathematics, what are the mathematical entities? They are the same as the symbols. There is not much of a distinction between what one is doing and the mind that does it. That type of distinction is insisted upon in the intuitionistic school. For them a logical formalization is only a tool. It can be a highly useful tool; it unifies isomorphic fields in mathematics; they concede all that to it. However, it is still a tool. Moreover, mathematics is essentially constructive. The business of the mathematician is not simply to talk in a logical fashion about mathematical entities, but to construct, that is, to show a manner in which you can arrive at the mathematical entities of which you are talking.

Because they regard mathematics as essentially constructive, they object to the indiscriminate use of the principle of excluded middle. What concretely does that mean?

Well, if you have some mathematical property D that you know about and can work on, establish, study, then if you use the principle of excluded middle you can talk about all the mathematical entities that are not-D and start discussing them as well. But according to Brouwer at that point you are just wasting your time. You are not doing mathematics at all in talking about these unknown entities that have the property 'not-D.' In other words, he refuses the automatic use of the principle of excluded middle.[17]

The fundamental point that they are raising has to do with the distinction that we make between *quid sit* and *an sit*.[18] If you are concerned simply with certitude, with necessary consequence from axioms, then all you are doing is trying to provide a check on the *truth* of your mathematics. And if you make your mathematics simply a consideration of truths of mathematics and do not do the intelligent work of determining *quid sit* in each particular case insofar as you can, then there is a tendency not to do mathematics at all. At least, that is the feeling of Brouwer, Heyting, and members of this school. It is not a large school, and the reason is that a rigorous application of their principles leads to lopping off large sections of classical mathemat-

---

17 Technical treatment of the principle of excluded middle by both Brouwer and Heyting (mainly by Brouwer, but with reference to Heyting's contributions) is available in Jean van Heijenoort, ed., *From Frege to Gödel: A Source Book in Mathematical Logic, 1879–1931* (Cambridge, MA: Harvard University Press, 1967, 1981).

18 The distinction is fundamental to Lonergan's perspective on human inquiry. Its historical grounding can best be appreciated from the directions of the first two chapters of *Verbum: Word and Idea in Aquinas* (see above, p. 13, note 14). Its psychological ground can be reached elementarily through part 1 of *Insight.* See the diagram below, pp. 322–23.

ics. The mathematicians do not want to let that go; they want to hang on to classical mathematics. For that reason the intuitionists have not had too much influence.[19] However, there is an influence that is seeping in, and it appears in accepting weakened axioms on various occasions.

3.5 *Gonseth and* Dialectica

In Switzerland, F. Gonseth founded or at least is the main name in an international union for the study of the philosophy of science. They publish a review, *Dialectica*.[20] Gonseth's approach is that mathematical logic is just a Euclidean avatar. Mathematics was done in this axiomatic fashion up to the time when mathematicians discovered something was wrong with Euclid. But some of them are trying to do the same thing still, and that is a mistake. The experience of the deficiencies in Euclid should have taught them not to rely so much on deduction. That is Gonseth's approach. In other words, he thinks there is something outdated about the attempt to deduce everything from strictly determined premises.

Gonseth and the people in this school are, one might say, more familiar than other schools with the general philosophic tradition. Gonseth himself speaks like an outright relativist, but the fact that philosophically he is a relativist does not mean that everything he says is going to be useless to us or that everything in his tendencies is going to be wrong. What he wants is a mathematics that develops. Mathematics has kept on developing, particularly in the last few centuries; there have been tremendous developments in mathematics. The developments arise not solely within mathematics itself but through the interaction between mathematics and the sciences, between mathematics and philosophy, between mathematics and general cultural movements. The idea that a mathematics has to be purely deductive, the attempt to set up a rigidly deductive system, is itself not a mathematical notion. It is just a methodological notion that happens to exist in the minds of certain groups of mathematicians who were not too up-to-date

19 Lonergan comments further on intuitionism in an interview of August 1978, with Paul Manning, 'Lonergan's Studies in Mathematics,' Toronto Lonergan Research Institute, File 947.

20 *Dialectica: International Review of Philosophy of Knowledge* is a Swiss journal founded in 1947 by Gaston Bachelard, Paul Bernays, and Ferdinand Gonseth. For a brief presentation of Gonseth's perspective, see F. Gonseth, 'Le problème de la méthode en philosophie,' *Proceedings of the XIth International Congress of Philosophy*, vol. 1: *Theory of Philosophy* (Amsterdam: North-Holland, 1953) 64–72.

philosophically and culturally. That is Gonseth's position, and the fact that he is able to publish a review and hold these international meetings – he has had at least two of them – shows that he has something of a following. Just how big it is and how influential, I do not know, but it is also not too important from the viewpoint of our present issue, simply because he does not think too much of this logical deductivism. He admits that it is splendid as far as it goes, but he is not putting all his eggs in that basket by a long shot.

3.6 The Bourbaki Group

In the sixth place, there is what is called the Bourbaki group. They sign their name in various mathematical articles, N. Bourbaki. You can find out something about them in an article by Ladrière. There was a meeting in Paris, in October of 1949, of members of these various schools, an international congress on the philosophy of science. The Gonseth group was represented, the Bourbaki group was represented, the Intuitionistic group was represented, and probably some others. An article by Jean Ladrière in *Revue philosophique de Louvain*, 1950, describes what went on at this meeting.[21] The Bourbaki group, represented by Cartan and Dieudonné, laid down the law from their viewpoint very clearly. First of all, no special parts of mathematics are of superior importance, as is implied by the intuitionistic school. In the second place, there are no scruples about the problems that caused people trouble at the start of the century with regard to contradictions in set theory. Again, all this business of meta-mathematics is all right for meta-mathematicians but not for mathematicians. Further, what is a mathematician's job? It is to work out the totality of properties contained in one's axioms. In other words, the Bourbaki people are high, dry, formal, closed in; and they came under attack, pretty well from all angles.

The general objection against them – and I think this would be particularly agreed to by Ladrière – was that there was no provision whatever in their position for any further development of mathematics. As someone said to either Dieudonné or Cartan, 'There would be very little to formalize if it had not been for the work of Lebesgue and Borel and Baire (major French mathematicians at the beginning of the century), and if we listen to

21 Jean Ladrière, 'Le Congrès international de philosophie des sciences,' *Revue Philosophique de Louvain* 48 (1950) 102–19. Lonergan added, 'We will be mentioning Ladrière again immediately,' and he referred to section 4 of the distributed notes; see below, p. 149).

you, we will have nothing more to formalize than already has been discovered. You are just staying within the logical consequences of a set of axioms that happen to exist at the present time. Just as mathematics has developed in the past, there is every reason to believe that it will keep on developing in the future. We are interested in that development as mathematicians. Your strictly deductive viewpoint is static. It fixes mathematics at a particular stage.'

So much, then, for a general outline of the tendencies in mathematics, tendencies to go right out into this logical movement, and on the other hand, tendencies to hold it at a distance, to make use of it insofar as it is helpful, and so on. The thing breaks down into several parts.

4   Gödelian Limitations

We now come to the extremely interesting point that I put under the title of Gödelian limitations. The book to get hold of on the subject is Jean Ladrière, *Les limitations internes des formalismes*, published both at Louvain and Paris this year.[22] I just saw it three or four days before leaving Rome, and for me it was a God-send. Before that I had been trying to piece this stuff together from various sources, and finally there came along a specialist in the field who pieced it all together for me and who did so with an authority that I could not attempt to aim at. Ladrière wrote the book and presented it along with a defense of fifty theses, to become a Maître agregé in the school of St Thomas at Louvain. He is *Licencié en mathématique,* and he also has a degree in philosophy.[23] His fifty theses roam through the whole of philosophy. There is a great deal of existentialism, or at least phenomenology, in his tone, in the way that the theses are presented. He is very up-to-date philosophically, very competent, and as far as I know – I do not know really but I have every reason to believe – he is completely competent on the mathematical side. His aim in the book is to present the whole question of Gödelian limitations in a way that a philosopher without special mathemati-

---

22 Jean Ladrière, *Les limitations internes des formalismes: Étude sur la signification du théorème de Gödel et des théorèmes apparentés dans la théorie des fondements des mathématiques* (Louvain: Nauwelaerts, and Paris: Gauthier-Villars, 1957). There is a recent edition in the series 'Les grands classiques Gauthier-Villars' (Sceaux: Jacques Gabay, 1992) with a complementary (and second) list of additions and corrections.

23 The following information appears after Ladrière's name on the title page: Docteur en philosophie, Licencié en sciences mathématiques, Maître agregé de l'Institut Supérieur de Philosophie.

cal training might follow. The reader needs an awful lot of good will, of course, and an ability to get down and learn particular things that he may not know, as the demand arises, but Ladrière's aim is to give you everything you need to be able to follow through this whole question of Gödelian limitations. The more mathematics you know, the easier it will be to read this book. And even if you do not know mathematics, you will need to have what I might call a mathematical mentality, an ability to follow complex trains of arguments that are expressed symbolically, if you are to be able to profit from the book. However, even for a person who does not know too much mathematics there is a good deal to be gotten out of it. For example, the bibliography is a complete bibliography on the point, and it goes up to 1956.

On bibliography I refer you to *The Journal of Symbolic Logic.* The first volume contains a complete bibliography of everything written in symbolic logic prior to 1936. The bibliography was prepared by Alonzo Church, the mathematical logician at Princeton, who has recently published his *Introduction to Mathematical Logic,* the first volume.[24] In the third number of the same review there is a supplement to that first bibliography, and in the subsequent numbers there is a complete bibliography of everything that appears in the field. Symbolic logic or mathematical logic is a field in which complete bibliographies of everything that has appeared at any time on the subject are available. But that is a little bit too much to swallow. A selective bibliography, up to about 1948, is available in one of the earlier fascicles, the third I think, of Bochenski's *Bibliographische Einführung in das Studium der Philosophie* (a bibliographical introduction to the study of philosophy).[25] It is a straightforward bibliography, and it is a systematic, selective bibliography. You may be thrown off by some of the abbreviations in that third fascicle, but if you go to the first fascicle, which has a list of all the general aids to studying philosophy – encyclopedias, periodicals, and so on – you will find explained any abbreviations that are used in the later numbers. Incidentally, in that same collection there is a very good bibliography of everything on Kierkegaard and on the French existentialists up to about

---

24 On Church's *Introduction,* vol. 1, see above, chapter 1, p. 19, note 19. Church's 'A Bibliography of Symbolic Logic' spans pp. 121–216 of *The Journal of Symbolic Logic* 1–2 (1936–37). Additions and corrections appear in the same journal 3–4 (1938–39) 178–92.

25 I.M. Bochenski, *Bibliographische Einführung in das Studium der Philosophie,* fasc. 3, E.W. Beth, *Mathematische Logik und Grundlegung der exakten Wissenschaften* (Bern: A. Francke, 1948).

1950, done by the head of the Catholic Institute of Lyons, Jolivet.[26] In the number on mathematical logic, you have selective bibliographies of the principal works prior to 1948, and the bibliographies are done by quite good people. They are very good, and they are selective; they do not give you everything, but they give you the principal things. For materials written after 1948 on the Gödelian limitations, you will find everything in Ladrière, who has a very full bibliography on the subject.[27] Also useful for finding the field, the present situation, on these things are the international congresses of philosophy. There was an international congress of philosophy at Amsterdam in 1948, one in Brussels in 1953, and there are other meetings of the various groups. When you get a number of people coming together, and each of them throwing out his little say, you begin to get a general idea of what the situation is at the time, and it provides you with leads around the names, and so on.

But we now come to what is more or less the sixty-four-thousand dollar question in this field, the Gödelian limitations.

The general fact is that there have been demonstrated a series of theorems setting limitations to the possibility of reaching the ideal of a rigorously deductive logical-mathematical system. The general form of the argument in such cases is approximately as follows.

4.1 Two Logical Formalizations

First of all, we consider any sequence of symbols – you will find the sequences of symbols in any book of logic – and we will call it an LF, a Logical Formalization. It is the sequence of symbols, what you see, p's and q's and x's and y's, signs for every quantifier, and so on, starting off from axioms, going on through derivation, and giving you the whole business. The discussion presupposes such a thing, a symbolic technique capable of representing a manifold of deductive sequences.[28] Moreover it admits, either in exactly the same symbols or in sufficiently analogous symbols, two sufficiently parallel systems, two series of symbols, LF_1 and LF_2. The two series are sufficiently similar, and they yield two interpretations, which we

---

26 In the same series, fasc. 4, R. Jolivet, *Kierkegaard*, and fasc. 9, R. Jolivet, *Französische Existenzphilosophie*.

27 The bibliography in the 1992 edition of Ladrière's book does not go beyond 1956.

28 The words from 'a symbolic' are taken from Lonergan's lecture notes. See below, p. 149.

will call LFL and LFM. In other words, the symbols will admit a logical interpretation and a mathematical interpretation. They need not be exactly the same symbols in the two cases, but there has to be a sufficient parallelism between them. That is the first idea.

4.2 Non-enumerable Sets

The second idea is that in mathematics there exist what are called non-enumerable sets. A finite group is something that you can finish counting. I can finish counting the number of people in this room. An enumerable set is a set you can keep on counting forever. It is very possible to count them. You can keep on counting them forever, and they are called countable sets, enumerable sets. For example, all the positive integers are countable. You cannot finish counting them; you can keep on counting them; they are called countable. Similarly, 1/2, 1/4, 1/8, 1/16, and so on. In general the criterion of an enumerable set, a countable set, is the possibility of a one-to-one correspondence with the positive integers or the natural numbers. So if you have a set, which may be in a set, that can be put in a one-to-one correspondence with the positive integers, you have what is called an enumerable or a countable set.

Such countable sets have various properties. For example, they contain parts of themselves. You can set a one-to-one correspondence of 1, 2, 3, 4, ... with 2, 4, 6, 8, ... all the even numbers. There is a one-to-one correspondence; for every positive integer you have to have a corresponding even positive integer, and so on. You have again a countable set in one-to-one correspondence.

But besides these countable sets, there are also non-countable sets. You can demonstrate that you have a non-countable set by showing that if it were countable you would be in a contradiction. Take, for example, the infinite decimals. Suppose you had all the infinite decimals down in order, arranged in a list. The first one is $a_1\, a_2\, a_3\, a_4\, ...$, the second one is $b_1\, b_2\, b_3\, b_4\, ...$, the third one is $c_1\, c_2\, c_3\, c_4\, ...$, and this goes on to infinity. Could you possibly say that all the infinite decimals are contained in that enumeration? You can demonstrate that you cannot by what is known as the diagonal theorem.[29]

---

29   The diagonal argument, generally attributed to Cantor, is found earlier in Paul du Bois-Reymond, 'Über asymptotische Werte, infinitäre Approximationen und infinitäre Auflösung von Gleichungen,' *Mathematische Annalen* 8 (1875) 363–414. There is an elementary presentation of Lonergan's perspective in chapter 3 of Philip McShane, *Wealth of Self and Wealth of Nations* (Axial Press, 2001).

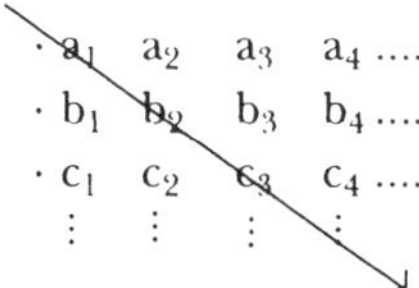

Pick out the numbers along the diagonal, and write down another decimal: $(a_1 + 1)$, $(b_2 + 1)$, $(c_3 + 1)$, ... and so on. This other decimal is not contained in any of the others. Why? Because it differs by at least one number from every one of the others, by definition. So there is no possibility of setting up the decimals in such a way that there would be a one-to-one correspondence between the infinite decimals and the counting numbers, the natural numbers.

This demonstration that certain sets cannot be enumerated is just a sample. There are various other types of non-enumerable sets, and the type of demonstration will vary in different cases, but this is the classical type, and it is known as the diagonal theorem.

4.3 Connection with Gödelian Limitations

All of the Gödelian limitations are tied in with this type of demonstration in mathematics, and the tie-in is our third step. The logical formalization is given a mathematical interpretation. If the mathematics is sufficiently advanced to include non-enumerable sets, the formalization will contain propositions in which you prove that such and such a number within the set, say k, is not enumerated. LF_2, as given a mathematical interpretation, will, if it is a sufficiently advanced mathematics, contain the proposition 'k is not enumerated.' The argument resembles the previous consideration. We considered the possibility of this list containing all the infinite decimals, and we constructed one immediately that was not contained. So we have the proposition, 'k is not enumerated among the others.' We have this true proposition, 'k is not enumerated.'[30]

30 Kurt Gödel's work of 1931 was entitled 'Über formal unentscheidbare Sätze der Principia Mathematica und verwandter Systeme.' It was published in *Monatshefte für Mathematik und Physik* 38 (1931) 173–98. It is available in English translation in Jean van Heijennort, ed., *From Frege to Gödel: A Source Book in Mathematical Logic, 1879–1931* (Cambridge, MA: Harvard, 1967) 596–616. German and English translation are given on facing pages in Kurt Gödel, *Collected Works*, vol. 1, *Publications 1929–1936*, ed. Solomon Feferman et al. (New York: Oxford University Press, and Oxford: Clarendon Press, 1986) 144–95. *The Journal of Symbolic Logic* 2 (1937) contains two useful

Now, consider these other propositions: 'The LF_1 is complete,' 'The LF_1 is coherent,' 'Some proposition is true,' 'Some term is definable,' 'Some problem is automatically soluble.' Because of the parallelism between the mathematical symbolism and the logical symbolism, it follows that where in the mathematics you can prove the proposition 'k is not enumerated,' in the logic you can prove that the system is either not complete or not coherent, or that a given proposition is true and is not true, or is both true and false. In other words, there are various ways in which, wherever you can prove that k is not enumerated, you can also prove that it is false or contradictory to say that the system is complete, that the system is coherent, that some proposition is true, that some term is definable, or that there is an automatic solution to a given class of problems. There is a parallel there. The argument involves a parallelism between the logical and the mathematical symbolism, such that proofs of non-enumerable sets in the mathematics will be paralleled by proofs in the logic that the system is either incomplete or inconsistent, or that all the problems of a given type cannot be solved automatically, or that 'true' becomes an impossible predicate under certain circumstances, or that 'definable' becomes an impossible predicate under certain circumstances.

So there you have the discovery that the project of finding an axiomatic system that will contain everything else is not a feasible business. The point of application of this is Hilbert's idea of proceeding on a finite basis, through a finite number of objects and a finite number of operations, to arrive at a logical justification of a mathematics containing an infinite number of objects and an infinite number of operations. The first demonstration concerns that position of Hilbert. Hilbert's plan regarding the finite basis of a logic against a mathematics dealing with infinities has been knocked out by these Gödelian limitations. That is certainly recognized by everyone. The ulterior implications of the limitations we can consider later on.

Another condition of this argument is, of course, that the mathematical meaning of the symbolism has to be sufficiently developed to give rise to the question of non-enumerable sets. However, you very quickly arrive at that.

presentations: Barkley Rosser, 'An Informal Exposition of Proofs of Gödel's Theorems and Church's Theorem' (53–60) and L. Chwistek, 'A Formal Proof of Gödel's Theorem' (61–68). A popular presentation (reprinted from *The Scientific American*, 1956) is found in Irving M. Copi and James A. Gould, *Contemporary Readings in Logical Theory* (New York: Macmillan, 1967) 51–71. The popular version is placed in a fuller context by the authors Ernest Nagel and James R. Newman, *Gödel's Proof* (London: Routledge, 1958, 1993). For the beginner, this is the best place to start.

In other words, if your mathematics is sufficient to deal with the theory of prime numbers – not arithmetic in the sense done in elementary school, but still very elementary mathematics – if your symbolism will deal with that in mathematics, you will run into these paradoxes in logic.

This conclusion to the non-automatically soluble is a limitation on the possibility of setting up automatic machines to solve problems. What is known as the functional calculus of the first order, a logic that represents subjects and predicates where the predicates are not variables that admit quantification, has been demonstrated to be complete and coherent. Every proposition in it can be demonstrated, and there is no contradiction that arises. However, there is no general solution, no automatic solution, to issues within the field of that calculus. That is a particularly interesting result. It was worked out by Church with regard to the functional calculus of the first order. The parallel problem directly in terms of machines was worked out by an Englishman in London by the name of Turing. He showed that if you tried to have a general solution you would have to be giving contradictory instructions to the same machine. In other words, the machine cannot solve it automatically. Partial solutions can be reached, but there is no general solution.[31]

This will give you some idea of the way these people in mathematical logic are able to discuss and study a deductive system as such, that is, the general idea of a deductive system as such and what kind of system can be mastered completely. For example, the propositional calculus can be mastered completely; the functional calculus of the first order, complete and coherent, resists an automatic solution. They are able to decide what problems can be solved automatically, what cannot, and so on, no matter how big the deduction is going to be.[32]

31 Alonzo Church, 'An Unsolvable Problem of Elementary Number Theory,' *The American Journal of Mathematics* 58 (1936) 345–63; 'A Note on the *Entscheidungsproblem*,' *The Journal of Symbolic Logic* 1 (1936) 40–41, with a correction at 101–102. Alan Turing, 'On Computable Numbers, with an Application to the *Enscheidungsproblem*,' *Proceedings of the London Mathematical Society*, ser. 2, vol. 42 (1936–37) 230–65, with a correction in vol. 43 (1937) 544–46. These articles are discussed in Martin Davis, *Computability and Unsolvability* (New York: McGraw-Hill, 1958).

32 A break was taken at this point. There is a gap in the tape of the lecture. Then Lonergan can be heard leading into the next paragraph of the text above with the words, '... deducing theology from it and you add in a few more premises from reason and then you get theological conclusions.' Evidently, he had digressed into broader considerations of deductive method, to which he returns later (see below, pp. 121–33). The text resumes with a summary of some of the major issues discussed to this point.

There is a conception of knowledge as deductivist, as essentially something that proceeds from premises to conclusions. All you have to do is find the right premises and draw the right conclusions and you will have your science. That is an idea of science and an idea of knowledge that produces an ideal. In other words, when we try to learn, when we try to understand something, when we try to build up a science, our thinking is guided not only by the immutable laws of thought but also by our preconceptions of what a science is, what knowledge is; and one of those types of preconception is the deductivist preconception. When I use the word 'preconception,' I am not using it in any pejorative sense: it may be good, it may be bad, but let us find out what it leads to.

One way of formulating the deductivist preconception is to say that there will be a set of n propositions. What they are we need not bother about, since we are just considering this ideal. Those n propositions will divide into two classes: some are not derived from any others, and some are. We will call those that are not derived from others 'primitive' and the rest 'derived.' The derivation will occur in accordance with certain rules. That is an ideal, what we are out to do, and we can conceive philosophy that way, we can conceive theology that way, we can conceive mathematics that way, or physics or any other science: the deductivist ideal. The question is, What is the value of it? What really can be done proceeding on the basis of that deductivist ideal? We considered different schools of mathematical thought, and we noted that some of them did aim at setting up mathematics as a purely deductivist science. They proceeded in various ways. Some presupposed the most difficult, most advanced part of mathematics that includes everything else, and it was not of too much interest because it was not too closely connected with the logic. Others, like Russell and Whitehead, took logical axioms, added a few more on, and deduced the whole of mathematics. Others, such as Hilbert, set out to do the mathematics on the basis of mathematical axioms that involve an infinity of objects and infinities of operations. But we will get the real elements of the thing when we start thinking about its logical properties, and particularly the property of consistency; and we will do that on a finite basis. Now, it was with regard to Hilbert's proposal that there came out the set of limitations discovered by various people, but the fundamental step and the fundamental method remains that invented by Kurt Gödel in an article that appeared in a German mathematical review in 1931. You will find it in Bochenski's bibliography under the name Gödel, and generally there are plenty of references to it.[33]

33 See above, p. 50, note 25.

4.4 Complexity of the Argument

I tried to give you some idea of what Gödel's argument is. That argument is enormously complex, because he has to be able to set up a symbolism for doing the theory of mathematics and a symbolism for doing the theory of logic, and he has to have the two symbolisms running pretty well parallel, or to find a way of communication between the two. That is where Gödelian numbers come in.

4.5 Transfinite Induction

We considered two sufficiently similar symbolizations, in which one has a mathematical interpretation and the other has a logical interpretation. Then, whenever the mathematical interpretation leads to such propositions as 'k is not enumerated' or 'Contradiction is involved in the assertion that k is enumerated,' you will get, because of the parallelism between the mathematics and the logic, logical affirmations to the effect that it is impossible to establish on a finite basis the coherence of Hilbert's exposition of arithmetic. However, shortly (three or four years) after Gödel's demonstration, a man by the name of Gentzen showed that by using transfinite induction in the logic, one could establish the consistency of the arithmetic.[34]

What is transfinite induction? It will show you just how far you have to go in logic to establish the arithmetic. The arithmetic does not normally involve transfinite induction. What is a transfinite induction?

In ordinary mathematics, there is a type of theorem in which you prove that if a theorem holds for n then it will hold for n plus one. For example, you want to prove de Moivre's theorem

$$\text{Cos } n\theta + i\text{Sin } n\theta = (\text{Cos } \theta + i\text{Sin } \theta)^{n}.$$

You first prove it when n is a positive integer. How do you prove it? Well, you can see right away that it is true if n is equal to one, because then the n's disappear and you have Cos θ + iSin θ = Cos θ + iSin θ. There is an identity.

---

34 Gentzen's work is available in M.E. Szabo, ed., *The Collected Papers of Gerhard Gentzen* (Amsterdam, London: North-Holland, 1969; see index, under Transfinite). Gentzen was Hilbert's last assistant, and no doubt benefited from Hilbert's interest, following Gödel's publication, in transfinite induction.

But how do you prove for the infinite positive values of n that that equation is true, that you have an identity there? Well, you show that if it is true for n, then it will be true for $n + 1$, and since it is true when n is 1, it will be true when n is 2; and if it is true for 2, it will be true for 3; if it will be true for 3, then it will be true for 4; and so on right down the whole line. Then you go on to negative integers and fractional values of n. That is the idea of an ordinary mathematical induction: if it holds for n, it will hold for $n + 1$, where n is an ordinary number.

Now we have seen already that there are enumerable infinities, infinities that you can set in a one-to-one correspondence with the positive integers, and non-enumerable infinities, ones that you cannot. There are different kinds of infinities. And the non-enumerable infinities have indefinitely more in them than the enumerable infinities. For example, consider a length of line. You can represent a half, a quarter, an eighth, a sixteenth, and a thirty-second, and so on, right on to infinity. And that is very many fewer than all the points in the line. It is just a few holes in the line, and still there is an infinity there. There is a much bigger infinity that is called the non-enumerable infinity that corresponds to the infinite decimals. The infinite decimals will just fill up the whole line, as if they do not leave any blanks; they will do this, at least if you include absolutely everything, all the numbers between zero and one including all the infinite decimals. You have a number for every point in the line. That is a different type of infinity, a non-enumerable infinity.

We can give those different types of infinity ordinal numbers; there is a first infinity, a second infinity, a third, a fourth, and so on. In a transfinite induction, you show that if it holds for any of the transfinite ordinals, it will hold for $n + 1$. That type of transfinite induction is what Gentzen used to establish the validity of Hilbert's arithmetic. Hilbert wanted to do it on a finite basis. Gödel showed that you could not. Gentzen showed that you could if you used a transfinite induction, if you used an induction that was just outside the field of arithmetic in any ordinary sense of the word. Gentzen's proof uses the transfinite numbers, but all he did was to establish arithmetic for the ordinary numbers and the first class of infinite numbers, not beyond the first class of infinite numbers. In other words, the logical problem is much more difficult than the original arithmetical problem. So far from making mathematics more evident by going into the logic, the logic is much more difficult. That is one way of putting the conclusion; it is a loose way of putting it, but it gives you more or less the idea.

4.6 Ultimate Significance of Gödelian Limitations

This brings us to the ultimate significance of such Gödelian limitations. We must note that there are a series of them. Gödel discovered one with regard to Hilbert's mathematics. Church, using pretty much the same general type of argument, shows that there is not a general solution to the decision problem for the functional calculus of the first order. Then Kleene and perhaps with him Rosser showed that there is a common root to both of those limitations, and Tarski went on to questions of truth and Mostowski to questions of the definable. (I think Tarski probably was involved in the problem of the definable too. You get full, detailed information in Ladrière.)

The ultimate significance, however, of such Gödelian limitations seems to be the same as that of inverse insight, on which you can see chapter 1 of *Insight*.[35] What is that significance? Well, the big moves forward in mathematics and science and, one might also add, to a certain extent in philosophy, lie not in understanding something but in understanding that there is nothing to be understood. That is one way of putting it. One example is the discovery of irrational numbers, such as root 2. The Greeks called them incommensurable numbers. The presumption is that, if you have a right-angled isosceles triangle, and if you make your measuring unit small enough (and you can make it as small as you please), you are bound eventually to find the right type of length that will go an even number of times into the side and an even number of times into the diagonal. Aristotle remarks that we think it is evident that, if one is free to make the measuring unit as small as one pleases, certainly one will be found that goes evenly into the side and into the diagonal. He goes on to add that, of course, the geometers would be much more surprised if anyone *could* find a unit that would go evenly into the side and into the diagonal. Why? Because they discovered the existence of irrational numbers. What does that mean? It means that where m and n are positive integers with no common factors, there is no such fraction that is equal to root 2. It just does not exist. You cannot express root 2, which is some number less than 2 and bigger than one, which you would expect to be some improper fraction, and the theorem of irrational numbers says that there are no two integers m and n to give you an improper fraction that is exactly equal to root 2.

You can prove there is not by a very simple procedure. Remove all the

35 See Lonergan, *Insight* 43–50.

common factors, so that m is prime to n. Square both sides. Then $m^2/n^2 = 2$. What does m^2/n^2 mean? It means $m.m/n.n$. If you have no common factors between m and n, then you have no common factor between the two m's and the two n's. So $m^2 = 2n^2$. Consequently, m^2 must be even. So, write $m = 2p$. Then, $4p^2 = 2n^2$. Take out 2, and you see that n^2 is an even number, which is contrary to our initial assumption that there are no common factors. Therefore, the fraction does not exist. The Greeks found another type of number besides the positive integers and the fractions you can make out of them. 'The incommensurables' was the name given to them by the Greeks because they proceeded geometrically. 'The irrationals' or 'the surd' is the name given by modern mathematics. The same sort of proof goes if you write in 3 or any other prime number. The m^2/n^2 will be equal to 9, and therefore m will be divisible by 3; and so on. You can prove for any surd that it is a surd, in other words, that it is not equal to any improper fraction.

Consequently, when you start talking about the arithmetization of the continuum, besides the ordinary numbers that you can get out of your integers and fractions, there are other numbers that are not that type at all, the irrationals. You can go on from there and discover another class of numbers called the transcendentals, which are not the solution to any algebraic equation.[36] They are a non-enumerable group. You can enumerate all the rational numbers and all the algebraic numbers, but you cannot enumerate the transcendentals. What are these transcendentals? We know a few like e and π; but since they are a non-enumerable infinity and they are all infinite decimals, it is a little hard to say what they are.[37]

That is a case of discovering that something is not the sort of thing you thought it was. When you discovered that root 2 is not a number in any ordinary sense of the word, you had an inverse insight. You start off with the notion root 2, root 3, and you think they each have to be some sort of fraction. After all, you have an infinity of numbers to choose from. But there is not any fraction. And you can go on to discover transcendental

---

36 One must add 'with integer coefficients.' Lonergan's regular illustration of irrational numbers, root 2, is an algebraic number, being the solution to the equation $x^2 - 2 = 0$. Irrational numbers have no regular repeating pattern in their decimals.

37 Cantor proved (1874) that the class of all algebraic numbers is enumerable. A proof for the transcendence of e was given by Hermite in 1873; the proof for π was provided by Lindemann in 1882. Kronecker, an ancestor of intuitionism, remarked to Lindemann, 'Of what value is your beautiful proof, since irrational numbers do not exist?'

numbers. Galois at the beginning of last century discovered that there is no general way of solving fifth-degree equations, though there is for quadratic and third-degree equations; and discovering that was an enormous development in mathematics.[38] So far from proving a block, it showed a dead-end street to be got around, and it revealed a whole new domain of mathematics. Galois died in his twenties, but he is a big name in mathematics simply from that discovery.[39]

In mechanics, Newton's first law, that a body continues in a state of uniform motion in a straight line as long as no force intervenes, is a case of an inverse insight. It is a matter of understanding that, while you need force for acceleration, you do not need force to account for uniform velocity. It is understanding that you do not have to postulate a cause. In other words, it is understanding that there is nothing to be understood. The uniform velocity takes care of itself without introducing causes. You come to something that demands an intelligible cause when you come to acceleration. Einstein's relativity, to my mind, is just a generalization of that first law of motion in Newton.

In such big movements, then, you open up new fields when you discover that something cannot be done. That is the significance of these Gödelian limitations. They started out with the idea that if you were to pick the right set of axioms, and if you had powerful enough techniques, you could deduce everything. They discovered that you cannot, and the fact that they discovered that you cannot is pretty much where the thing is at present. Still, there is a terrific significance to that discovery of limitations: the history of mathematics and of science has been a matter of discovering that not everything is understood; not everything is an object of intelligence, something to understand.

38 P. Ruffini (1765–1822) and H.N. Abel (1802–1829) produced independent proofs of the impossibility of solving by radicals the general equation of degree greater than the fourth. Abel's proof (1824) inspired Galois's deeper view of insolvability, which introduced to mathematics the technical use of the word 'group': an algebraic equation is solvable by radicals if and only if the symmetric group of its roots is solvable. See also Lonergan, *Topics in Education*, vol. 10 in Collected Works, ed. Robert M. Doran and Frederick E. Crowe (Toronto: University of Toronto Press, 1993) 93 and note 47.

39 A readable presentation of Galois's legacy, relevant to the drive of the present volume, is I.M. Yaglom, *Felix Klein and Sophus Lie: Evolution of the Idea of Symmetry in the Nineteenth Century*, trans. Sergei Sossinsky, ed. Hardy Grant and Abe Shenitzer (Boston: Birkhauser, 1988). The twentieth-century legacy within physics is the topic of the two works by O'Raifeartaigh cited at note 31 of p. 33 above. See also below, note 13 on p. 356.

The philosophic illustration of this is, Why is one point different from another?[40] If you say that it is the distance between them, then let us take a third point in an equilateral triangle, and I will ask you why are the three *distances* different? Well, you may say, they are different directions. You cannot say they are unequal: they are not different because they are unequal. And you cannot go back to the points, and say, 'The distances are different because the points are different,' because what you want to explain is the difference of the points. If you say, 'It is the directions,' then we can just move on to another question: Why are the directions different? Finally, you will say, 'Well, we have to presuppose something.' Of course, and that is precisely what I am trying to show. There is such a thing as material individuation. There is the fact of difference without a reason for difference. Angels are not of the same species; they are intelligibly different; there is a reason for their difference. But things that differ materially differ as a matter of fact, not in nature. Material difference is a case of inverse insight. It is a case of hylomorphism. What does hylomorphism mean? There is a form that you know by understanding, but there is something else there that you know as a matter of fact, not by understanding it. This gives us a way to introduce philosophy through the very things that mathematicians and scientists are familiar with.

5   The Transcendence of Gödelian Limitations

Now we come to the transcendence of Gödelian limitations. What is going on that seems to suggest the development of a mathematical-logical system that will get around these Gödelian limitations? When you discover these limitations, the real significance of them is that you know that such-and-such is a dead-end street and that you have to find another street. What are the implications of this looking for another street?

When I start speaking on this topic, I am in a much more speculative field. My presentation of the other material has been just the barest outline of the sort of thing that is very elaborately developed. But on this topic, I would not say that the mathematical logicians are altogether aware of the

---

40 Lonergan repeats this brief task description, with slight modifications, in *Insight* 51, 527–28, and in *Understanding and Being*, chapter 2, conclusion. Its brevity belies its importance for foundational work. One can glimpse this by taking the axiom '*x* or *x* implies *x*' mentioned below in note 2 on p. 69, placing the three *x*'s in an equilateral triangle, and pursuing a more fundamental axiom that meshes with the psychology of this-saying.

significance that I am suggesting exists in the series of points in this section. One might say that the *nota theologica* here is pretty weak![41]

5.1 Mere Avoidance

First of all, there is mere avoidance of Gödelian limitations. An example of this can be found in an article by J.R. Myhill in *The Journal of Symbolic Logic* for 1950.[42] He avoids the implications of these limitations simply by using a logic in which there do not occur quantification and negation. Without quantification and negation, you do not get into those difficulties. But this is a matter of avoiding the difficulties by stepping into something that is less rich. It is not a satisfactory type of solution.

5.2 Indefinitely Large Stratifications (Analogy)

The next example is the use of indefinitely large stratifications, which we would call analogy in this present case. This can be found in Church's 1935–1936 work, his mimeographed lectures published at Princeton in 1936, pp. 45–50.[43] I do not know whether it comes into his *Introduction to Mathematical Logic*, a book (first volume) published in 1956.[44] I glanced hastily through it, but did not notice it.

What we have here is an instance in which Church is avoiding difficulties by saying, 'Let us not say that implication always means the same thing.' Supposing we symbolize 'implication' by the arrow: $x \to y$, x implies y. The arrow takes on different meanings according to the level on which you are working. The same can be done for 'quantification.' This is what he means by a stratification, and it is the sort of thing that we mean when we speak of analogy. The meaning of the quantification or the meaning of the implication varies with the level. This gets us into a logic that uses analogy as a principle of structure.

The same type of thing can be found in Curry, who is the chief name in combinatory logic. He is among the people who study the possibilities of symbolism as such. His basic notion is canonicity. He has a series of levels

41 A *nota theologica* was a regular feature of theses in older manuals of theology. It indicated the degree of certainty of the thesis, ranging from certainties of faith to the tenuous certainties associable with disputed issues in theology.

42 John R. Myhill, 'A Complete Theory of Natural, Rational, and Real Numbers,' *The Journal of Symbolic Logic* 15:3 (1950) 185–96.

43 See note 19 on p. 19, above.

44 See the same note.

for the meaning of canonicity, and that series ultimately turns out to be transfinite. A reference can be found in *The Journal of Symbolic Logic*, volume 7, 1942, pp. 49–64, where Curry, to avoid contradiction, introduces an eventually transfinite hierarchy of levels of canonicity.[45]

5.3 Skolem Paradox

We start getting down to the real problem in what follows. The Skolem paradox brings us right into the nub of the matter.[46] If you remember, the trick had to do with the enumerable set. The negative proposition is in the mathematics: it is contradictory to assert that this number k has been enumerated. You are dealing with the term 'enumerable,' and you are using enumerable sets. What the Skolem paradox reveals is that 'enumerable' has to take on different meanings. There is an ordinary meaning: one-to-one correspondence between the positive integers and the things that you are lining up in a set. After that initial meaning you will use a more elaborate syntax to describe your one-to-one correspondence, and when you use the more elaborate system of expressing your correspondence, then what previously was non-enumerable now turns out to be enumerable. An analogy of the term 'enumerable' makes sets that from one viewpoint are non-enumerable into sets that are enumerable.

Zermelo's account of set theory is in first-order predicate logic. It is in the functional calculus of the first order. Set theory includes Cantor's proof of non-enumerable sets, and Skolem showed that the functional calculus of the first order sufficed to handle Cantor's theory of sets. There seems to be

45 Haskell B. Curry, 'The Combinatory Foundations of Mathematical Logic,' *The Journal of Symbolic Logic* 7:2 (1942) 49–64.

46 A good deal of the volume cited in note 6 above focuses on the significance of the Skolem paradox. The original work of Skolem is reproduced in English translation in the work cited above in note 17 of this chapter, p. 46. More elementarily, W.V. Quine, *Methods of Logic* (3rd ed., New York: Holt, Reinhardt and Winston, Inc., 1972) chapter 32, 'Löwenheim's Theorem,' pp. 177–80. 'In a word and in general, the force of the Löwenheim-Skolem theorem is that the narrowly logical structure of a theory – the structure reflected in quantification and truth functions, in abstraction from any special predicates – is insufficient to distinguish its objects from the positive integers' (p. 178). Less elementary but readable is William Kneale and Martha Kneale, *The Development of Logic* (Oxford: Clarendon Press, 1978), chapter 12. 'In short, axioms designed for the purpose of focusing attention on a non-denumerable infinity must inevitably fail to achieve that purpose.' Ibid. 712. This page meshes nicely with Lonergan's discussion.

no doubt about this paradox of Skolem, and it introduces the idea that the notion of enumerable is what we would call analogous.

5.4 Leon Henkin

There was also work done by Leon Henkin, not exactly along that line, but studying the relation between Logical Formalizations, LFL, and their models, their interpretations, what they represent. He distinguished between regular models and irregular models. Because of the difficulties he ran into he introduced the notion of categoricity. He has a theorem or proof to the effect that, when a logical formalization admits non-regular models, then the terms in the logical formalization will have to be what we would call analogous. What he says is that they lack categoricity; they do not have an absolute definite meaning. That work of Henkin is in *The Journal of Symbolic Logic*, volume 15, 1950, pp. 81–91, and it is treated by Ladrière, pp. 371–73.[47]

These points I have given so far are straws in the wind. They are seeing the necessity of what we call analogy: analogy of enumerable, analogy of canonicity, analogy in the sense of the correspondence between a formalization and what it represents, analogy in the notions of implication and quantification.

5.5 Hao Wang

Hao Wang, as far as I can see, seems to have exploited this business in a way that provides a strategy for getting round the Gödelian limitations. His work appears in *The Journal of Symbolic Logic*, volume 19, 1954, pp. 159–60, 241–66. The description is given in Ladrière, pp. 393–95.[48] What does he do? His mathematical-logical system consists not just of one system but an indefinite series of systems. Each system is given a transfinite ordinal, as far as I can see. So you get an idea of the sort of level on which he is working. That is the first thing, then: it is not a simple mathematical-logical system in

---

47 Leon Henkin, 'Completeness in the Theory of Types,' *The Journal of Symbolic Logic* 15:2 (1950) 81–91. For this and the next note Lonergan gave the page references in Ladrière as they are repeated here, and his references were correct.

48 Hao Wang, 'The Formalization of Mathematics,' abstract, *The Journal of Symbolic Logic* 19:2 (1954) 159–60; full paper with the same title, ibid. 19:4 (1954) 241–66. Hao Wang's work was already referred to. See note 21 on p. 19 above.

any of the previous senses; it is a series of systems. And they go on indefinitely; it is indefinitely open. On each level, there are new means of construction that arise, new resources; each level is more resourceful and has resources not on the previous level, and on each level 'enumerable' takes on a new meaning in view of the new resources of construction. What he showed was that the consistency and the theory of truth for any level i is demonstrable at level $\mu + 2$.[49] If you want to do the thing thoroughly what you have to do is set up an indefinite series of strata, in which on the second level up you are able to handle the logical problems of two levels lower down.

5.6 Significance

In other words, the human mind, as St Thomas says, has a natural desire for the beatific vision; it is infinitely open.[50] That openness is something that upsets this effort, the initial, naïve logical ideal of starting out from a whole set of axioms from which you deduce everything that is to be known.

6   Conclusion

The material today has been very abstruse, first of all in the sense that I have been talking about logical formalizations without attempting to give you concrete ideas of what is being dealt with. I have not attempted to do that because you could spend a month on it. If people are mathematicians they are able to do it for themselves without difficulty, and if they are not it is quite difficult to teach it. But what I hope to have given is some suggestion, and if I can give any further help by answering questions, I will only be too

---

49 At least several sentences at this point were not recorded because of a change of tape.

50 See Lonergan, *De ente supernaturali*, class notes of 1946, 83 pp., translated as *On Supernatural Being* by Michael Shields, 1992, 151 pp. Both works are available at the Lonergan Research Institute, Toronto. Lonergan's comments here on the openness of intellect correspond to remarks of Wang on pp. 261–62 of the article referred to in the text and in note 48 above. For example, 'From this point of view, the fact that no enumeration can exhaust all sets of positive integers is explained not by the existence of indenumerable sets but rather by the impossibility for our intellect to have a clear and distinct idea of the totality of *all* sets or laws defining enumerations, for unless we are able to contemplate such a totality, it is quite senseless to ask whether there exists any set which is indenumerable in the absolute sense.' On the theological issue, see below, pp. 349–55.

glad to do so either immediately or later. What I wanted to communicate is the idea of rigorous deductive system and what an impasse that notion leads to at least when applied to anything as complicated as mathematics. We will see later that it is probably even more complicated when you get down to more concrete things than arithmetic.

We have now finished what we are going to say about mathematical logic as such. From now on, we are going to be discussing questions like, What about it for us? What is the truth of it? What can we get out of it? We will be discussing the foundations of logic in general, and the relation between this logic and Scholasticism. So you need not worry about tomorrow being as much worse than today as today was worse than yesterday!

3

The Truth of a Mathematical-logical System[1]

Yesterday we attempted to obtain at least a *docta ignorantia* with regard to the difficulties that arose in the effort to set up the whole of mathematics within a single deductive system. The results were, roughly, that to embrace the whole of mathematics and to make room also for the development of future mathematics – although this was not an operative fact in reaching the conclusions – it seems necessary to proceed not on a single level, from a single set of axioms, no matter how big, from which to deduce along a single plane all the conclusions of mathematics, but to have a series of levels. Further, on each higher level one needs richer resources of construction and conceptualization for setting up axioms and principles. In other words, the notion [of a single axiomatic basis] is naïve. Even in a science such as mathematics, which by and large does not consider concrete matters of fact as a decisive factor, there is not a single set of axioms from which all the propositions of the science can be deduced. What you have to have is a series of levels, each with its own set of axioms, and each higher level richer than the level below.

That development of mathematical logic was due to its close linking with mathematics. Logicians were, in general, mathematicians, and they took as

---

1 The third lecture, Wednesday, 10 July 1957. The tape begins with Lonergan in mid-sentence, no doubt talking about the first lecture: '... implications that run counter to Scholastic practice in proving theses and answering objections, and the general possibility of a mathematical logic that lies in what is called isomorphism and consists in a structural identity or similarity of relations.'

the control in their investigation of deductive systems the question, How much mathematics can be deduced from what premises?

The question we are asking today is non-mathematical. We are considering simply systems of symbolic logic, systems of mathematical logic. They are purely logical in intent and in content. They are related to mathematics externally just as logic is related to any other subject. But they are not specifically mathematical. Since there are many such systems and they are not all compatible, we are asking what kind of truth they have. It is a somewhat subtle question because the type of axioms assumed by the logicians in drawing up their list of axioms and primitive propositions are absolutely trivial, the sort of thing that no one would dream of denying. They get everything they want for their system from an absolute minimum, and that is their ideal: to start out from the sort of thing that no one would dream of denying.

For example, in Hilbert's system there are four axioms, of which I remember two.[2] 'x or x implies x': where x always denotes the same proposition, which may be any proposition at all. If it is true that you cannot deny both x and x, then you are going to be left with x. It is a little hard to imagine any possible situation where you can get around that. Another one is, where x and y are any two propositions, then 'x or y implies y or x.' What they aim at in their premises, in their axioms, is triviality, the sort of thing that no one would dream of denying. Yet they come up with incompatible systems, and so there arises a question of truth.

1   The Truth of What?

First of all, we ask, The truth of what? The truth in question is not the truth of what may happen to be written down. What is written down is a set of symbols. What do you see? What you see may or may not be 'wf,' well-formed.[3] When you write down a set of these symbols, the thing may make sense or it may not. It may be just a casual jumble, the sort of thing that would happen if someone dreamily hits a typewriter key. That would not be of any significance whatever. You have to know a certain amount of math-

---

2 My suspicion is that Lonergan is referring here not to Hilbert but to *Principia mathematica* (2nd ed., vol. 1, pp. 95–97). 'x or x implies x' would be equivalent to *1.2, and the other axiom he notes is equivalent to *1.4. In *Principia mathematica*, there are five, or perhaps six, primitive propositions.

3 See below, p. 151. Lonergan is simply explaining the shorthand used in his notes, standard in mathematical-logical texts.

ematical logic to decide whether any given set of symbols really has any meaning in the mathematical logic. Consequently, the symbols alone are not the sort of thing about which you ask the question, Is it true? but the symbols as objects of a necessary measure of understanding, at least a measure of understanding necessary to determine whether the symbols are well-formed, whether you have an expression that could mean something.

Moreover, not only is there the question of the expression being well-formed, but also the expressions that are written down are only part of the system. The system is the totality of propositions. And when you ask about the truth of a system, you ask about the truth of all the propositions that constitute the system, even though all of them may never have been written down and never may be written down. When you are asking about the truth of a system, you are asking about the truth of the lot.

Finally, truth does not reside in expressions. Expressions may be adequate or inadequate. What is true or false is the judgment. *Veritas formaliter est in solo iudicio.*[4] We are asking about truth in our meaning of the word 'truth,' and so we are asking about the truth of a judgment that accepts the system. We are not talking about the symbols as not understood, and we are not talking about the symbols that are understood and written down. We are talking about all the symbols that might be written down and understood. And we are not asking about simply the expressions of symbols as symbols even when they are understood. We are talking about what is meant by the totality of expressions that might be written down. And to know all that might be written down, we have to postulate a knowledge of mathematical logic sufficient to manipulate the symbols and bring out all the virtualities of the axioms.

The truth of a mathematical-logical system is the truth of a virtual totality of propositions. What do I mean by a proposition? I mean by a proposition what may be expressed in several languages. 'The king is dead.' 'Le roi est mort.' 'Der König ist tot.' There we have three different sentences, in three different languages, but they all mean the same thing. There is only one proposition there, though there are three sentences in three different languages. That is an easy way of distinguishing propositions from sentences.

Again, you can distinguish sentence from utterance. I can say, 'The king is dead, the king is dead, the king is dead.' That is one sentence, but three

4 'Only in the judgment is there formally truth.' See Lonergan, *Insight* 298–99, 578–80.

utterances of the same sentence. We are talking about the proposition: what is common to all the sentences you can get by using various languages and what is common in all the repetitions of the same sentence in the same language.

2   The General Character of Such Truth

What is the general character of the truth we are asking about? In book XIII of the *Metaphysics*, which Thomas unfortunately did not write a commentary on with the result that very frequently it is unknown, in the tenth chapter, Aristotle distinguishes between science in potency and science in act. He says that science in potency is indeterminate and universal and of the indeterminate and determinable, while science in act is determinate and of the determinate, of this particular thing.[5] So you have, on the one hand, science that is in potency and of the universal and of what can be determined, and, on the other hand, science in act, which is of the concrete and particular. Clearly, then, when we ask of the truth of a mathematical logic, we are asking about the truth of something that is universal and indeterminate and to be determined. In other words, we are asking about the truth of a system in which there are no particular propositions but just blank x's, p's, q's, and r's in which propositions could be put, as long as you always put the same proposition wherever there is a p and another same proposition wherever there is a q, and so on.

Again, Aquinas distinguished between abstraction of the universal from the particular and abstraction of form from matter.[6] The abstraction of a mathematical-logical system is of form from matter. What one is talking about when one speaks of a mathematical logic is a structure of relations: what the contents are is indifferent. They may be limited in certain generic fashions, but what is significant is the relational structure. And a relational structure is what is meant by a *forma artificialis*. Consequently, what we are inquiring about is the truth, not of a universal abstraction from particulars,

---

5 Aristotle, *Metaphysics*, XIII, 10, 1087a 16–20. Lonergan quoted here freely from the Greek text (from memory). The English in the Loeb text reads as follows: 'Knowledge, like the verb "to know," has two senses, of which one is potential and the other actual. The potentiality being, as matter, universal and indefinite has a universal and indefinite object; but the actuality is definite and has a definite object, because it is particular and deals with the particular.'

6 See Lonergan, *Verbum*, chapter 4.

such as man from these men, but the truth of a relational structure. For example, there is the *unum per accidens* (improperly so called), as opposed to substance, but there is also *unum per se* that consists in an intelligible unity of parts within a whole.[7] Thus, a motor car can be called an *unum per accidens* insofar as it is made up of many substances, but it can be called an *unum per se* because it is one machine, a realization of one idea.[8] Well, the *forma artificialis* is the one idea in the one machine. Similarly, *mutatis mutandis*, a mathematical logic is a relational structure, a *forma artificialis*. The object of our inquiry is a form abstracted from matter. We are asking about the truth of a relational structure into which you can substitute any particular that you wish.

3   What Is Meant by 'Truth'?

The third point is, What is meant by 'truth'? We have to distinguish between the definition and the criterion of truth. Truth as defined is *adaequatio intellectus ad rem*, but the criterion of truth is commonly expressed as the *perspicientia evidentiae sufficientis qua sufficientis*.[9] That is a rather woolly notion. How do I know it is true? Because I have sufficient evidence. How do I know that I have sufficient evidence? Well, it is evident. And how do I know it is evident? Well, it is manifest. And how do I know it is manifest? Well, it is obvious. And how do I know it is obvious? Well, no sane person will say anything else. And where does that get you? Well, you do not have anything precise on which you can operate and through which you can argue in a systematic fashion. So in *Insight*, in chapter 10, I attempted to work out exactly a formula that would enable one to say just what one gets hold of when one grasps the sufficiency of evidence for the truth of a true proposition. The expression I used was 'the virtually unconditioned.'

The unconditioned may be divided into the formally unconditioned and the virtually unconditioned. The formally unconditioned has no conditions whatever; it is God and God alone. The virtually unconditioned has condi-

---

7 See Lonergan, *Insight*, chapter 8.
8 In response to a question at the break, Lonergan said, 'I spoke of the *unum per se* and the *unum per accidens*, and distinguished two meanings of the *unum*. In one case, the *unum per se* means what is substantial, and in that sense a man is *unum per se* and a machine is not. In another sense, an *unum per se* is an intelligible unity, and in that sense a machine is an *unum per se* and a heap of stones is not. And the relational structure is *unum per se*, like the machine, but not like a man.'
9 See Lonergan, *Insight*, chapters 9 and 10.

tions but the conditions are fulfilled. So de facto it is an unconditioned. It is a de facto unconditioned. God is *de iure* unconditioned. He has no conditions whatever. Consequently, there is no relevance to talking about the fulfilment of the conditions. But the virtually unconditioned is a de facto unconditioned. It has conditions but de facto those conditions are fulfilled.

If you analyze the notion of the virtually unconditioned you will find three elements: a conditioned, a link between the conditioned and its conditions, and the fulfilment of the conditions. I argue in chapter 10 of *Insight* that that will provide the answer, whenever there is a true judgment. If you analyze what you have whenever you make a true judgment, then you will find that you have grasped the virtually unconditioned.

I note briefly some examples. First, there is deductive inference, any inference you please. If *A*, then *B*, where *A* and *B* are propositions or sets of propositions.[10] You have a minor premise *A* and a conclusion *B*: If *A*, then *B*; but *A*; therefore *B*. If Socrates is a man, he is mortal. He is a man. Therefore, he is mortal. We have here *B* exhibited as a conditioned: if you have *A*, you have *B*. You also have a link between the conditioned and its conditions. If *A*, then *B*: the hypothetical premise gives you the link between the condition *A* and the conditioned *B*. In the minor premise, you have the fulfilment of the conditions, and so *B* is exhibited as virtually unconditioned. The point to syllogism, on that analysis, is to exhibit the conclusion to intelligence as a virtually unconditioned. The point to syllogism is not the point given it by Kant. Kant says that in a syllogism you have two premises, and to prove the first premise you need two more, and to prove the second one you need two more, and so you have a regression to 2^n premises, where *n* is as big as you please. In other words, for him the syllogism never really gives you anything. Well, it does not if you are thinking of proving the premises. But what does syllogism do to the conclusion? It exhibits the conclusion as a virtually unconditioned, or it exhibits it in a fashion in which it might be seen to be a virtually unconditioned. That is one example of the virtually unconditioned.

Another example is the concrete judgment of fact.[11] You come back to your room, and you find water on the floor, the window broken, chairs smashed, and smoke in the air. You don't make any rash judgment such as,

10 See ibid., chapter 10, §1. See also Lonergan, 'The Form of Inference,' in *Collection* 3–16. I take the opportunity to note the brief comments of Alonzo Church on 'The Form of Inference' in *The Journal of Symbolic Logic* 8 (1943) 48.
11 See Lonergan, *Insight*, chapter 10, §2.

'I may have left a cigarette in the wastebasket.' No, you say, 'Something happened.' That is your judgment: 'Something happened.' It is your *B*. You want now to get the link between your *B*, 'Something happened,' and the conditions of true judgment. How do you know something happened? Well, you have to have a link between your *B*, 'Something happened,' and its conditions, and the link is found in the meaning you give to the word 'happened.' If you say that something happened, you mean that the sensible data now are not the same as the sensible data before. You needn't *say* that, but whenever you say that something happened, that is what you mean: the sensible data now are not the same as the sensible data before. If the sensible data now are not the same as the sensible data before, then something happened. Only you do not say all that. It is something in the back of your mind. You do not have any explicit proposition, any prior judgment there, but you have that meaning in your mind of what you mean by saying, 'Something happened.' And you have the fulfilment of the conditions in your present experience of the state of your room and your memory of the state of your room when you left it. In the present experience and the memory, you have the fulfilment of the conditions, and so your judgment, 'Something happened,' rests upon a grasp of the virtually unconditioned.

So we have taken a concrete judgment of fact, on the one hand, and on the other hand syllogism, and if you read chapter 10 of *Insight* you will be able to fill in all the intermediate stages. All I am concerned to do here is to give you the general idea of the analysis of a judgment. It fits right in with Thomist psychology, at least as I worked it out in the *Verbum* articles. You have an *intellectus possibilis*, a *verbum incomplexum* with which you answer the question, *Quid sit?* and a *verbum complexum* with which you answer the question, *An sit?* A *verbum* is preceded by an act of understanding. In every direct act of understanding you grasp some intelligibility in phantasm and express it in a definition: a circle is the locus of coplanar points equidistant from a center. And then you ask, 'Are there any circles?' And you cannot tell, so you say, 'I do not know whether any circles exist.' And when you say, 'I do not know whether any circles exist,' you mean, 'I have no means of telling whether the conditions are fulfilled so that I can judge that de facto something is absolutely round.' Your negative judgment depends upon the absence in you of a grasp of the virtually unconditioned.

So reflective understanding issues in the judgment. Just as direct understanding grasps form in matter, *intelligibile in sensibilibus*, reflective understanding grasps the virtually unconditioned and says, 'It is' or 'It is not.' The virtually unconditioned is an absolute, and so on the level of judgment you step into the order of what is absolute, into the order of *ens*. You know being. You are in an absolute order.

So much for the criterion of truth, the virtually unconditioned, and the definition of truth, the *adaequatio* of the judgment *ad rem. Verum est medium in quo cognoscitur ens, id quod est.*[12]

When you ask about the truth of a mathematical-logical system, you can consider it in two different ways. Those two ways correspond to the difference between an analytic proposition and an analytic principle.

Now, that may be a startling distinction. By an analytic proposition I mean the sort of thing that can be formed *ad infinitum, ad libitum.*[13] If you define A as what has a relation R to B, then there cannot be an A without a relation R to B. You have an analytic proposition. Whether in this world or in any other world there is such a thing as an A with a relation R to B, you do not know and you do not care. But you have something that is necessarily true. If you define A as what has the relation R to B, then you can say with absolute certainty that there cannot be an A without a relation R to B because that is what you mean by A.

What do we have there? We have an analytic proposition. Is it a virtually unconditioned? Yes. First of all, the fulfilment of the conditions is the set of definitions. A is what has a relation R to B. The definitions of terms provide the fulfilment of the conditions. The link between the conditioned and the conditions lies in the laws of syntax, the way the defined terms come together to form a proposition. If you define A as what has the relation R to B, then you must say there cannot be an A without a relation R to B: you are grasping syntactical laws, the relations between defined terms and the propositions in which the terms occur. So the link between conditions and conditioned is syntactical. The fulfilment of the conditions is by definition, and so the analytic proposition is virtually unconditioned. You can form an infinity of such analytic propositions. 'Analytic proposition' in that sense is quite closely related to – it is not exactly the same as – what is meant by the logical positivists when they speak of tautologies.

The analytic principle goes further. It adds on concrete judgments of matters of fact. There *exists* an A that possesses a relation R to B. When you bring in that concrete judgment of fact, you move from the mere analytic

---

12 'Truth is the medium in which being, that which is, is known.' See Lonergan, *Verbum* 73, note 53.
13 'Endlessly and freely.' See Lonergan, *Insight*, chapter 10, §7.

proposition to the analytic principle. The difference between the analytic propositions that you can form indefinitely and the analytic principles is enormous. There is an analytic principle in metaphysics, for example, that a contingent event proceeds from an efficient cause. We are not merely defining our terms and in virtue of those definitions getting the analytic proposition; we can also say, 'This is a contingent event. This is the sort of thing I mean by "proceeding from an efficient cause."' In other words, the terms in the sense in which they are defined are found *in the existential order.* There is a big difference between the merely analytic proposition and the analytic principle. The fulfilment of the conditions in the analytic proposition occurs simply in a linguistic world. If I use the term A in your sense, in which A is defined by a relation R to B, then I am apt to agree with your principle, but there is no reason in the world why I should ever agree to talking of A in that sense. Consequently, pure analytic propositions with all their implications are fine, but they do not bear any relation to any knowledge that I possess or ever will possess. On the other hand, analytic principles, because their terms in the sense in which they are defined also occur in concrete judgments of fact, are not merely propositions but also principles. They refer to the real order, because with the term you are knowing something that is virtually unconditioned, something that is absolute with regard to the real order. And so the analytic principle belongs to an entirely different sphere of judgment from the analytic proposition.

With regard to analytic principles, one can distinguish three levels: the absolute, the provisional, and the serial. The absolute is such as occurs in metaphysics: every contingent event proceeds from an efficient cause. The provisional: for example, when a physicist works out Boyle's law, the relation of pressure and volume; he presupposes Boyle's law when he goes on to work out the relation between pressure and temperature, and when he gets that correlation he further presupposes both. He uses those empirical laws as if they were definitions of what the relation is. In other words, the empirical law can be transformed into the equivalent of an analytic principle. What I mean by temperature is what varies in a certain way with pressure and volume. And what I mean by pressure is what varies in a certain way with volume and temperature. You start using empirical generalizations in an empirical science as means of definition, but those empirical conclusions of science are subject to revision. And insofar as they are subject to revision, they are not analytic principles in the full sense but revisable analytic principles. If physics were to develop beyond the level represented by the successive formulations de facto of Boyle's and Charles's and Gay-Lussac's law, then you would be changing your analytic principle. That matter has been treated by Arthur Pap in a book referred to in *Insight*.[14]

So there is a first type of analytic principle, the absolute, as in metaphysics; a second type, the provisional, as in the principles and laws that are accepted in physics and made the basis, as it were by definition, of further investigation. And finally, there is the serial type. The mathematician knows de facto that there are instances of 1, 2, 3, but how far does he go? At least his concrete experience does not take him on to 10^n, where n is as big as you please. That number, 10^n, where n is as big as you please, is not the number of something that he has counted and knows exists. But 1, 2, 3 are something he has counted and knows exist. That idea can be extended throughout the whole of mathematics. In mathematics you deal not merely with analytic propositions but with analytic principles in some serial sense.

However, these are minor subdivisions. The basic issue is the virtually unconditioned, which applies to syllogism and to concrete judgments of fact, and which can be applied to yield the distinction between analytic propositions and analytic principles. Analytic propositions are the sort of thing you can form as many of as you please, just by using more symbols in a more and more complex fashion. You can say, 'I mean by A what is related R to B, and is in a relation R_1 to C, and in a relation R_2 to D,' and so on. You can form as many of those as you please. And they yield analytic propositions. But they bear no relation to knowledge of an existing reality until you can make concrete judgments of fact in which occur all the defined terms in the sense in which they are defined. And when you can do that, then you move from the analytic proposition to the analytic principle.

4   A Mathematical-logical System Is by Postulation a Virtually Unconditioned

In the light of that distinction, we will attempt to answer the question of the truth of a mathematical-logical system. In the first place, we will say that every mathematical-logical system that is respectable, that gets credit and is respected by logicians and mathematicians, possesses the truth of the analytic proposition. In the second place, we will go on to ask, In what sense do the mathematical-logical systems possess the truth of analytic principles?

14 Arthur Pap, *The A Priori in Physical Theory* (New York: King's Crown Press, 1946; referred to at the end of *Insight* 334).

The first step is quite easy: a mathematical-logical system is by postulation, or by definition if you wish, a virtually unconditioned. A mathematical-logical system is a virtual totality of propositions that result through explicitly stated rules of derivation from primitive terms and propositions. The mathematical-logical system, then, the totality of the propositions, is the conditioned; it depends on something else. Both the something else on which it depends and the link between the something else and it are by postulation accepted, just as the definition of *A* as what has the relation *R* to *B* is by postulation or by definition accepted. Consequently, a mathematical-logical system has the truth of an analytic proposition; every mathematical-logical system satisfying the definition of such a system has the verbal type of truth that pertains to the analytic proposition. If you are going to speak in the manner prescribed by a given mathematical-logical system, or if you are going to think in that manner, then *eo ipso* you are committed to accepting the system as true for you as long as you think or speak within that system. That is putting it hypothetically; that is envisaging the possibility that this system should not be simply something set down by postulation but should be actually employed. When it is actually employed, you step into the existential order. But when we say that it has the truth of the analytic proposition, we are talking about the system as such, whether or not it is employed. If it is actually employed, then you have someone actually using it. If he is using it in its defined sense, then it is occurring in a mind, and the mind will be linking. Thus, if you accept the premises and the link between the premises and all the conclusions, you accept all the conclusions. Consequently, once you start operating within a mathematical-logical system, you get beyond the question of the merely analytic proposition. You are raising the further question whether a mathematical-logical system has *factual* truth, whether it has anything of the aspect of the analytic principle. The analytic principle, again, arises when there is truth in the sense of a concrete judgment of fact, such as we made when we went back to our room and said, 'Something happened.' If the fundamental axioms and relations and terms occur within concrete judgments of fact, then the mathematical-logical system has not merely the truth of an analytic proposition. It steps into the order of the analytic principle, and then we have to consider further questions. Consequently in the fifth section we say that various types of mathematical-logical system contain fragments of factual truth.[15]

15   Lonergan took a break in his lecture at this point, and questions were
     raised. Some of these, with Lonergan's responses, are given below,
     pp. 333–36.

5   Various Types of Mathematical-logical Systems Contain Fragments of Factual Truth

Can a mathematical-logical system be said to possess factual truth? In other words, do there exist concrete judgments of fact such that mathematical-logical systems can be brought out of the airy realm of mere analytic propositions into the field of analytic principles? That question we answer by saying that the various types of mathematical-logical systems contain fragments of factual truth.

This assertion stands with the movement from the analytic proposition to the analytic principle. It is not simply that: it is movement from a judgment on the system as analogous to analytic propositions to a judgment on the system as analogous to analytic principles.

In our next lecture, on the foundations of mathematical logic, we will raise the question whether mathematical-logical systems can yield absolute analytic principles such as you have in metaphysics. The question of foundations, then, will be the question of putting logic on the same absolute basis as we conceive metaphysics to have. But we will not touch that question today because today we want to know just where these people stand, and they certainly have no idea of getting themselves into metaphysics. One of their reasons for going into this is not to have to bother their heads about metaphysics. The interest in metaphysics is ours. What we want to know is where they stand in virtue of just what they say.

We have divided our analytic principles into three types: absolute, provisional, and serial. Whether there is a possibility of having absolute analytic propositions as the basis of a mathematical logic is a question that we defer for the moment. One might think from the name 'mathematical logic' that the thing to do would be to pick on the third type, namely, the analytic principle of the serial type such as you have in mathematics. And there is a possibility in that direction. But there is also a difficulty. The difficulty is this: that the mathematical logic has been introduced to help the mathematicians decide exactly what on earth mathematics is. And they are nearer an answer to the logical foundations of mathematics than to settling their minds on what mathematics is. One of the big names in the field is Paul Bernays. In the discussion of the nature of truth at the 'Colloque de logique' connected with the International Congress of Philosophy at Brussels in 1953, he stated that, while the choice of a method and of a deductive framework for the investigation of the foundations of mathematics may soon be settled – he did not say it was settled yet – still the solution of the problem of the *Auffassung*, of the notion of mathematics, seems to belong to the indefinitely removed future, on account of the different schools among mathematicians.[16] We saw something about that yesterday: the different approaches, the different ideas of what mathematics is. So the question of the nature of mathematics is awaiting a solution in the distant future, and while we could conceive possibly a mathematical logic as the generalization of mathematics and in that way say that mathematical logic possesses serially analytic principles, still, until we know exactly what mathematics is, we do not have any foundation for proceeding along that line.

If (*if*, I say, I do not say absolutely), if that triple division is adequate, so that analytic principles either are absolutely so, as in metaphysics, or provisionally so, as in empirical science, or serially so, as in mathematics, then we are left with the second alternative. What has been going on in mathematical logic has been an empirical inquiry. They are very far from thinking of it that way, of course. But still, this is what has been happening: just as in the sciences you have a series of hypotheses that are tested, so in mathematical logic there has been developed a series of systems. These people have been heading to a determination of what precisely a mathematical-logical system is or has to be. In other words, we have to consider the various systems that have been proposed so far as provisionally analytic, as having the type of truth that is found in the provisionally analytic principle. They started out with an ideal of a deductive system. They were out to determine its properties. Their investigations have led to various discoveries, particularly to the discovery of Gödelian limitations of various types, and so they are gradually trimming down the ideal from which they started into accordance with logical fact or logical possibilities.

Now this suggestion is not just an a priori conclusion from the fact that they are not thinking of metaphysics, and that they do not have any determinate notion of mathematics to say that the logic is just a most general form of mathematics, and consequently that we are left with the third alternative. It also seems confirmed by the fact of the succession of mathematical-logical systems. As I remarked before, you will find in A.N. Prior's book *Formal Logic*, in an appendix, a brief exposition of the axioms and

16 Bernays delivered no paper at this Congress, but to the article referred to in chapter 6 below, p. 153, one may add P. Bernays, 'Die Erneuerung der Rationalen Aufgabe,' *Proceedings of the 10th International Congress of Philosophy*, Amsterdam, 1948 (Amsterdam: North-Holland, 1949; Liechtenstein: Kraus Reprint, 1968), vol. 1, 42–45.

rules of derivation of thirteen mathematical-logical systems.[17] And besides those listed there, there is also the combinatory logic developed by Curry and others, which is of a more general type and of a different type. And finally, there are new systems coming out, for example the work of Hao Wang mentioned yesterday, in which there is a series of levels.[18]

Now, if we consider those systems enumerated by Prior, we find that some are equivalent systems. For example, Łukasiewicz's system is equivalent to the classical propositional calculus that we find in Russell-Whitehead and in another form in Hilbert. So there are a number that are equivalent, and they just differ in the axioms through which they proceed. On the other hand, there are different types that are mutually incompatible, and if they are mutually incompatible they are not all the one and only logic. Consequently, there arises the problem of truth, and what we mean when we say that they contain fragments of factual truth is that each one, as far as it goes, is true enough, but we cannot consider any one of them as exclusive unless it is one that embraces all the others.

We treat this in considering four different types.[19] First of all, with regard to what is called the classical propositional or sentential calculus, there is no doubt about the existence, and so the factual truth of the occurrence, of propositions, of negative propositions, and of compound propositions in the truth-functional sense. We have, 'There exist propositions,' 'There exist negative propositions,' and 'There exist compound propositions such as *Apq* or, in the Russell and Hilbert notation *P* or *Q*, not neither *P* or *Q*.' If you find that propositions of that type exist, then everything in Russell and Whitehead's logic and everything in Hilbert's logic and everything in Łukasiewicz's logic follows, and since such propositions occur, since you cannot talk at all without using propositions and negations and some compounding propositions, then that whole logic follows.

However, there is this catch: in those logics, there is built in 'and nothing else,' and that is why I say 'fragments.' In other words, there are other things that are not taken account of in that logic. So if you want to proceed from one of these *CPC*,[20] in any equivalent formulation, as an analytic proposition, to an analytic principle, you can admit what they assert, the

17 See note 17 on p. 15 and note 34 on p. 35, above.
18 See the text above, at note 21 on p. 19, and notes 48 and 50 on pp. 65–66.
19 See below, pp. 154–55.
20 *CPC*: 'Classical Propositional Calculi,' as, for example, listed by Prior, *Formal Logic* 301–303.

relevant existential statements, that there are propositions, there are negative propositions, there are compound propositions in the truth-functional sense. But, insofar as the system exhibits as a system, or implicitly supposes, that as far at least as this logic is concerned, there is nothing else that is relevant, nothing that cannot be had out of functions of that type, you hold back. There is something else, and that is why other logics arise. Consequently, I say that that system contains fragments of factual truth. It is true, as far as it goes, in an absolute sense. But it does not give you the whole truth; it is not a complete logical system. When I say 'not complete,' I mean that I do not want to confine things to the propositional calculus: for the '*p*' you can substitute functional statements, and quantify them, and put in or treat relations or classes in all their complexity, but still you are not dealing with a complete logic. You accept it *in sensu aienti*, as the theologians or philosophers say, not *in sensu exclusivo, in sensu negante*, in which anything that is not contained in their axioms and rules of derivation is ruled out of consideration.[21]

Again, there is no doubt about the factual truth of strict implication, and there is no strict implication in that system. You have a strict implication when your antecedent has an intelligible nexus with your consequent. A strict implication is a complex sentence: If the flag is at half mast, the mayor is dead. There is an intelligible connection between the two parts, and it is because of the intelligible connection as verified that you assert that proposition. But in that calculus, all you have is juxtaposition. You have no intelligible nexus, and there is no way of expressing a strict implication within that calculus. All the combinations and propositions that occur within that calculus are of the type 'not neither *P* or *Q*' or 'not *P* without *Q*.' 'Not *P* without *Q*': if you have a *P*, you will have to have a *Q*. But their sense of that is not the sense of an intelligible nexus. This proposition, *Cpq* in the Polish notation, or *P* ⊃ *Q* in Hilbert, does not mean for them that there is an intelligible nexus between *P* and *Q*. They mean a new whole, a new proposition composed of those elements, in which you never have *P* true and *Q* false. So, 'If bananas grow on trees, Eisenhower is President of the United States' satisfies that. It does not give you any intelligible nexus between bananas growing on trees and Eisenhower being President of the United States. But that is a satisfactory interpretation or statement, to substitute for the symbols '*P* implies *Q*' in the sense of that calculus.

---

21 These distinctions were familiar to Lonergan's audience from their training in Scholastic debates. '*Aienti*' is from the present participle of the incomplete verb *aio*, to say, to affirm.

Consequently, C.I. Lewis developed a modal logic that includes strict implication. In fact, he developed several systems, eight in all, each adding more on in the way of the axioms.[22] Now, there does exist strict implication in the sense of 'If the flag is at half mast, the mayor is dead,' in the sense of an intelligible nexus. For example, in any mathematical logic, in any such system, the conclusions follow from the premises in virtue of the rules of derivation. Any mathematical logic is an example of a strict implication. So you cannot affirm a mathematical logic without affirming a strict implication. Therefore, it is also true that there exists strict implication. Consequently, insofar as there are mathematical logics that can represent strict implication, as that first class of systems cannot, you have systems that contain an element of factual truth, and *in sensu aienti* they are to be affirmed.

However, Lewis's systems are not all compatible. He has eight: 4 and 5 are not compatible with 6, 7, and 8. There are different ways of developing the basic idea, so you cannot accept the whole eight of Lewis in an absolute sense. You can accept them provisionally for different purposes, but you do not have a universal factual truth in the whole eight. You can have it only in some of them, namely, those that are not incompatible.

But to go into discussing the differences between the different systems in Lewis is beyond our point. We make our point sufficiently by comparing the main divisions of mathematical-logical systems. We have considered a first type, which comes under the classical propositional calculus and its extensions. We spoke of a second type, systems that include strict implication. Under the name 'modal systems' there are also included systems that involve propositions: 'it is possible,' 'it is necessary,' and of course, when this is put into symbols, the logicians need to know about propositions, 'that it is possible,' 'that it is not impossible,' 'that it is necessary,' 'that it is not necessary,' and they find it hard to give a meaning to the compounds of possible, impossible, necessary, not necessary. That forms something of a paradox for them, but we need not go into this. I just note the fact that they come under the name 'modal logic.'

22 See note 37 on p. 35, above. Lonergan may be referring especially to appendix II (492–502), written by Lewis, of Lewis and Langford, *Symbolic Logic* (see the same note). Appendix II is called 'The Structure of the System of Strict Implication.' The 1959 edition has an appendix III, written by Lewis in January of that year (503–14), 'Final Notes on System S_2.' See also William Tuthill Parry, 'Modalities in the *Survey* System of Strict Implication,' *The Journal of Symbolic Logic* 4 (1939) 137–54, and the articles by J.C.C. McKinsey in the same journal 1940–41.

In the third place, according to Aristotle the statement, 'Tomorrow there will be a naval battle,' is neither true nor false.[23] It is the contingent future. And Aristotle's reason for this is that, if I assert this proposition now, then I must now have sufficient evidence such that it is impossible that tomorrow there could not be a naval battle. And if there is sufficient evidence now for it to be impossible not to have a naval battle tomorrow, then the naval battle will not be contingent but necessary. Consequently, either the event is not contingent or it is not future. But if there are propositions that are neither true nor false, then a logic that is two-valued, that says every proposition is either true or false, is not adequate. So you have three-valued logic, in which propositions are true, false, or indeterminate, unknown. The classical propositional calculus, logics of that type, logics of strict implication – in general, the mathematical-logical systems – are two-valued. They have not room for propositions that are neither true nor false. Consequently, if you hold with Aristotle and St Thomas that events either are not contingent or not future, then you have to acknowledge that there are propositions that are neither true nor false. So, if you want a complete logic, then you have to say that there is an element of factual truth in three-valued logic. Thus, you get another fragment of mathematical-logical system that you have to consider true.

Finally, as long as human knowledge is *in potentia*, as long as it is in genesis, as long as it is developing, you do not have a fixed set of concepts that are not open to revision. For example, in physics, in any empirical science, scientists will all tell you, 'We are not propounding this as definitive; we are propounding it as the best opinion available at the present time; we're doing pretty well, but we are not saying that we have the last word in this field.' Similarly in matters of common sense: men of common sense are apt to be very pertinent, downright: 'that is all there is about it.' But when a Socrates comes along and wants them to define their terms and say exactly what they mean by this and that, they find that their only way of providing a solution is by inviting him to drink the hemlock. Common sense does not go into accurate definition, and precisely because it does not commit itself to very accurate definition, it is able to be so certain. In other words, it has a general idea, and it is much more certain when it is not trying to say exactly what it means, but rather remaining content with having its own general ideas on the subject. When it gets down to analyze specific things, it is not done accurately. In general, then, you cannot admit a blind application of the principle of excluded middle insofar as (1) your concepts are open to

23 See note 39 on p. 36 above, and p. 340 below.

revision, that is, your basic concepts are not hard rules, regular little nuggets that are not going to change a hair's breadth in any direction down the ages, and insofar as (2) principles, primitive propositions, are not simply immutable, so that no one is going to come along with a further distinction at any stage in human development that will throw a little more light on this subject you are handling at the present time. Insofar as you are not in that position of gazing at the Platonic Ideas, in which you know just what everything is, insofar as your knowledge is in process and you have half an idea on the thing, and you could get an awful lot more light on it (and that is the normal human situation), the principle of excluded middle cannot be applied blindly. In other words, it is true ultimately that you are going to affirm any given proposition or deny it. But until you know the accurate formulation, the formulation that covers all the facts with perfection, you are not committed to selecting either one of *these* contradictories. You may prefer to formulate it in a different way when you get a little more light further on. The classical example is the loaded question, Have you stopped beating your wife? According to the principle of excluded middle, you have to say yes or no. And there is no possible alternative. But everyone wants to reformulate the issue because the question has a false presupposition: if you have a wife, you have been beating her. If you posit excluded middle as a law that is built into an automatically operating logic, then you are presupposing that all your concepts have received a definitive formulation, that there are no further distinctions to be added to them at any time in the history of human thought.

Various logical systems contain elements of factual truth. There is factual truth to the negative and compound proposition, and in that sense the classical propositional calculus possesses truth. There is factual truth to strict implication, and in that sense Lewis's systems contain factual truth. There is factual truth to the indeterminacy of the contingent future, and in that sense the three-valued logic contains factual truth. There exist states of knowledge in which everything is not all chrome and steel, all fixed up with sharp edges, absolutely determinate, and until you get things in the perfect situation, you cannot apply blindly the principle of excluded middle, and therefore there is factual truth in logics such as the intuitionistic logic and the logic of Fitch that weaken excluded middle or restrict it in some fashion.[24] This series of distinctions, this nuanced answer to the question of

---

24 Frederic Brenton Fitch, *Symbolic Logic: An Introduction* (New York: The Ronald Press, 1952).

the truth of systems of mathematical logic, fits in with our previous general assertion that the truth of systems of mathematical logic is like the truth of the principles reached in empirical science: they are provisionally true. It is not exactly the same situation, because it is not in an empirical field such as the empirical sciences, but what they have been doing has been gradually determining with greater accuracy and greater fullness what the properties of deductive systems are. For example, each of the thirteen types enumerated by Prior contains a significant measure of truth, but if you want a complete and adequate logic, you have to combine them all together. And I think that you have to add the series of levels such as are suggested by Hao Wang in the article already referred to.

4

The Foundations of Logic[1]

We have considered, first of all, the general idea of a mathematical logic; secondly, what happened in the development of the idea; and thirdly, the general question of the truth of the various systems that have been propounded. Today we come to the question that I have called the foundations of logic.

1   Traditional Logic

Originally logic was the clearheaded assertion of human rationality against sophistry, against the abuse of language to deceive mind. Again, logic meant a distinction of rationality as apart from rhetoric. Rhetoric is rational enough, of course, but it is concerned with persuasion with regard to the contingent. That was the Greek cultural context within which logic arose.

Logic was not a *pure* assertion of rationality, however. In other words, it was connected with techniques. For example, there were techniques involved in figures and moods of syllogisms, like the reduction of the second and third figures to the first and the various other things that appear

---

1 The fourth lecture, Thursday, 11 July 1957. A question and discussion session was announced as scheduled for the evening, in addition to those sessions that were included with the lectures. No recording of the evening session has been found.

It seems clear from the tapes that Lonergan rearranged some points, so that the lecture follows a somewhat different order from that indicated in the notes that he distributed. We will call attention to this at the appropriate places.

traditionally in manuals on logic. For this reason, the affirmation of human rationality that you have within the Greek context, in which logic is something different from rhetoric and opposed to sophistry – it appears in an earlier form in the Platonic dialogues as the distinction between eristic and dialectic – was not a pure affirmation of rationality, a pure affirmation of rational discourse. It was connected with techniques for the more or less automatic securing of the end at which it aimed. However, up until twenty or thirty years ago logicians did not stress the technical aspect to any great extent.[2] No one was worried, either, whether you thought the fourth figure of syllogism was just so much nonsense; no one paid much attention to the issue. Occasionally, someone would go into such technical questions, but in general these questions did not seem a matter of any concern.

There is another aspect as well. The phenomenon of philosophic difference that existed in the Greek world and in the Medieval period and that became most pronounced in the fourteenth century and has continued to grow in modern philosophy traditionally was contained, as it were, within a field called major logic or epistemology. Logic, minor logic, was considered a sacred little preserve in which all men of good sound sense were in entire agreement. There were no disputes in that field. This unquestioned and conventionally unquestionable status enjoyed by minor logic probably was one of the principal motives that moved the mathematicians, when they were looking for foundations of their science, to seek to base it all on logic. In other words, most of philosophy was just a jungle in which wild beasts lived and no one safely ventured, but there was a preserve in which everyone seemed to be agreed. Such agreement was taken as a mark of something in the way of scientific thinking. Consequently, it was thought, it would be a good thing to base mathematics on something more general than mathematics itself, that is, on logic.

2   The Changed Situation

The development of mathematical logic has changed that situation. We are not where we were before. We cannot conceive logic simply as it was conceived by the Greeks or as it was conceived in the Medieval period or as it was conceived, say, fifty or even thirty years ago. The reason is that there is a new fact. There exists a new complex of logical techniques. We cannot say

---

2 Lonergan added in an aside in the middle of this sentence, 'It depended, of course, on the authors.'

that we are entirely unaffected by it simply because there have existed technical elements traditionally within our own logic. In my opinion we have to concede that mathematical logic as a technique is far superior to traditional logic considered as a technique. In other words, traditional logic as a matter of jingles, such as 'Barbara, Celarent, Darii, Ferioque prioris,'[3] and so on, traditional logic conceived as a technique, has been supplanted by a new technique that admits infinitely more in the way of refinement, of precision, and of capacity to deal with large, complicated questions in logic such as the properties of deductive systems. I think that is a fact. I may be wrong, but that is what seems to me to be so.

Another point has to do with the question of the name 'logic.' The question of a name among philosophers has been traditionally considered a minor point, and quite rightly. Still, the use of a name can generate an awful lot of heat, and the fact is that in a broad and steady stream of articles and books the word 'logic' now does not mean any concern whatever with the immutable laws of thought, as logic used to be conceived; it means rather familiarity with some or all of a set of symbolic techniques. We may take different attitudes with regard to that use of the name 'logic.' We may resist its usurpation, or on the other hand consider it not worth the fuss: let them keep the name. There is a problem there, and it is a problem of naming. It would be a considerable difficulty to convince a number of these mathematical logicians that there was anything in traditional logic except the techniques. The issue that arises between mathematical logicians of a somewhat closed mentality and Scholastic philosophers generally on the matter is something that we will come to later. We are just noting the existence of a problem with regard to the name 'logic.' They have taken over the name. They use it to denote their set of techniques. Are we going to let them keep it? Or are we going to challenge their taking it? I do not think that it is an important question, but the fact is that the use of words generates an awful lot of heat very easily.

In the third place, there are differences between the techniques in Aristotelian logic and the techniques in some of the mathematical-logical systems. One can think that those differences are a matter of high concern, and of course one is moved to that judgment all the more readily when people make the differences an occasion for ridiculing traditional logic. But I think it would be a strategic blunder to pay much attention to those

3   Part of a Scholastic jingle used to recall the classification of syllogisms; the jingle would be familiar to Lonergan's audience.

quite minor differences. No matter what your logic is, if it is coherent, you can develop a technique to express it. As far as Aristotelian logic is conceived in terms of classes, it has already been provided with a logical formalization. The issues do not lie simply in technical differences: that is not the important point.

What is essentially new in the present situation is the invasion of the field of minor logic by philosophic differences that formerly were contained by convention, by common consent, outside logic. Formerly, no one was surprised if people held widely different views on epistemology or ontology or psychology or natural theology. We had become accustomed to the fact and recognized it. But it was taken for granted that everyone would agree upon logic. That was a transitory situation, from the nature of the case. If there are profound philosophic differences between people, those differences are bound to come in and influence their logic. So, at the present time, it is not possible to treat logical questions adequately – I mean by that, fully, going to the roots of all differences – in the old fashion in which it was previously possible. In other words, fundamental philosophic differences give rise to differences in the conception and use of logic. That fact is simply an instance of the interdependence of all provinces of human knowledge. The fact of that interdependence and its manifestation in the present case should not be unwelcome to any serious philosopher. It is the interdependence of all departments of knowledge and their relations with one another that provide the philosopher with one of his principal modes of entry and the possibility of his having something to say on all fields, at least to some extent.

Connected with this development of mathematical logic, there have arisen two philosophic movements. The first is logical atomism, from Russell, and the second is logical positivism, from Wittgenstein.[4]

For Russell, the discovery of mathematical logic was a philosophic liberation. He could not get round Meinong's position, in which everything you spoke of had to be something; otherwise, you were not talking of anything.[5] If you said that a square circle was a contradiction in terms, well, what was

4 In Lonergan's notes (see below, p. 157), these two movements are mentioned at the end of the section on the symptoms of technical ambivalence. In the lecture Lonergan includes this discussion as part of his treatment of 'the changed situation.' This new order is continued in the next section, on the question of foundations, which in the notes followed the treatment of the ambivalence of technique rather than preceding it. See below, at note 10.

5 The early Russell was sympathetic to Meinong. See Russell, 'Meinong's Theory of Complexes and Assumptions,' *Mind* 13 (1904) 204–19, 336–54,

this square circle that you were talking about? You may say that the square circle is impossible, but to be talked about it somehow had to be possible. Russell discovered that you could say that there is no *x* where *x* is both square and circular, and in that way express the impossibility of a square circle. He eliminated the phrase 'square circle' and said, 'Where *x* is square and *x* is circular, there is no *x*, such that *x* is square and *x* is circular.' He eliminated the expression 'square circle,' and that for him was a philosophic liberation.

The fundamental idea in logical atomism is the logical calculus.[6] You can substitute ordinary words for the variables, and ordinary sentences for the propositional variables, and the logical connectives are all that you need to link together atomic experiences: white, here, now. All that you need to build up knowledge out of such atoms as white, here, now are the connectives provided by a logical calculus.

In both these movements, it was pretty well taken for granted that the type of logic found in *Principia mathematica* was the last word in mathematical logic, that there was nothing new to be done in the field, that it was done very satisfactorily in that work. In general, that position presupposes that a language is equivalent to a logical calculus plus a vocabulary. That presupposition need not be quarreled with, provided the logical calculus is sufficiently complex and nuanced. But it was also presupposed that no questions were to be asked about the logical calculus itself, and that is the important presupposition for us to challenge if we want to argue with these people. They are pretty well off the map now, of course, but what does that presupposition mean? It means that all the fundamental things like affirmation and negation and so on are taken care of by the calculus. It is from those fundamental elements that you get all your connectives. Now it is in those connectives that all the fundamental elements of philosophy, of a metaphysics, come to light and are contained, insofar as they are represented in a language. Just to accept the logical calculus and ask about the tailpieces, the 'white,' 'here,' now' that you substitute for your variables, is to eliminate philosophic problems or to kid yourself that you do not have to bother your

---

509–24. In Russell's introduction to the second edition (1937) of his *The Principles of Mathematics*, he speaks about the steps of his withdrawal from a type of Platonism, and says (p. x), '[The] theory [of descriptions] swept away the contention – advanced, for instance, by Meinong – that there must, in the realm of Being, be such objects as the golden mountain and the round square, since we can talk about them.' On Meinong and his place in European and British philosophy, see J.N. Findlay, *Meinong's Theory of Objects and Values* (2nd ed., Oxford: Clarendon Press, 1963).

6 See below, pp. 165–66.

head about them at all. The real philosophic problems are hidden in the logical calculus and in the supposition that no man of sense is going to ask us any questions about that. 'Either x or x implies x': the triviality of the axioms of the logical calculus was taken as a guarantee that there are no philosophic questions to be raised of anything connected with the calculus. And if there are no philosophic questions to be raised with regard to that, then logical atomism becomes a detractive program. Logical positivism arises – there will be something more explicit on that tomorrow[7] – when they posit the verification principle. Either logical functions have no conditions of their truth, and then they are tautologies; or under no condition can they be true, and then they are contradictions; or if one or more of the variables are true or false, then the function will be true or false, and to decide which it is you have to have some extralogical means of deciding for or against the proposition. That led to the affirmation of a verification principle, and the problem on which logical positivism went on the rocks was the question, Is the verification principle a tautology, or is it something you have to verify? And how do you get a verifiable verification principle?

On both these movements, there is an excellent little work by Urmson, *Philosophical Analysis*.[8] Note also the statement by a Fellow of Magdalen College, on the third program of the BBC in 1956, 'I am not, nor is any philosopher of my acquaintance, a Logical Positivist.'[9] Those movements in their original forms are dead at the present time. They still hold for philosophical analysis, but it is hard to see that they are doing more than going through the old motions having renounced the old objectives. However, that is a further matter; it is a little difficult to say what is going on in a movement at the present time unless you are right on the spot.

3   The Question of Foundations

This brings us to the question that we want to treat today, the question of foundations.[10] There are many mathematical-logical systems, and while

---

7 See below, pp. 165–66. In fact, Lonergan did not treat the issue in the fifth lecture.

8 J.O. Urmson, *Philosophical Analysis: Its Development between the Two World Wars* (Oxford: Clarendon Press, 1956).

9 In Lonergan's notes (see below, p. 157), the statement is attributed to G.J. Warnock, and reference is made to A.J. Ayer et al., *The Revolution in Philosophy* (London: Macmillan, 1956), where the statement appears on p. 124, in Warnock's essay, 'Analysis and Imagination.'

10 See below, p. 157, the section '5 The Question of Foundations.' Note that the order of presentation here differs from the order of the outline. See above, note 4.

some are equivalent many are not. This fact posits as a problem the theory of choice of any given system either absolutely, or relatively to a particular task.

In the second place, when the field to be formulated is as complex as the theory of arithmetic, the mathematical-logical system becomes extremely complicated. You have to go into transfinite induction to handle arithmetic, so that the logic is not a source of greater evidence but of bigger problems. What is the basis from which these larger problems are attacked?

Thirdly, while it is the problem of the infinite that makes the logical formalization of arithmetic so complicated, there seem to me to be parallel problems when you are dealing with the finite domain of ordinary matters of fact. You come upon a sort of infinity there that is far less regular than the infinity you are dealing with in arithmetic. It is not an infinity that you can foresee and work out; it is not predictable. But the mere fact that human knowledge is progressive, that we start out with a *tabula rasa*, that each individual grows in knowledge and is continuously capable of advancing further in knowledge, that societies advance in knowledge, and so on, provides a continuously changing background and context to any type of knowledge in the concrete sphere. That gives rise to a complexity that is the source of difficulties just as serious in the concrete domain of matters of fact as the difficulties you have in arithmetic.

The reasons for that I put down under two headings. First of all, ordinary concepts are not the simple, smooth, regular, homogeneous nuggets needed to conform to a mathematical-logical system. In other words, the symbols have to denote the same thing right through, and you have to know exactly what that same thing is, but ordinary concepts are simply not of that type. You can get a lot on that from the book *Insight*, and we will have something more to say about it tomorrow.

Secondly, there is the matter of forms of thought. There was a man by the name of Leisegang[11] who was concerned with forms of thought as exhibited in languages. He set about studying the way in which the expression of thought occurs in Heraclitus, in St Paul, in St John, in Hegel, in Aristotle, in Plato, and so on, and became interested in what he called *Denkformen*. The book has marked deficiencies, and we cannot discuss it in detail at the present time. But it has proved extremely stimulating. Another relevant writing is by Hellmut Stoffer, 'Modern Attempts at a Logic of *Denkformen*,'

---

11 Hans Leisegang, *Denkformen* (2nd ed., Berlin: Walter de Gruyter, 1951). The first edition of 1928 was followed by studies of Hegel, Kierkegaard, Dante, Goethe, and others. Hellmut Stoffer draws his inspiration from this 'Pionierarbeit.' See the work cited in the following note, pp. 442, 621.

in *Zeitschrift für philosophische Forschung.*[12] He presents an extremely fully documented pair of articles, with references all along the line to different works. There is a terrific bibliography connected with these two articles, because they are review articles of the situation. He argues that you need to distinguish six types of logic if you are going to deal with *Denkformen*, thought patterns, and he distinguishes them as plane, dialectical, existential, magical, mystical, and hermeneutical. All mathematical-logical systems would be contained within the first, as would traditional logic and perhaps a bit more. The discussion of the first five would fall under the sixth. You will find in *Insight*, chapter 17, in the section on the truth of an interpretation, some basis for the ideas connected with hermeneutical logic. Such a logic is pertinent when you are understanding why so and so understands someone else in such and such a manner, and why someone else interprets him in another manner, and so on.[13] It provides a logic for theories of interpretation. A friend of mine teaching philosophy at McGill in Montreal, who did a degree in philosophy in Berlin, told me that in his time there were fifteen professors expounding Kant. Fourteen had each his own different interpretation of Kant, and the fifteenth refuted the other fourteen! Now, when you want to deal with the logic of that situation, you are in a hermeneutical logic. There is the question of what Kant thinks, and the question of why each of the fourteen thinks what he thinks, and the relations of them to the fifteenth fellow who disagrees with them all and talks to you about the other fourteen and about Kant. As there is a succession of such professors in most universities of the world, the general problem is considerably more complex. That is the question of hermeneutical logic. We run into the same problem in regard to St Thomas, Aristotle, the scriptures, and so on. The problem has to do with the ultimate context in which to discuss the differences between the different logics.

This direction of thought seems confirmed by the English experiment as described by Urmson. They discovered that ordinary language follows patterns that do not fit in with the logical calculus. The patterns of thought in commonsense discourse have their own fundamental laws, and whatever they are, they are not something you can reduce to a couple of trivial

---

12 Hellmut Stoffer, 'Die modernen Ansätze zu einer Logik der Denkformen,' *Zeitschrift für philosophische Forschung* 10 (1956) 442–66, 601–21. Lonergan added: 'At the present time, that is the official organ of the association of all German professors of philosophy. *Kantstudien* retains its old position, but since the war this journal is among the first journals in Germany.'
13 *Insight*, chapter 17, §3.1

axioms, as has been the ideal in some of the mathematical-logical systems. Just as the mathematical logicians step in and take over the technical element in Aristotle and improve upon it enormously and then someone claims that this is the basis of language, so others study language, which is, after all, the ordinary technique of communication, and find in it thought forms that are not the forms of mathematical logic but that pertain to entirely different types that fit under these six headings. Hermeneutical logic has been developed in Germany mainly in connection with Heidegger and existentialist presuppositions of that type.

My fourth point under this heading of the question of foundations is that developing intelligence, on the one hand, and perfectly transparent expression, on the other, form a dialectical couple.[14] To go all out for either one is to sacrifice the other. If you want to make people understand, the way you go about it is not by giving them formal proofs, but by figuring out how much they understand of the question already and determining what signs you can make so that they will catch on and get the next point and then what signs you can make so that they will go on to the further one, and so on. You gradually build up their understanding. It is a matter of person-to-person communication. I figure out from someone's reactions, from when he looks dumb and when he smiles, just how much he is getting of what I'm trying to say. From my familiarity with his background, which may be greater or less, I am able to throw out the signs, the hints, the clues, to use the tone of voice, to employ the gesture, and so on, that will lead him to the further act of understanding. That type of communication of understanding is something such that after the understanding has been achieved you can say, 'Now let us put this all down in perfectly rigorous logical order.' That can be done when you are both on the same level of understanding. On the other hand, if you go entirely out for the rigor, what happens? If, for example, you do Euclidean geometry in a purely formalized fashion, you can go right through Euclidean geometry in virtue of the axioms, but when you get to the end of it and want to go on to something else, you have to take a terrific leap. But if you do Euclidean geometry by adding on casual insights as you go along, then you are not moving along on a flat plane. There are all sorts of little steps upwards that you are taking, and when you come to something, to a new field, it is not a terrific leap, it's just a larger step upwards. So there is an incompatibility, there is not an exact concomitance, between development of understanding, on the one hand, and

14 See below, p. 158. See *Insight*, chapter 7, §5; chapter 17, §2.4.

perfectly rigorous and transparent expression, on the other. You can put your language into the armor of Saul, but it will not have the agility of David. On the other hand, if you have David without the armor, he will be open to difficulties and objections that he would not be vulnerable to if he had the armor on. Still, he will be able to move about and get things done in a way that would not otherwise exist.[15]

4   Symptoms of the Ambivalence of Technique[16]

4.1 The Ambivalence of Technique

We come now to the fundamental point. These previous things have been approximating to the issue. What is the real issue? There exist logical techniques that are virtually independent of any particular mind. They are equivalent to computers. After you get so far in a mathematical deduction, there is no question of anyone coming along and saying, 'The way I feel about it, this should not be x and y, but something else,' or 'The way I see it,' or 'The way I understand it,' or 'Really if you were to ask my opinion.' No one asks you what you feel about it; no one asks your opinion. The thing runs out automatically. That is the meaning of the logical technique that is independent of any particular mind, independent of the strokes of genius or the aberrations or deficiencies of any particular mind. It goes along like a steamroller. It can handle problems too complex to be considered previously. But it also raises the question whether the techniques are simply a tool for mind or rather a substitute for mind.

That way of putting the question is a little too bald. We can put it more accurately as follows. One may, for one's own personal satisfaction, work out a theory of truth as we did yesterday and come to the conclusion that in certain senses these mathematical-logical systems are true. On such grounds, because one considers them true, one becomes rationally committed to the acceptance of the systems, either hypothetically (namely, if one were to use them) or, with restrictions that vary with various systems, absolutely. Now that is the attitude that we have spontaneously taken. We have wanted to know whether these mathematical-logical systems are true and in what

15 Lonergan added, in an aside, 'That is just a throw-out.'
16 See below, pp. 156–57. This section includes material from Lonergan's '3. The Ambivalence of the Technical Achievement' and '4. Some symptoms,' both of which preceded the section on the question of foundations in Lonergan's distributed notes. See below, p. 157.

sense they are true, in what sense they are factually true, totally so or fragmentarily. But that is only one possible attitude. One may consider, and many do find it more pleasing to consider, the rational commitment to truth as a private and minor matter. Then what alone counts is the external fact that you employ some language and that any language presumably can be reduced to a logical calculus plus a vocabulary. What really counts is not anyone's private and internal dedication to truth, which is an extremely complex and abstruse matter, but the public and external fact that he talks or ventures to write. The ambivalence of the technical achievement is an ambivalence that arises when you consider the human attitude towards the achievement. Thus, it seems to offer the alternatives of either, on the one hand, a rational commitment to truth, and on that ground an acceptance with restrictions of the mathematical-logical systems, or, on the other hand, a nonrational, pragmatic acquiescence in the fact of talk.

We can put the matter differently. Bronislaw Malinowski, in his *Magic, Science and Religion,* did some very fine work from the viewpoint of the anthropology of primitives.[17] The book is something of a classic. He shows that while the minds of the Trobriand Islanders are full of myths and all sorts of magical practices, still this nonsense does not interfere in the least bit with their plowing and preparing the soil, planting the crops, taking care of them, reaping them, and so on. The magic and the myths deal with spheres in which they have no concrete everyday evidence to guide them. But there is also the sphere of light, and that is guaranteed to them by the result that if they do not do this about their crops, they starve. They have concrete factual evidence with regard to that, and there they do not make mistakes. Where the magic and myth creep in is in spheres outside that.

What is the point? Modern scientists and modern mathematicians exhibit a very strong tendency towards pragmatism in their attitudes. There is a conception of science that makes science not something based upon a conquest of mind from which myth and magic have been eliminated, but something based upon an expansion of the everyday pragmatic field. Is science right? Well, it works. 'It works' is the ultimate answer. It is just as if you asked these primitives why they planted their crops in a certain way; they would have their equivalent to 'it works,' and that would be, as far as they could express it, the ultimate answer.

---

17 Bronislaw Malinowski, *Magic, Science and Religion and Other Essays,* selected, and with an introduction by Robert Redfield (Boston: Beacon Press, and Glencoe, IL: The Free Press, 1948).

Yesterday we set up a criterion of truth in terms of the virtually unconditioned, but that is a difficult notion to grasp. It is much easier to say, 'It is obvious,' and when you are asked, 'What do you mean by "It is obvious?"' to say, 'It is manifest.' Unless people have been brought up within an intellectualist tradition, they find it extremely difficult to give any meaning to the notion of evidence other than 'it works.' People with that pragmatic mentality at the present time inevitably are extremely numerous, because they have never studied philosophy at all, and in their cultural traditions there is lacking the intellectual element that exists in ours. They are led to think of the ultimate criterion as 'what works.' That is the ambivalence of the technical achievement. So the question we put yesterday of the truth of mathematical-logical systems is a question the raising of which leaves them wondering what on earth you could possibly mean. What you mean by it would be a real problem for them. Insofar as they could give a meaning to an answer to it, their answer, due to their horizon, their limitations, their cultural traditions, and so on, would be, 'It works.' While we disagree with them, we must have a profound sympathy for them. After all, on the whole, we have not been too brilliant, even with our traditions and opportunities, in emphasizing the importance of the purely intellectual elements. As I see it, that ambivalence of the technical achievement is the fundamental issue raised by mathematical logic. When I said earlier today[18] that there has been an invasion of the field of logic by philosophic differences that formerly had been contained outside the field of logic, I was referring to this ambivalence of the technical achievement. We have the technical achievement, it works wonders in its own line, and it supersedes the purely technical element within Aristotle. But people can accept it, not because of some rational commitment to an ideal of truth. That may be operative within them, of course, but what they will be able to express and manifest is usually a great deal less than what really is effective in their lives. We must not judge the individuals. But there seems to be a definite tendency to say, 'This works, and that is all there is to it.'

4.2 Symptoms

Now I will give you some straws in the wind that seem to confirm that interpretation. In the Proceedings of the Second International Congress of the International Union for the Philosophy of Science, held in Zürich in

18 See above, p. 90.

1954 – that is Gonseth's group – Heinrich Behmann said that strict implication was natural only in the sense that the notion of flat space is natural.[19] He was speaking of 'If, then' in the sense of an intelligible nexus that is an element in logic as it is found in Lewis's modal systems. He said that this is the kind of idea that people have before they get down to the scientific study of logic. It is like the idea that people have before relativity that space has to be flat, or the idea that people have before the rotundity of the earth is established. (I do not know if he used the latter example, but it will serve to bring out his point.) When you first hear that the earth is round, you might say, 'Well, do the people on the other side down in Australia walk with their heads hanging down?' Spontaneous notions of that type are corrected by the advance of science, and among them is the notion of strict implication. What has happened? The development of the technique becomes the criterion of whether or not strict implication is a notion to be taken into consideration.

Again, at the Colloque de Logique held in connection with the Brussels International Congress of Philosophy in 1953, as reported by Feys in *Revue philosophique de Louvain* and published in 1954 in the *Revue internationale de philosophie*, there was a head-on collision between Gonseth and Tarski.[20] Gonseth represents the cultured European who has had a rather broad philosophic formation. As I remarked already,[21] his attitude to the issue of the foundations of science exhibits more familiarity with philosophic problems and the possibility of solutions to philosophic problems, with general philosophical culture, than you usually meet in these writers. At this Colloque de Logique, on the topic of proof, Gonseth set forward his very nuanced views, and Tarski rose and said that he did not see anything rational in them whatever. Gonseth got up and answered every possible thing he could see

19 *Actes du deuxième Congrès international de l'Union internationale de philosophie des sciences*, Zürich 1954 (Neuchatel: Éditions du Griffon, 1955), vol. 2, pp. 97–108.
20 Robert Feys reports on the exchange in 'Le Colloque International de Logique (Bruxelles, 18, 19, 18 et 19 août 1953),' in *Revue philosophique de Louvain* 51 (1953) 594–624, at 618. There is no account of the exchange in the 14-volume publication *Proceedings of the XI International Congress of Philosophy* (Amsterdam: North-Holland Publishing Co., 1953), but vol. 1 contains the paper by Gonseth already referred to above, in note 20 of chapter 2. Gonseth also presented at the Colloque the paper 'La preuve dans les sciences du réel,' *Revue internationale de philosophie* (1954) 25–33, and this sparked the exchange. The discussion mentioned by Lonergan is reported there, pp. 46–64; the Tarski-Gonseth exchange is on pp. 49–51.
21 See above, p. 47 and the references there.

that would be guiding Tarski's position, and Tarski rose again and said that all he could say was that he could not see anything rational in Gonseth's views. As far as one can judge from the report, it would seem that Tarski gave an exhibition of Edwin Schroedinger's remark, 'The specialist is a barbarian.' What did he mean by 'rational'? He was not able to say. He was not able to pick up Gonseth's answers and meet them or discuss them. He was simply able to repeat his position. What he meant by 'rational' probably was 'symbolic and technical.' In other words, the meaning of the word 'rational' has become the *symbolicum technicum*. I will not say that this is so in Tarski, but that seems to be the interpretation that arose. There was an obvious clash between the two men at that meeting, and to my mind Gonseth had it over the other man in the discussion.

But I suspect that at the root of Tarski's objection was the question, What do you mean by 'rational'? Well, unless you have an extremely accurate analysis of intellect, unless you have a grasp of the nature of intellect that enables you to conceive with accuracy the psychological analogy of the trinitarian processions – and that is not too common among Catholic theologians! – you will not be able to state with accuracy what you mean by 'rational.'[22] You have your rationality, and you had it since the age of reason, but to state just what it is, is something that is extremely difficult. When one becomes immersed in these symbolic techniques, the meaning of rationality can become simply the technique. You have the contrast, the opposition, between mind and mind's own creations. With the objectification of mind, what counts, what is ultimate, is the technique.

A third symptom of the dominance of technical pragmatism is the fact that, within mathematical logic, the classical propositional calculus and its derivations hold a dominant position. I should say that it does so, not because of any merits of a logical character, but because it effects an assimilation of logic to mathematical techniques. In mathematics, there is the functional symbolism, the symbolism for functions, and logic can be put within that technical expression – a function has two arguments, true and false – insofar as you drop out the parts of logic, such as the strict implication, that do not fit. The classical propositional calculus holds what seems to be a dominant position within the field because it has the maximum assimilation to techniques that are mathematical, not because it is the best

---

22 Lonergan's *Verbum* drives toward that accurate conception. The *pars systematica* of his *De Deo trino* (Rome: Gregorian University Press, 1964), which will appear in translation in the CWL, establishes the psychological analogy.

expression of what is logical. In fact, I would say that it is the weakest expression of what is logical. It has no room at all for the complex sentence or for strict implication.

In the next session, we will go on to samples of what we mean by foundations of logic.[23]

5   Samples of Foundations of Logic

The technical part in our traditional logic, I would be inclined to concede, has been superseded by the techniques available in symbolic logic. However, in our traditional logic, there are more than techniques. There are ideas in it that supply foundations for logic in a fashion that both reveals the inadequacies of a purely symbolic technical approach and also provides the means of estimating or judging when a logic will be adequate. These ideas also indicate the means by which you can construct a logic that takes into account the spontaneous dynamism of the human mind.[24]

Now I will give some samples of what I mean by foundations of logic, under a series of headings: middle term, subject term, predication, judgment, the levels to Aristotle's explanatory syllogism, and the properties of such foundations of logic. We are going outside the symbolic, technical idea of logic and asking the question of foundations. Logic traditionally has contained two elements, one of which, the technical side, is expounded in mathematical logic; but the other side, the side of foundations, of grounds of ultimate meaning that lie behind traditional logic and that in Aristotle live in symbiosis with the techniques, also has its logical value. And it is by penetrating on that side of Aristotle that we can retain what is good in our tradition and use it to give the proper rational approach to logic in the full sense of rationality. What is our answer to the ambivalence of the symbolic techniques? It is to emphasize elements in Aristotle and the Aristotelian tradition that in the past have been eclipsed to a great extent, and by that emphasis to have some sort of an answer to the people who tend – I just say 'tend' – to speak as though logic were the symbolic technique and nothing else. If they want to call the symbolic technique, and nothing else, logic, perhaps there is no point in disputing about the name. We can call this other material 'foundations of logic' and settle the question of names that

---

23 A break was taken at this point.
24 The diagram on p. 322 provides a useful context for the reflections to follow.

way or in any other way that happens to please people. That is not an issue into which I would care to go.

5.1   *The Middle Term*

What is the real significance of a middle term in Aristotle?[25] In Aristotle a middle term, in its fundamental scientific significance, is a matter of knowing the cause. Aristotle started out from the common Greek name *epistēmē*, which is roughly our 'understand,' and to say what understanding is he used the expression 'knowing the cause.' To make that still clearer and more precise, he invented his doctrine of middle terms. If you interpret Aristotle by saying, 'We will think of middle terms and forget about knowing the cause and understanding,' you are transforming Aristotelian logic into a mere technique and omitting what in Aristotle is the soul of the technique.

What is a middle term, knowing a middle term, as a technical expression of a logical statement, of what psychologically or scientifically would be called knowing the cause? Take the case of the moon and its phases. If this room were darkened and we had a basketball here and a flashlight to illuminate one side of it, then by walking around the basketball or moving the flashlight around the basketball, you would see on it a succession of illuminated shapes. When such illuminated shapes occur on the moon, they are called 'phases.' In other words, there is a nexus between the fact that the moon is a sphere and the shapes that appear night after night in the moon, according as the illumination of the moon comes from the sun at a different angle relatively to us. There is an intelligible connection between phases and sphericity in the moon. That intelligible connection can work in two ways. First, the phases are *causa cognoscendi*, the cause of knowing, the reason that we know that the moon is spherical. Inversely, in the line of *causa essendi*, the cause of being, the sphericity is the cause of the phases of the moon. So, you can argue syllogistically, if the moon goes through these phases it must be spherical; it goes through these phases; therefore, it is spherical. You have an explanatory syllogism in which the middle term is the cause of the predicate in the subject. That cause is the *causa cognoscendi*: the phases are the reason that we know that the moon has to be spherical. Inversely, if you argue that, if the moon is a sphere, it must go through phases of this type, then you are arguing from the *prius quoad se*, what is first in itself, the *causa essendi*, to the *prius quoad nos*, what is first for

---

25 See Lonergan, *Verbum*, index, under Syllogism.

us, what we see, the phases of the moon. That is the significance, ultimately, of the middle term in Aristotle. Besides the technique in which middle terms have certain properties within syllogisms from a purely technical point of view, there is to Aristotle's middle term the notion of knowing a cause, the notion of understanding, and a distinction between *causa cognoscendi* and *causa essendi*. The same distinction occurs in another way as *prius quoad nos* and *prius quoad se*; the phases are *prius quoad nos*, and the sphericity is *prius quoad se*.

What, though, is the status of such notions from the viewpoint of modern science? Are those notions of Aristotle completely out of date? Well, turn to books on modern science. Take a history of physics or a history of chemistry, and what do you find? You find that the history is a development from sensible data, and at every step forward the decisive factor is the uncovering of sensible data. The process is not any deductive process, but a process of gradually moving towards the chemical or the physical system. On the other hand, if you open a textbook of chemistry, what do you find at the start? You do not find the sensible data that led the chemists to discover their chemistry. You find a periodic table, and as the book develops you find that there are over 300,000 compounds, quite apart from mixtures, that can be explained in terms of about 100 elements. The process in the textbook of chemistry is from the *priora quoad se*, and the student, in experiments guided by the theory expounded in the textbook, is confronted with the *priora quoad nos*. You have two movements represented, on the one hand, by the history of a science and, on the other hand, by the systematic exposition of the science in a textbook. The history of the science starts from the *priora quoad nos* and gradually mounts to the *priora quoad se*, to the chemical elements or to the mathematical theory of the various departments of physics. But the textbook will start from the *priora quoad se*, and gradually work to the *priora quoad nos*. So the notion that Aristotle expressed in his middle term undergoes a terrific expansion in the modern sciences; but it is not a notion that is irrelevant to their structure; it is an element that appears within a much larger context.

5.2 *The Subject Term*[26]

We have considered the relation between the middle term and the predicate. Where does Aristotle get his subject term? He distinguished four

26 The central difficulty in what follows is the reading of the word 'question.' The recent surge of interest in the topic *questions* may not exhibit the same

questions at the beginning of the *Posterior Analytics* – I think it is the second book.[27] He says that all questions can be reduced to four: *An sit A? Quid sit A? Utrum A sit B? Propter quid A sit B?* (Is there an *A*? What is *A*? Is it true that *A* is *B*? Why is it true that *A* is *B*?). All questions can be reduced to those four, and in every case the question is a question for the middle term. *An sit A?* (Is there an *A*?) and *Utrum A sit B?* (Is it true that *A* is *B*?): in this pair you want a middle term that will show you the existence of *A* or of *B* in *A*. *Quid sit A?* (What is *A*?) and *Propter quid A sit B?* (Why is it true that *A* is *B*?): in these questions you want a middle term that will account for *A* or will account for *B* being in *A*. And he raises the question, What on earth do you mean by the *Quid sit A?* The questions, *An sit A?* and *Utrum A sit B?* are existential, and *Propter quid A sit B?* is a question for explanation, for a cause. But what is the meaning of the question, *Quid sit?* In the *Posterior Analytics* he is able to offer a partial answer. He says that frequently it happens that by changing the verbal expression you can reduce this question, *Quid sit?* to the question, *Propter quid?* Take, for example, the question, What is an eclipse? You can change that to the question, Why is the moon darkened in this fashion? Instead of asking, 'What is an eclipse of the moon?' you are asking the same question when you say, 'Why is the moon so darkened?' When you get a question in the form, Why is the moon so darkened? you transpose this question to this form.

However, in his *Metaphysics*, book VII, the last chapter, 17 – he has been hammering away at the notion of substance from all angles for the previous sixteen chapters – he comes up with the ultimate answer to this problem of the question, *Quid sit?*[28] There are cases in which you cannot effect that simple transformation from what to why. What is a man? How do you get that into a 'why' question? What is a house? How do you get that into a 'why' question? Aristotle's answer is as follows.

---

horizon that Lonergan found in Aristotle and developed in his own terms. At any rate, see David Sandberg, 'Mathematical Explanation and the Theory of Why-Questions,' in the *British Journal for the Philosophy of Science* 49:4 (1998) 603–24, where an up-to-date English-language bibliography is provided. See also Aron Ronald Bodenheimer, *Warum? Von der Obszönität des Fragens* (Stuttgart: Philipp Reclam jun., 1984); Tadeusz Kubinski, *An Outline of the Logical Theory of Questions* (Berlin: Akademie-Verlag, 1980); and Andrzej Wisniewski, *The Posing of Questions: Logical Foundations of Erotetic Inferences* (Dordrecht, Boston : Kluwer Academic Publishers, 1995).

27 Aristotle, *Posterior Analytics*, II, 2, 89b 36 – 90a 34. See Lonergan, *Verbum* 26 at note 53.

28 Aristotle, *Metaphysics*, VII, 17, 1041a 9–32, 1041b 4–8. See Lonergan, *Verbum* 28 at notes 60 and 61. What follows in the text above is treated more adequately in the subsequent pages of *Verbum*.

You ask the 'why' not with regard to the concept 'man,' since that is what you want to arrive at. You ask the 'why' with regard to *this*. Why is *this* a man? And the 'this' may be, as Thomas remarks, either this supposit or these phenomena. The question, What is a man? is equivalent to the question, Why is this a man? And why is this a man? It is because of the form, the *eidos*, the *aition tou einai*, and in a man that is the soul. Why are these phenomena the phenomena of a man? Because he has a human soul. So you have the material cause and the formal cause; and the two together give you 'man': *this*, the materials, and the form, the *aition tou einai*, the *ousia*, of a man, in the meaning of the seventh book. (In the eighth book, he goes on to show that in material things *ousia* is a compound of matter and form. But in the seventh book, form is taken as substance.) In discussing form, substance – the language is not too steady because he is dealing with a very subtle idea – he has *this* and the form, and the form is the middle term between the *this* and the man. The form is *why* this is a man. Just as you understand that the moon has to be a sphere because of the phases – understanding in the sensible data – so in these sensible data before me, a man who is talking, presenting the phenomena of a man, I grasp in the sensible data a rational soul. That rational soul is the form, the *aition tou einai*, the *causa essendi*, the middle term between what is presented and what is conceived, and man is the subject term. The subject term then comes in the same way as do the predicates. It is just as when you ask, *Propter quid A sit B*, Why is a man mortal? It is because he is compound, composite, and what is put together can be taken apart; that is why mortality is in a man. The notion 'man' itself is generated by a similar process. There is a middle term. You grasp in the act of understanding the form in the sensible data. In Thomas there is phantasm, act of understanding, and concept, definition, where understanding grasps form in sensible data, and understanding expresses itself in a definition. So in the real order, there is in the man the matter and the form, and the form together with the common matter is the grounding of defining a man, and the answer to the question, What is a man? Those are things in Aristotle that reveal the coming-to-be of the subject term. Just as the connection between middle and predicate is given in the notions of *causa essendi* and *causa cognoscendi*, so there are parallel notions that account for the genesis of the subject term.

5.3 Predication

Next, consider predication. When you say that one thing is another, what is the ground from which that activity of predication arises? You have before you phenomena of a man, let us say. If you consider those data as individual, what you arrive at is the unity within those data, the spatially extended and temporally extended series of data in which I grasp a unity, one man, an intelligible unity, an *unum per se*. Consequently, when you consider data as individual you arrive at a notion of the subject, the notion of the substance. When you consider the same data as of a kind, then you arrive at properties. You see that these data are all black, these men are dressed in black, and you have the predicate 'black' or 'dressed in black' from considering data, not as the particular 'black' that there is in a given case, but as of a kind. When you consider data as of a kind, you arrive at properties. These properties can be on the simply descriptive level, as black, round, heavy, hot, and so on. Or they can be explanatory: for example, you speak not of weight but of mass, and mass does not mean the same as weight. Again, you speak of temperature and not of heat, and temperature is not the same as heat.

5.4 Judgment[29]

While what we grasp is absolute and what we know through that grasp may itself be strictly absolute, as God, or relatively absolute, as an analytic principle, nonetheless that occurrence of the judgment in us is a contingent event. There is no process of mental chemistry that runs parallel to a symbolic technique. If the subject uses the symbolic technique to express, to sum up to himself, the fact that there is there a virtually unconditioned to be grasped and grasps it, then he will go on to make the judgment. But unless that occurs in *him*, there is no occurrence of a rational judgment in *him*. That point is of some moment in the analysis of faith, where there is a confusion between the merely verbal expression of the process and what actually goes on psychologically in the individual.[30] What is affirmed, the

---

29 There is a bit of a gap in the tape at this point. When the tape resumes, Lonergan has moved on to the fourth of his 'samples,' namely, judgment, but the first two paragraphs that appear in his notes are not reflected here. Those paragraphs read:
   'A Judgment: virtually unconditioned; if A, then B; but A; therefore B, where A and B are propositions, sets of propositions; where A is simply experience; where "if A then B" is merely implicit insight, invariant of thought, invariant of expression.
   'Significance of syllogism is not Kantian regress to $2''$ premises but manifestation of conclusion as virtually unconditioned.'
30 This is a major point in Lonergan's 1952 work, 'Analysis fidei.' A translation by Michael G. Shields is available at the Lonergan Research Institute, Toronto.

content of the judgment with regard to the existence of a necessary being or with regard to the affirmation of a necessary principle, is absolute. Still, because that event occurs contingently in us, there is a radical fallacy in waiting to be necessitated. People will take the attitude, 'Well, if this is scientific it can be demonstrated, and I can be forced to assent.' And they will wait until the force comes over them, and it does not come, and they become skeptics. To secure the occurrence of the contingent event that is a judgment, there is necessary in the individual a personal commitment, a sense of personal responsibility, where that commitment is intrinsically rational, namely, motivated. The virtually unconditioned within rational consciousness gives rise to the necessity of rational consciousness acting, and that element within rational consciousness is fundamental in judging, and it is fundamental in all philosophy.[31] In the fourteenth century, we have within the Scholastic experience a search for absolutely rigorous arguments. That is what the fourteenth century fundamentally was concerned with. And we have a succession of theologians coming along and gradually whittling down what you really could demonstrate with absolute certainty. What you could demonstrate with absolute certainty all along the line became less and less and less and less, and they ended up with complete skepticism. Why? Because they were expecting to be necessitated. And they were not paying attention to the necessity of an effort to understand and the necessity of assuming personal responsibility for their own judgment when there happened to arise in them what de facto was the virtually unconditioned. Unless one is aware of those fundamental facts of the human mind, what happens? If you are waiting for someone to force you to understand and not making an effort yourself to understand, not systematically striving to understand, what happens? Insofar as you are understanding, you grasp a nexus between subject and predicate, and the more you understand, the greater the number of nexuses you will see all the more clearly. But the less you try to understand, the fewer nexuses you will see, and the less clearly you will see them. And when you drop altogether all effort to understand, what happens? There are fewer and fewer nexuses that you see to hold between terms, and the fewer you see, the fewer principles you find in the evidence. In other words, the evidence of the principles, the clarity of the principles, is not something absolute. It is something relative to a mind, and something relative to my mind. People will say, 'It is not clear,' and that is just a polite way of their saying, 'I do not

31 See Lonergan, *Insight*, the index, under Unconditioned.

understand it.' If they understood it, it would be perfectly clear. They do not understand it yet, and if they are waiting until *you* make it clear, they are expecting the impossible. What is needed is an effort on *their* part to understand. You can help them to understand; you cannot force them to understand. And it is only when they do understand that they will see a nexus. These are a few remarks connected with judgment.

5.5 *Three Levels to Aristotle's Explanatory Syllogism*

This brings us to what seems to be a fundamental point in the matter of foundations. I call it 'three levels to Aristotle's explanatory syllogism.' Aristotle discussed explanatory syllogisms, in which the middle term is the ground, the reason for the predicate in the subject. For example, the phases of the moon are the ground for sphericity in the moon as far as our knowing goes; and compositeness is the ground of mortality in Socrates in the order of the *causa essendi*. Aristotle distinguished syllogisms of that type from syllogisms of other types in which you are not really explaining anything.

Now, in that syllogism we can distinguish three levels. The first level is the level of the sensible signs in which the syllogism is expressed: the words, the symbols, in which it is expressed, or the sensible data to which you can point when you are asked what you are talking about. You are talking about the moon, and if someone asks what you mean by that, you can point to the sensible data in the evening sky. So there is the whole level of the sensible.

The second level is that of understanding. There is grasping the subject (as in the discussion of the subject term, the *quid sit*). There is grasping predicates and the nature of predication, as in our discussion of predication. The same data understood as individual and then understood as of a kind give rise to the predication of what is grasped, uttered, as a result of one understanding and what is uttered as a result of the other. Again, there is grasping the ground of the predicate being in the subject, or being known as pertaining to the subject, as in our discussion of the middle term.

Therefore, you have first of all a level of sensible data, of words, symbols, and things you can point to. The second level is that of understanding, the understanding that goes from sensible data to the subject term, the understanding that goes from sensible data to the predicate term, the understanding that gives you predication, namely, understanding the same thing in two ways, as individual and as of a kind, and the understanding that puts predicates in order, that makes one predicate the ground of the other predicates being in the subject. That is the level of understanding.

The third level is the level of judgment, represented by the syllogism insofar as the syllogism is the expression of a virtually unconditioned. There are, then, three quite different aspects to the syllogism. There is the sensible level. There is the level of understanding, in which you grasp subject, predicate, middle term, and the connection between middle term and predicate in the subject. And there is a third level, when you consider the syllogism as a whole. The syllogism as a whole exhibits the conclusion as a virtually unconditioned. Insofar as it exhibits the conclusion as a virtually unconditioned *to you*, you have a *perspicientia evidentiae sufficientis qua sufficientis*, and you are rationally bound to affirm the conclusion. And in that 'rationally bound to affirm the conclusion' there is a submission in you to an immanent rational necessity.

Now it is insofar as one goes behind ordinary expressions of terms, propositions, inferences, to these intellectual facts that one finds the real meaning of what Aristotle was trying to convey when he was talking about syllogism. And it is insofar as you go behind the terms, propositions, inferences, to experiencing, understanding, and judging, that you come to the foundation on which you build a logic and from which you can judge whether a logic is adequate or not.

However, to recognize the existence of those three levels in oneself is in general not a simple matter. As a matter of fact, you can set up the fundamental differences in philosophy in terms of the measure in which the subject finds those levels in himself. What do I mean by that? Hume presented human knowledge as a succession of impressions linked together by customs and habits, and Hume presented human knowledge to be that and nothing but that in an extremely acute, intelligent, convincing fashion. In fact, there is an opposition between the knowledge that Hume described and the knowledge that Hume used in his describing. The knowledge that Hume was using to put across this doctrine of knowledge was something far more intelligent and rational than the mere succession of impressions linked by customs and habits that he said that knowledge was. What we have here is a phenomenon that recurs regularly in the history of philosophy: the subject is beyond his own horizon. He finds in himself most easily the sensible part of his experience. He cannot pin down exactly what his intelligence is, although he is extremely intelligent, and so his intelligence and his rationality are omitted in his account of knowledge. Then you have an account of knowledge as though it were simply the sensible, most easily found elements in knowledge. That gives you a first class of philosophers, who are materialists, sensists, phenomenalists, positivists, pragmatists.

You can go beyond that and discover the fact of understanding. You can not merely be intelligent but also know what it is to be intelligent. You can have experience of intelligence, experience of your own intelligence; that experience can come into your own thinking, and you can attend to it. Then you are in a new world. Philosophers at that level are an entirely different type. Here we are dealing with the idealists, the relativists, the immanentists, Kantian criticism, and so forth. These people are not aware merely of the sensible order; they are not just intelligent without having any notions of what intelligence is. Idealists and relativists have a very accurate and full notion of what it is to be intelligent. They can tell you a lot about it that perhaps you did not know. They are extremely competent with regard to the level of understanding. On the other hand, they do not have a satisfactory account of judgment. They may be rational, but the rational element in themselves does not appear in their account of knowledge and reality.

When you move to including the third level, the level of rationality, namely, the asking of 'Is it so?' and 'Is it true?' and the demanding of the virtually unconditioned, grasping the virtually unconditioned, and affirming, 'It is so,' when you introduce into the theory of knowledge and reality the third level, then you are on the level of Thomist realism as, among certain Thomists at least, Thomist realism is understood.

The effective distinction of those three levels occurs in the subject, and just what its occurrence is I attempted to describe in the eleventh chapter of *Insight*. But it is insofar as each individual philosopher grasps in himself the facts of his own intelligence and his own rationality that he will be capable of providing foundations for logic and of setting up reasons, demands, regarding what a completely adequate logic can and must be. In other words, the subject, each subject, in his self-appropriation, in his personal appropriation of his own intelligence and rationality, is the real foundation of a logic, of an adequate account of logic, of an ultimate judgment on logic, and of the wisdom that is needed to choose between different symbolic techniques and to decide that this is the technique to be used now and that is the technique to be used at another time.

It is an extremely difficult matter, a matter of making the subject *leap*, to move him from the first level, the level of sensism, materialism, and so on, to the second, to bring him up to the level of the idealist. And it is another leap to bring him from the idealist position to the realist position.[32] The

32 Lonergan's realism is axiomatized in *Insight* 413. To reach his view of faith one must supplement his treatment of belief in *Insight* 728–40 with his 'Analysis fidei,' nineteen pages of class notes from Toronto, 1952. See above, note 30.

realist position is common among Catholics, principally because of their faith. That faith is an affirmation of truths: you know that there are three persons in one God because that proposition is presented to you and you assent to it. It is through your assent to a truth that you know the Trinity and the other mysteries of faith, and because from childhood we have been brought up to know reality through assenting to truths it is easy for Catholics to move to this level of realism. But it is extremely difficult for people brought up in other traditions to move to that third level.

5.6 Foundations of Epistemology, Metaphysics, and Ethics

This foundation that I have offered for logic is a foundation not only of logic. I believe it will supply a foundation of epistemology, as I have suggested by the implications that it yields of three classes of philosophers. It also supplies the foundation of a metaphysics. For just as the subject is experiencing, understanding, judging, so the proportionate object of his knowledge will be a triple compound of potency, form, and act. And insofar as you distinguish substance and predicates as we have done or in a more complex fashion, you will have substantial and accidental forms; and since potency and act are defined relatively to form, you will have substantial potency, form, act, and accidental potency, form, act. You step right into metaphysics from that analysis. And you could go on to an ethics by adding will to knowledge. Related points are developed in *Insight*, chapters 14 to 18.

5.7 Foundation Dynamic

This foundation is dynamic. We found that the attempt to set up symbolic techniques seems to head to an open series of levels. On two levels up you can establish the theory of truth and the consistency and coherence of the symbolic techniques that are two levels down. The series is open; you can go right up.[33]

Is human intellectual development a matter of marching along one level, and then coming to the end and popping up and starting to march along another? Is that the way the mind works? When I say that the mind is dynamic, I mean that it does not move *along* those levels, it moves up *through*

---

33 Lonergan's notes (see below, p. 160) indicate that he has in mind here such work as that of Hao Wang. See above, pp. 65–66 and notes 48 and 50 there.

them. In other words, a logical symbolic technique defines an ideal plane that actual thinking is tending to at any stage. You do not have complete coherence and exact definitions of all your concepts yet. You are moving to that. And symbolic technique or logic forces you to seek ever more accurate definitions of concepts and presentations of arguments. But it is an ideal, not an attainment. Logic expresses an ideal of the mind, not a law of the way in which the mind de facto works. It expresses a goal towards which the mind strives; it does not express or describe a process. The description of the process is a matter of the genesis of subject terms, of middle terms, of predicate terms, the grasp of the virtually unconditioned, and making the judgment. The logic is not a description of that process. The description of that process is found to be psychological. What does the logic formulate? It formulates an ideal. You *should* have all your terms accurately defined. All your arguments *should* be rigorous. All your premises *should* be true. But what have you got? You have a number of opinions and you are looking for proof, you are trying to find out just how you could define this more accurately and make that more clear. That is the situation in which the thinker actually finds himself in philosophy and in the sciences. So, while there is a series of ideal planes, still the mind is never *on* one of these planes, but is moving *towards* a plane. And if you get there, what happens? New questions arise, the plane for which you are striving is simply transposed, and you are aiming at another plane. So the logical formalization, the strictly rigorous presentation of conclusions from premises that are accurately defined, is an ideal of mind, not a statement of actual process in mind. The business of drawing the conclusions from the premises is essentialist trifling; the real labor of mind is the movement up to a plane that you have not reached yet; and that is what I mean by the dynamism of the human mind.

It is insofar as one has foundations such as we have attempted to describe that one can think and deal with and more or less cope with a mind that is essentially dynamic. If you try to think of the mind as moving along one of these planes, you will be putting a straightjacket on it; you will not be dealing with a reality. A good deal of the difficulty experienced in the development of mathematical logic arose because it started out with the presupposition that this ideal was a reality, that concepts were perfectly and neatly defined, that propositions were formulated with perfect accuracy. It did not take into account the fact that concepts are in genesis, propositions are in genesis, and, above all, systems are in genesis. As soon as you arrive at the polished system, you get further questions and move on to something else, as is clear from the history of science. Because, then, we exhibit the coming-to-be of the subject term, the coming-to-be of predication, the coming-to-be of the middle term, the coming-to-be of judgment, the foundations of logic are dynamic.

5.8 Further Points[34]

Furthermore, this foundation grounds a grand-scale analogy. Experience stands to understanding stands to judgment as potency stands to form stands to act, for every stage in the development of understanding. While the content of particular sciences is shifting and is developing, still, just as you remain a being that is experiencing, understanding, and judging, so you have the same structure throughout science. Because of the immutability of the structure in your mind which is experiencing, understanding, and judging, you have a basis for a logical structure, a foundation of logic, that is similarly invariant despite the continuous variations in the content of understanding and expression as the sciences develop.

I illustrate that from the notion of fire. First of all, there is the experience: you have the phenomena of fire. Your heuristic structure is the question, What is fire? The question has been answered in three ways. Aristotle says that it is one of the four elements; chemists prior to Lavoisier said it was phlogiston; and subsequent chemists say that it is a chemical activity, a type of oxidization.

If fire is phlogiston or if it is a type of chemical activity, if that is what you mean by fire, then there is no comparison between the two, and there is no possibility of saying that one is right and one is wrong. It is possible to say that one is right and that others are wrong only insofar as you have some fundamental idea of fire that takes on successively the particular forms of one of the four elements, or of phlogiston, or of chemical activity. You can compare these three with that fundamental idea of fire. You consider these three relatively to your heuristic concept or your heuristic structure, What is fire? where by 'what' you mean 'the nature of,' what is to be understood in sensible data similar to these. The fundamental open heuristic structure that may be represented by the question, What is fire? provides the constant through which you compare the successive explanations of fire that are offered in the history of thought. In general, the successive explanations, the efforts towards them, the attainment of them and going beyond them,

---

34 See below, pp. 160–61, points *h* to *j*.

imply an ongoing variation in the content of concepts. But though concepts vary historically, that structure remains the same. 'What is fire?' meant exactly the same thing to Aristotle and to Lavoisier's predecessors as it does to us, though the three answers are totally different. You can see the interest of this from the viewpoint of development of dogma.[35]

Finally, there is the relevance of those foundations to the different types of logic that have been mentioned. But that is rather extensive. And you can get more help from *Insight* if you want it.[36]

35 The topic becomes explicit in the final ten pages of *Insight*, but it is evidently central to Lonergan's Latin works. The focus of *Method in Theology* is on a paradigm shift in 'methodological doctrine' (295).
36 The deeper communal help, of course, was still in the future. The manner in which it restructures both the phenomenology of logic and the logic of phenomenology is sketched, against the background of Husserl's early essay on geometry (an appendix in *Krisis*: see p. 247, note 2 below), in chapter 3 of Philip J. McShane, *PastKeynes Pastmodern Economics: A New Pragmatism* (Halifax: Axial Press, 2001). This will be developed in some detail in the work mentioned on p. xxi, note 6, above.

5

Mathematical Logic and Scholasticism[1]

My subdivisions are, first, a new factor in the problem of method, a point that we have said something about already; second, is Scholastic thought to be cast in the form of an axiomatic system? third, does mathematical logic eliminate existence? fourth, does mathematical logic eliminate substance? fifth, what is the relation of mathematical logic, first to logical atomism, and secondly to logical positivism?[2] These two movements had considerable influence in England. They more or less dominated Oxford after the Second World War. They had been building up, though. Logical atomism starts with Russell at the time of the First World War, and logical positivism is Wittgenstein in his earlier phase, the phase of the *Tractatus Logico-Philosophicus.*

1 A New Factor in the Problem of Method

The first point regards a new factor in the problem of method. I pointed out yesterday that traditionally and by convention there has been recognized among philosophers a preserve out of which disagreements were kept, namely, the field of logic, in which philosophic differences were not significant. That situation is not really changed by the discovery of mathematical logic and the use of the symbolic technique. Of itself, the symbolic technique is philosophically indifferent, and that indifference is recog-

---

1 The fifth lecture, Friday, 12 July 1957.
2 Lonergan did not in fact treat the fifth question. See below, p. 138.

nized and maintained by the better writers. However, even the better writers find it difficult in their Prefaces not to put in a few slams here and there to sell their own goods, and that may annoy Scholastics. And the writers who are not first class, while they present themselves as logicians, very frequently, because of an unfamiliarity with philosophic positions, can be caught out, I think, in fallacies. They are not making distinctions that are fairly simple from a philosophic viewpoint. They are not aware of their own presuppositions.

This unawareness of presuppositions is a topic that will come up next week when we discuss horizons.[3] But an example that struck me may help here. Max Black, in the first essay of his *Language and Philosophy*, attacks C.I. Lewis, the man who wrote on mind and world order and on modal logic.[4] Lewis had advanced the following position. Suppose that, when everyone else sees blue, I see green, and when everyone else sees green, I see blue. But I use the word 'blue' to denote my green and 'green' to denote my blue. Then with suitable adjustments all along the line, there is no way of testing out the situation, and any view is perfectly right on it. For Black that is pure skepticism. In his attempt to refute Lewis's contention that what is significant is the relational structure, Black seems, if I remember correctly, to regard the relational structure as simply equivalent to another perception like the blue or green. And of course, I do not think that would be Lewis's meaning, especially since Lewis is a sponsor of strict implication. So I think it is well, when reading these people, to be aware that, while they may be experts in symbolic technique, it does not follow that they are aware of their own philosophic presuppositions. And the less they are aware of them, the more their philosophy will obtrude into their expositions. Consequently, while of itself mathematical logic is indifferent, still it tends to be exploited in an empiricist-pragmatic direction. The exploitation increases as the quality of the writer decreases, until you get to the popular level where it is claimed that everything that happened before mathematical logic was discovered was just nothing at all.

That creates a problem in method. What is to be done about it?

I think you have to distinguish between practical procedures and the theoretical viewpoint, and again between practical procedures on an el-

---

3 See below, pp. 198–215, and chapters 13 and 14.
4 Max Black, *Language and Philosophy: Studies in Method* (Ithaca, NY: Cornell, 1949) 3–22; Clarence Irving Lewis, *Mind and the World Order: Outline of a Theory of Knowledge* (New York: Charles Scribner's Sons, 1929; 2nd ed., New York: Dover, 1956); for more on Lewis, see p. 35, note 37, and p. 83, note 22.

ementary level, a very elementary introductory course in logic, and such procedures in more advanced courses in logic. A good deal depends upon the judgment of the man on the spot, what he can do in his situation. But I think that there is considerable value to studying logic as the introductory step to the study of philosophy. And I would say that to continue to treat logic as philosophically indifferent would be a sound procedure, because, strictly, mathematical logic is philosophically indifferent. You can make students aware, when you are simply teaching the logic, that philosophic questions can intervene and will be treated adequately only at a later stage of the question. In other words, the problem essentially is not too different from the problem that existed in previous times when there were nominalists and empiricists, people who wanted a merely class interpretation of syllogism, and so forth.

From a theoretical viewpoint, one's attitude towards the issues raised by the invasion of logic, by this new technique and, associated with it, by an empiricist and pragmatic attitude in philosophy, depends entirely upon one's notion of the foundations in philosophy. For example, on a view associated with the name of Étienne Gilson, one can claim that one can begin philosophy with metaphysics. There is no doubt that this view has a solid foundation in tradition. Metaphysics was discovered, I would say, simultaneously with a satisfactory psychology and epistemology, but it was much easier to express the metaphysics, and then to express the psychology and epistemology in terms of the metaphysics, than to express the psychology and epistemology in a manner that was independent of metaphysics. For that reason, you have in Aristotle and in St Thomas the expression of psychology and similarly of epistemology, the treatment of epistemological questions such as existed at their time, on a metaphysical basis. St Thomas presupposes a theory of accidents, potency, form, act, before he says anything about psychology. The *intellectus possibilis* is a metaphysical potency; the *intellectus agens* is a potency in a metaphysical sense; the *species intelligibilis* is a form, or a part of a form, in a metaphysical sense; the *actus intelligendi* is an *actus* in a metaphysical sense; and so on, all along the line. Metaphysics is presupposed throughout. Again, the word 'intentional' as employed in Aquinas (not as employed by Husserl) is a metaphysical term; an entity is intentional insofar as it is a means of knowing some other entity. Therefore, the word 'intentional' presupposes a metaphysics.

In St Thomas the notion of *sapientia* is the fundamental notion.[5] In the

5 See Lonergan, *Verbum,* chapter 2, §§3 and 5.

Summa theologiae, 1–2, q. 66, a. 5, ad 4m, St Thomas is asking whether the intellectual habit *sapientia* is superior to the intellectual habit *intellectus.* He says that by *intellectus,* if you know the terms, you will grasp right away the principles, and therefore, it would seem that, since it supplies you with first principles – through the grasp of terms you arrive at the principles – *intellectus* is the fundamental intellectual habit. St Thomas says, 'No.' What you need to guide *intellectus,* to govern the habit of *intellectus,* to pick out the right terms, is *sapientia.* You need good judgment with regard to the fundamental terms that you link together to supply yourself with first principles. Otherwise, you will be going wrong.

That can be put more concretely by saying that there are several ways of conceiving *ens,* from Parmenides through Plato, Aristotle, the Arabs, the Scholastics, and the idealists with Hegel. The meaning of fundamental principles changes in each case with the meaning of being, of *ens,* the most fundamental notion. Consequently, a judgment on what the right notion of being is determines the meaning of the fundamental notion. And this fundamental judgment on the right notion of being that is going to be presupposed and that is going to be determinative of the meaning of any fundamental principle depends upon *sapientia.* Accordingly, for Thomas, the fundamental intellectual habit is not *intellectus.* It is not *scientia,* which deduces conclusions from *intellectus.* It is *sapientia,* which picks out fundamental terms in their correct meaning and on that basis, by determining the terms, determines what intellect lays down as the first principles.[6]

The question obviously arises, What is *sapientia?* And St Thomas has two answers to that question. The first answer is that *sapientia* is a gift of the Holy Ghost. Just what that gift is he explains by quoting Dionysius the pseudo-Areopagite, to the effect that by the gift of the Holy Ghost, you are *patiens divina* rather than active, you are suffering divine things. That is the full sense of this first meaning of *sapientia.* But there is another meaning of *sapientia,* and in that case it is Aristotle's metaphysics, in which judgment is a matter of reducing this particular affirmation to its ultimate foundations. One is of good judgment, one's judgment is good, insofar as one correctly reduces a given affirmation to its ultimate foundations. And the ultimate foundations lie in one's view of the universe, in *sapientia,* which is Aristotle's metaphysics. For if it is a view of the universe, it is also consequently an ability to place any particular entity in its proper place within the universe.

---

6 For more details on the relation of Thomist wisdom to principles, see ibid. 78–87.

That is the Thomist position, historically. So, within the natural order, where do you start from? You start from Aristotle's metaphysics.

Now in that sense I say that Gilson's position, that metaphysics is fundamental, is solidly traditional. However, in Aquinas, while you have a distinction between natural and supernatural, between faith and reason, you do not have a separation. St Thomas never did just philosophy. To do just philosophy was pagan; he was a Catholic, and he did theology. And when you work at the Thomist philosophic synthesis, what are you doing? You are picking parts out of his theology that you think you can get from natural sources. In Thomas there is not a separation between philosophy and theology. That separation comes out clearly when you get to Descartes, where you have someone attempting to set up a philosophy simply on the basis of reason. And in Descartes, while you have a sharp, clear, effective separation between philosophy and theology, you do not have such a separation between philosophy and science. Descartes deduces the principle of the conservation of momentum from the immutability of God. Descartes believed that it was easier to learn all the sciences than to learn any one of them properly, and consequently, he was not in favor of separating philosophy from the sciences. He had his philosophy as a basis of his science, and his philosophy is a means of putting across a new idea of science. That intimate connection between philosophy and science is still present in Newton. His fundamental work is named *Principia mathematica philosophiae naturalis*. The principle of universal gravitation is presented as a natural philosophy. It is only later that there came the same separation between science and philosophy that you have between philosophy and theology in Descartes.

In other words, all development is a matter of drawing distinctions. The distinctions that the theologians drew in the Middle Ages between grace and nature, faith and reason, natural and supernatural, involved a distinction between philosophy and theology. Still, it was a distinction that at that time was drawn by the theologians within the theological schools. It is only later that you get philosophy setting up on its own, apart from theology. And it is still later that you get science setting up on its own, apart from philosophy.

Consequently, we are not in the historical, cultural position that Aquinas was in in the Middle Ages. It is within the context of the Cartesian separation between philosophy and theology and of the later separation between philosophy and science that we have to live and operate and teach. It is within the context of the condemnations of fideism and traditionalism that, again, we have to operate.[7] Finally, it is within the context of our own English-speaking traditions, which are largely non-Catholic cultural traditions, that we have to do our job. So basing the whole of our philosophy on metaphysics and our metaphysics on Aristotle's say-so or on 'Well, it is the best there is' or on some other summary meaning does not seem to me, for that reason, to be satisfactory.

However, it is quite a job to express Scholastic psychology and epistemology in a form that will provide a basis for metaphysics without presupposing the metaphysics. You have to go from Aquinas's expression in terms of potencies, forms, species, acts, and so on, to the description of the psychological events if you are going to put on the Thomist edifice a portal through which people of the present time can enter and arrive at Aristotelian and Thomist metaphysics. You do not want them to say, 'Well, you accept the whole thing or nothing.' If you want to lead them on into it, what you have to do is have an expression of the epistemology and the psychology that is independent of, and does not presuppose, the metaphysics.

That was the aim of my book *Insight*: to outline an ascent of the mind through understanding and judgment towards the metaphysics without presupposing the metaphysics, but proceeding from the psychological, epistemological side towards the metaphysics. It is in terms of that that I would say that when you ask for the foundations of logic, you are asking for the same thing as the foundations to, the solution of, the epistemological problem, the foundations of metaphysics, the foundations of ethics, the foundations of natural theology. And those foundations lie ultimately in the concrete individual, in the light of his intellect and in his sensible experience. St Thomas says that judgment reduces ultimately to its principles, and what are the principles? The principles are not propositions. The principles are real causes, namely, sense and the light of intellect. That is the approach followed in *Insight*.

So much, then, for the new factor in the problem of method. It is the same old problem of method that we have been having all along: do you start by presupposing metaphysics or do you build up to it from another basis? It is a point disputed among Scholastics. I happen to have an opinion on the matter. But the new factor in the problem of method arises insofar as the field of logic has been perfected on the technical side to an astounding degree. If we are going to judge its product and say, 'Well, I use this now in

---

7 The condemnations belong to the nineteenth-century oscillations between fideism and rationalism, dealt with during the papacy of Pius IX (1846–78).

that sense,' or 'I prefer this to that type of mathematical-logical system,' then we stand in need of foundations for our judgments and preferences. If you are to have connections between the technique and ordinary thinking you have to go to more ultimate points, and that was the point we gave samples of yesterday in dealing with subject, middle, and predicate terms, judgment, and so on. Those foundations, I suggest, lie in the concrete, individual, experiencing, understanding, judging subject. If a person wants to follow the position that metaphysics is the basis of everything, well, just how he would work out his foundations of logic I do not know. Very possibly, he could do so. But just what they would be, I would not be able to say.

2   Is Scholastic Thought an Axiomatic System?

The second point: Is Scholastic thought to be cast in the form of an axiomatic system? Again, we have a question that Scholastics will answer differently according to their school, their tendencies, their conceptions of what Scholasticism is. When we ask the question, Is Scholastic thought to be cast in the form of an axiomatic system? that question is not the same as the question, Is Scholastic thought to use deduction, Is it to use syllogisms? The two questions are entirely different. I believe that if you take any Scholastic manual in any department of philosophy, you will find a series of theses. Where do the proofs of the theses come from? One argument comes from working up one way, and another one from working up another way; one has one set of premises, and another one has another set of premises. The arguments are deductive arguments but they come from all over the map. When Scholastics come to meet objections, other points are introduced, just brought in. So what do you have? You have a sequence of positions. One position will depend on another, but it will also depend upon further evidence drawn from other fields or from other aspects of the matter. We are dealing with a series of issues, picking decisive points as we go along. That is to my mind the logical structure of the Scholastic manual, the Scholastic presentation of most departments. In natural theology, there is much more deduction than there would be, for example, in psychology. In natural theology, there is more of a chain of arguments. On the other hand, in a science such as ethics, particularly special ethics, there is an appeal to arguments and considerations from all over, because the good is the concrete, and you know the concrete by considering all aspects of the matter. But, by and large, that is the structure.

Again, St Thomas in the *Contra Gentiles*, book 2, from chapter 46 to chapter 94, develops the idea of soul. He asks, What is an intellectual substance? and then, What is an intellectual substance in a body, What is a soul? There you have a systematic development of an idea. Now, if you care to make the experiment, study the logical structure of those chapters in the second book of the *Contra Gentiles*, and ask yourself, 'Is Thomas setting up a series of positions in which he draws all the arguments that are relevant to the case, all those that he thinks will serve to bring out his point to his readers, with his premises coming pretty well from all over? Or is it an axiomatic structure in which there are at the beginning a certain number of premises laid down and the whole expansion consists in rigidly deductive procedures from that closely well-defined set of premises?'

When we ask whether Scholasticism is to be cast in the form of an axiomatic system, *that* is the question we are asking. Is the Scholastic at the beginning of the whole course of philosophy, or is the professor of a particular section of philosophy at the beginning, to lay down a set of principles, premises, and say nothing in the whole course that he does not deduce with strict rigor from those premises? That is the question. And the point of the question is that one eliminates a great deal of the tendency to think that, since Scholasticism is deductive, therefore it is an axiomatic system. An axiomatic system has a much more precise meaning than simply the use of deductive argument. Theology uses deductive argument; it argues from scripture, it argues from the Fathers, it argues from the councils, it argues from the opinion of the theologians, and it argues from papal documents, all of them deductive arguments. But theology is not the exposition of an axiomatic system. The speculative part of a given treatise may, more or less, be something like an axiomatic system. But that is only one element in theology.

For example, take the treatise on the Trinity. There is the psychological analogy. You posit two processions in God. You deduce from the two processions four relations. You show that the relations are identical with the substance and although identical are rationally distinct from it, and that three of them are really distinct from one another, and that they are subsistent. And you show that the subsistent relations in God, divine subsistent relations, are persons. So, there are speculative procedures for setting up the treatise on the Trinity, and the sequence of notions is somewhat deductive. Is there a rigorous deduction from processions to relations? Well, theologians offer a *ratio theologica* in that direction, but at the same time they argue from scripture: the names 'Father' and 'Son' are relative names. They also argue that it is only if the persons are relations that you can have real distinctions in the absolute being. They argue perfectively from the name 'person.' In other words, the arguments they bring in to affirm the thesis come from all over. There is the fundamental line of development in the thought, but the trinitarian speculation does not rest simply on that fundamental line. It rests on considerations coming from all over.

I believe that there is a similar structure, by and large, in the presentation of the Scholastic doctrine in any department of philosophy.

However, in different Scholastic schools, there is more or less of a tendency towards a deductivism. A Scholastic may say that Scholastic philosophy is the deduction of necessary conclusions from self-evident principles and that the theology is the deduction of further conclusions from the Word of God, where, sometimes, you add on premises from reason. However, that view of philosophy and theology is much easier to substantiate from Scotus than from St Thomas, and it is much easier to substantiate from St Thomas when you are using your own mind to interpret St Thomas's general statements about matters than when you get down and see exactly what St Thomas does in a series of chapters treating a concrete point, the way he argues.

Consequently, to form some notion of the structure of Scholasticism is perhaps an important point, something worth discussing. And I think that it is useful at this stage to elaborate distinctions between nonempirical, empirical, and comprehensive types of inquiry. Nonempirical types of inquiry are found, for example, in mathematics; empirical in natural sciences; comprehensive in philosophy and theology.

In the nonempirical type of inquiry, there is little or no appeal to concrete matters of fact. This is clear, for example, in mathematics, which is also a markedly deductive science. However, if you want to understand the possibility of deduction in mathematics, you have to go to a more fundamental point, namely, that it is *constructive*. The mathematician starts out from very simple notions: one, two, three, point, line, plane, and he constructs infinities of further entities on that basis. Because the further entities that he discusses are constructed from the simple things from which he starts, it is possible for him to deduce the relations between any two or three or four or any number of these further entities. Suppose that you have any three mathematical entities: x, y, z. What is the possibility of deducing the relation of x to y and of y to z? How is it that the mathematician can rigorously deduce what the relations between those three key terms are or, on the other hand, show that you cannot deduce, that it is an insoluble problem? What is the possibility of that feat? It is that he starts out from simple elements, say, *a, b, c,* and proceeding along one route he constructs *x,* along another route he constructs *y,* and along a third route he constructs *z.* By knowing the way in which these entities are constructed, he is able to transform the relations by which *x, y,* and *z* are constructed into relations between *x, y,* and *z* themselves. That is the transition from the constructive side of mathematics to its deductive side. Consequently, you have in mathematics, due to the constructive side of mathematics, the possibility of a rigorous deduction.

The possibility of deduction that exists in the nonempirical type of inquiry exists to some extent, but not at all to the same extent or in the same manner, in the empirical inquiry, because, as we mentioned yesterday, in the empirical type of inquiry there is a double movement. There is the movement from sense data to principles. The chemists eventually arrive at the periodic table. From the periodic table, they are able to construct over 300,000 different types of compound. Once they arrive at that structure, then they are able to start deducing. The more accurately they know what precisely it is that makes the compound what it is, the more accurately they are able to make systematic investigations, to predict what exactly will happen when you combine certain compounds. Similarly in the whole history of physics, in the development of physics, the decisive factor is always the sensible matter of fact. But if you examine a textbook on physics, you find that they start you off with a lot of preliminary mathematics, and they lay down theories, draw conclusions, and so forth. In that second movement, from the *priora quoad se* to the *priora quoad nos,* they are deductive. And they can be deductive because by proceeding systematically from the data, they have been able to construct a system. Insofar as they have constructed a system, that construction makes deduction possible. And insofar as the constructed system is accurate and adequate, the deductions that are made will hit off exactly concrete matters of fact. In physics and chemistry, the construction is not in complete control of the situation as in mathematics, because in mathematics what is constructed is simply just what the entity in question is. There is more security in the mathematical deduction than in the deduction of the empirical science.

There is a further point to be made that is common to both mathematics and science. Mathematics, because it is constructive, is deductive; science is deductive insofar as it is constructive, namely in its second movement, when you have risen from the *priora quoad nos* to the *priora quoad se.* You become deductive, because you have constructed a system. However, does it follow that in these two fields one has a single axiomatic system, a single set of rigidly defined initial terms and propositions from which everything else within the mathematics or within the physical science is to be deduced?

The discovery of the Gödelian limitations of which we spoke is to the effect that you do not have a single set. My claim here involves an inference, and I noted regarding this claim that the *nota theologica* was considerably weaker than the *existence* of the limitation. For example, in the formalization of arithmetic offered by Hao Wang, what you have is a series of levels that is open at the top. And on any level $\mu + 2$, you can establish the theory of truth and the consistency of the level μ. So what you have is an open series of levels with richer means of construction, in particular, a different meaning, effectively, for the name 'enumerable,' that is, for determining enumerable sets as you move upward. In mathematics, on that formalization of arithmetic, which at least directly avoids any Gödelian limitations, you have rigid deductivism, but not from a single set of premises but from an open series. You have an open series of systems.

You get a similar phenomenon in the empirical sciences insofar as the science develops on one basis as far as it can. When it runs into difficulties, it moves to a higher viewpoint. There is the shift from Newton to Einstein, and you complement Einstein with, or shift from him to, quantum mechanics.[8] In the higher viewpoint, fundamental concepts are different, one is starting from a new basis, and the physicists do not say, 'We have arrived at the last word about everything.'[9] They do not exclude the possibility of some further systems later on. Because mathematics is constructive, it can be utterly deductive. Empirical science becomes constructive insofar as it becomes science of the explanatory type, not merely science of the descriptive type such as the old type of botany. The latter consisted of endless lists of definitions of what the different species and sub-species and so on are. But, insofar as a science becomes explanatory as in physics, chemistry, and biology as dominated by a theory of evolution, it becomes constructive. Insofar as it is constructive, the possibility of deduction arises. But while the deduction can be, theoretically, from a rigorous set of premises and per-

---

8 For Lonergan's context, see *Insight*, chapter 4.

9 There is a drive in contemporary physics for a unified theory. Popular presentations tend to exaggerate the significance of Toes and Guts, 'theories of everything,' and 'grand unification theories.' The recently developed fields of fractal geometry and chaos theory draw attention to the complexity of secondary determinations.

formed rigorously, still the scientist never excludes the possibility of an ulterior higher viewpoint, and his deduction is more or less rigorous. My friend Fr Eric O'Connor in Montreal remarked to me, 'The physicists all want to be rigorous as long as it does not interfere with their getting results.' In other words, the physicist cannot afford to be rigorous in the same sense as the mathematician, because his construction has not the same efficacy as that of the mathematician, who more or less creates his own entities.

We now turn to the question of the logical structure of the third type of inquiry, the type that is comprehensive, that deals with the totality of reality as it may be known naturally as in philosophy, or in the light of faith as in theology.

What we have to consider is this. If philosophy is to include a philosophy of science, if in some sense it is to be a *regina scientiarum,* not merely a constitutional monarch – you do no wrong because you can do nothing at all! – but an effective monarch that exerts a real influence within the fields of the sciences, then, as a philosophy, it will have to be something fixed. But it cannot have the fixity of a monolith, one big stone, solid and homogeneous throughout. Its fixity has to be the fixity of an invariant form in which the sciences are included; but what are included are not fixed sciences but sciences free to develop.

This is an extremely important notion, and it can be put graphically.[10] We want a philosophy that will be or will include a philosophy of science. If our philosophy, as including a philosophy of science, is just a solid block, with no give anywhere, then, when the sciences start moving, they move away from it, and we may continue to claim to be *de iure* the queen of the sciences, but our subjects will pay no attention to us whatever. If our philosophy, as philosophy of the sciences, is an invariant form and the sciences are, as it were, free-flowing through the whole, we can be the whole of a certain determinate shape that remains that shape all the time, and in that way exert a legitimate philosophic influence on the different departments of science, and at the same time, let these sciences develop according to their own nature. Do you see the difference between those two attitudes? Are we setting down laws as to what sciences have to be, or are we offering an invariant form within which the sciences can develop freely and yet be legitimately kept in their place? The distinction between those two attitudes of philosophy towards science is a distinction that is somewhat pragmatic,

---

10 Lonergan moved to the board at this point to draw a graph.

putting things on the lowest level. It is not appealing to the structure of the mind and what things have to be from a priori principles; it is viewing the thing from a practical point of view. If you offer a philosophy of science that is going to tie the sciences down, you will not succeed. They will pull away from you and do as they please; they will be in revolt. If you have a philosophy of science that provides the sciences with the foundations that they are seeking, and at the same time in no way limits their legitimate freedom of inquiry, then you are doing your job. You make philosophy not merely a queen of the sciences in a constitutional sense but in an effective sense. You will be doing a service to the sciences. In physics, and so on, at the present time, they are wondering what on earth they are dealing with: at least, the older men are; the younger men are out to make discoveries, to make a name for themselves. You will be offering them something that we can give and providing them with means to go on to something further. That seems to me to be a fundamental point.

How is that worked out? It is connected with hints that I have been throwing out throughout the week. As I said at the start, to arrive at the notion of mathematical logic we had to set aside a number of very familiar Scholastic ideas. For example, we set aside drawing distinctions as you go along. We had to see that paradoxes are not just fallacies that you have to find a distinction for. The whole approach of the mathematical logicians is something different. What is a paradox to them? It means that the logical machine has jammed: it is not working properly. So, we had to build up a new mentality to understand this view and to get into the type of thinking that the mathematical logician is doing.

Now let's come back to our present case. I said that a Scholastic treatise is not an axiomatic system at the present time. De facto you may find occasionally among Scholastics attempts at a rigid deductive expansion.[11] And of course you will find in various parts, for example, in deducing the attributes of God, a liberal deductive expansion. But the general picture is a series of theses with arguments coming from where they are most useful and most effective, and the selection of arguments according to the mentality of the people with whom one is dealing. That is not a rigorous deductive system; it is not an axiomatic system.

Moreover, that structure of the Scholastic treatise – cosmology, epistemology, psychology, ethics, and so on – is related to the Scholastic attitude of a deductive procedure in which you draw distinctions as occasion arises.

11 Lonergan made a reference to 'a doctrine attributed to Scotus.'

A difficulty comes up that was not provided for in my initial definition of terms, but, 'If I see the distinction now, I make the distinction now.' In other words, as I said at the beginning, the Scholastic uses the spontaneous dynamism of the human mind. The Scholastic deductive procedure is not a deductive procedure from a limited, sharply defined set of premises, but from premises that de facto are true, wherever you can get them. It is based upon a set of fundamental conceptions, but those conceptions are open to development, and when one gets into a difficulty and finds a distinction that shows the way out, one is not inhibited from using that distinction. All the better if it makes the matter clearer. In other words, Scholasticism de facto is something spontaneous, dynamic, vital, developing.[12]

So I think we can say that Scholastic manuals of philosophy as they de facto exist are not axiomatic structures proceeding from a closely defined set of premises and consisting in nothing but a series of deductions from those premises. They do not have even the structure of Euclid's *Elements,* and Euclid's *Elements* does not present a satisfactory axiomatic structure. The attempt to construct a philosophy along the lines of Euclid's *Elements* is found in Spinoza, in his *Ethica more geometrico,* but the structure of Spinoza is not a Scholastic structure. Moreover, one can doubt – as I do in *Insight,* in chapter 14, where I discuss deductivist methods – whether a deductivist method could possibly yield a satisfactory philosophy.[13] In a deduction the conclusions necessarily depend upon the premises. What kind of a universe could you know through a deductivist system? You could know a monist universe in which there is just one being with a series of attributes that are exhibited by the conclusions from the premises. You could know a universe such as Leibniz's, in which God was morally bound to create; the other beings necessarily proceed from the necessary being with a moral necessity. If the universe proceeds contingently from God by a free act, then the first principles logically cannot be the first principles ontologically, because then the logical deduction with the necessary consequences would be in contradiction with the ontological order in which there is contingency. In other words, Scholasticism is a much more complex structure than a concrete deduction in which the premises correspond to reality.

---

12 A break was taken at this point. There was a final sentence before the break, but the last part of it is blurred on the tape. What is clear is the following: 'The difficulty in the historical study of a man like St Thomas Aquinas is that for seven centuries we have had professors of philosophy and theology who ...'

13 See Lonergan, *Insight* 427–33.

If you attempt a deductivism of the abstract type, then you have a philosophy of the kind that is implicit in Scotus, Ockham, and so forth in the fourteenth century. What happened then? They studied what was necessarily so in every possible world, and their theory was abstract, involving a set of abstract, necessary principles from which they could deduce whatever they knew. They saw the necessity of effecting the transition from that abstract philosophy to the concrete. To effect that transition, they introduced the notion of an intuition of the existing and present as existing and present. Since God is omnipotent, since the omnipotence of God is restricted only in the sense that God cannot do what is contradictory, they were naturally led to the question, Would it be contradictory to have an intuition of the existing and present where what was intuited as existing and present was not present and even did not exist? And they found no contradiction in it. It would be contradictory to say that you intuited and did not intuit. It would be contradictory to say that it existed and did not exist. It would be contradictory to say that it was present and it was not present. But there is no contradiction in saying that you intuit the existing and present where what you intuit neither exists nor is present. And there you have the skepticism of Nicholas of Autrecourt, which follows automatically from the Scotist attempt and the Ockhamist attempt, the fourteenth-century attempt, to set up an abstract deductivism. If there is no contradiction in having an intuition of the existing and present where what is intuited is not existing and present, there is some possible world in which that occurs. And how do you know that it is not this world? Not from an abstract deduction, and not from the fact that you have intuition. So, you arrive at fourteenth-century skepticism, as a result of an abstract deductivism.

You can consider a deductivism of the Kantian type. That is, there are concrete deductions of this world, and your philosophy consists in determining the possibility of such a concrete deduction. That is a third way of setting up a deductivist philosophy. Kant asked the question, What is the possibility of a world in which Newton's *Principia mathematica philosophiae naturalis* is possible? He answered that question by his transcendental analysis, and the Kantian answer to that question is not a satisfactory deductivism. Certainly, it is not a Scholasticism. So deductivism alone does not seem to provide a satisfactory philosophy. However, that is a question that is beside the point.

We have said, first, that de facto Scholasticism is not an axiomatic system. It is a dynamic procedure in which you deduce, you have all sorts of deductions, but they come from premises, from truths, things that de facto are true, and there is no limitation on where you take them. Secondly, it is dynamic in the sense that it is developing. New distinctions are being added all along the line. Whenever you hit off something correctly, you are entitled to introduce a further distinction. De facto the history of Scholasticism has been the history of the introduction of further distinctions, refinements on them, and so on. Thirdly, so far is Scholasticism from being an axiomatic system in which you have exactly defined terms and a limited and exactly determined set of primary propositions from which everything else is to be deduced, that de facto Scholasticism as it exists tends rather to be a little too loose. It uses terms in almost any meaning, as I attempted to illustrate from common usage of the terms 'act' and 'potency.' They come to mean almost anything that happens to occur to anybody at this moment, instead of having strictly determinate meanings, such as I have suggested are based on Aristotle and St Thomas. There you find first potency and first act: as the eye is to sight and the ear is to hearing, prime matter is to the soul. You find second potency to second act as the soul to its existence, the sight to seeing, the habit of intellect or will to the act of intellect or will.[14]

We have, then, the fact that Scholastic philosophy as it has existed hitherto has not been presented as an axiomatic system. We have a general doubt that that sort of thing is desirable from the history of the rationalist philosophies exhibited by Spinoza and Leibniz, from the skepticism in which the attempted deductivism resulted in the fourteenth century, and from the unsatisfactoriness of a transcendental deduction such as is offered by Kant.

One may ask the further question, Granted that de facto we have not had an axiomatic system, still should we try for it? Here is where there becomes relevant the point about a philosophy that includes or is a philosophy of science. I said that if we want our philosophy to be queen of the sciences, to set their legitimate boundaries and frontiers while at the same time leaving them their legitimate liberty, then our philosophy as philosophy of science cannot be a monolith. It has to be an open structure that determines the shape of things in which the sciences can develop and allows them freedom to develop within those shapes. Could a philosophy so conceived be presented in an axiomatic structure? On that question, I hedge. On the one hand, I do not think that it is impossible. The thing to be attended to would be to be sure you have room for analogous terms. Analogous terms in mathematical logic hitherto have resulted in a series of strata. But if you

14 See ibid. chapter 15, §1, especially pp. 458–60.

have an explicit concept of the analogous term at the start, it should not be impossible to develop a system that would take care of them. It would require for Scholasticism a much more complex type of logic than I think has hitherto been developed. But let us suppose that Scholastic thought did achieve the consistency, the coherence, the accuracy necessary for going about working out an axiomatic system. Let us suppose that a symbolism of the type that was adapted specifically to Scholastic needs was worked out. There is no a priori difficulty about that from the side of mathematical logic, because every one of them, when he has a special problem, and he is a really bright fellow in the field, develops a symbolism that is suitable to his purposes. For example, in the normal system developed by Post,[15] for certain purposes all rules of derivation can have a form such that, when all the changes in the initial symbols conform to the pattern, then one can foresee that no matter how many are the changes, the ultimate result is bound to be of a certain type. He has conceived the logic in such a way that he is able to predict the possible conclusions that could follow within the system. There you see the power of this type of logic for dealing with systems as a whole and the potentialities of systems. That was a particular type of symbolism worked out for a specific end. And there is no difficulty in Scholastics, if they have their philosophy worked out with sufficient accuracy and coherence, developing a symbolism specifically for the purposes of expressing that philosophy.

So, on the one hand, I say, I do not think that it is impossible, and on the other hand, I am not certain that it would be very useful. The reason is that the problems in philosophy at least at the present time are not problems of exploitation. They are problems of getting people to the starting point, problems of opening up people's minds and bringing them to fundamental truths on which the system would rest. When you have got them there, you have practically got them the whole way. Deduction looms as something enormous in a constructive type of inquiry such as mathematics, where the

15   In the next few sentences Lonergan condenses pp. 254–65 of Ladrière. Ladrière is drawing on Emil L. Post, 'Formal Reductions of the General Combinatorial Decision Problem,' *American Journal of Mathematics* 65:2 (1943) 197–215; 'Recursively Enumerable Sets of Positive Integers and Their Decision Problems,' *Bulletin of the American Mathematical Society* 50 (1944) 284–316. Post is recognized as one of the originators both of truth tables and of many-valued logic, the fruit of his doctoral dissertation at Columbia University in 1920. That work, 'Introduction to a General Theory of Elementary Propositions,' is available in Jean van Heijenoort, *From Frege to Gödel* (see above, p. 46, note 17) 264–83.

construction implies right away the possibility of deduction. But you do not have that same significance to deduction within a Scholastic or, in general, within a philosophic system. The problem in philosophy is to start off from the average naive realist and bring him on to something that involves a fuller grasp of all the issues and a more profound understanding of what his real basis is. The problem is not having people repeat with Augustine that 'The real is not a body, it is what you know when you know something true.'[16] The problem is to get people to *mean* as much as Augustine meant when Augustine spoke about truth. And that is a transformation of the subject. It is bringing the subject up to the level of thought of a Plato and an Aristotle and an Augustine and an Aquinas. And that is a terrific development in the subject. There is no automatic, surefire means of bringing that about. You push people along the way as far as you can. It is just as in teaching any particular point, putting any particular point across: the teacher can utter the sounds and make the gestures and draw the diagrams but he cannot guarantee that people are going to understand. He can repeat himself over and over again, but their understanding of it is an immanent activity of their own, and still more so their arriving at truth, at a true judgment, and a true judgment in which the implications of the judgment are something vital, something permanent in their living. This depends upon them. All the teacher can do is something external, and still more all that a symbolic technique can do is something external.

Therefore, with regard to the question, Is Scholastic thought to be cast in the form of an axiomatic system? we can say the following. First of all, the question is useful insofar as it enables us to get a clearer, more exact idea of the way in which deductive arguments are employed within Scholasticism. Secondly, it is important to distinguish between the significance of deduction in different types of inquiry: the non-empirical or mathematical, the empirical type of the natural sciences, and the comprehensive type of inquiry that deals with everything, as do, from different viewpoints, philosophy and theology. Thirdly, we do not want an axiomatic system that would be just a monolith, because then we would be setting up a philosophy that would sooner or later disassociate us from, and lose all relevance to, the sciences that develop and concrete historical cultural situations that change. If we are to have a philosophy that is relevant not only to our own situation but to the whole succession of subsequent situations, and that is a key to understanding the past history of situations, then we need a philosophy

16 See editorial note *a* to the Introduction to *Insight* 778–79.

that is of the type of an invariant form within which developments can occur within their proper limits. A philosophy of that type is attempted in *Insight*. The fundamental thing is the development of the subject, a grasp of the subject as experiencing, understanding, judging. Then there is a grasp of the isomorphism between the subject and proportionate being com pounded of potency, form, and act in the substantial and accidental orders. This provides a basis from which you ascend to God and from God return to the world. That is an open structure. It is not impossible that a symbolic technique could be invented ad hoc to express it, but the invention and formulation of such a technique is not the fundamental problem in philosophy. A philosophy is a personal question to each student of philosophy, to develop, to rise to the level of a Plato, an Aristotle, an Augustine, and an Aquinas, and to face the problems that have been raised by later philosophers. After all, the questions raised by modern philosophy are the questions that have their roots in the development of Scholasticism and were not fully treated by the Scholastics. In a sense they are *our* problems. We must not think of modern philosophy as simply an aberration. It is also asking further questions that were not solved in the medieval period. The fact that they were not solved in the medieval period is the fact of the fourteenth century when Scholasticism just went to pieces.

So much, then, for that second and main question, the main question perhaps of this week, Is Scholasticism to be expressed as an axiomatic system?

3   Mathematical Logic and Existence

I have added on a few other questions. First, does mathematical logic eliminate existence? With regard to that, there are references in the notes.[17] There is a notion of existence implicit in the quantifiers, particularly the particular quantifier such as '*some* men are white.' A nuanced discussion of the existential import of the quantifier can be found in an article by J. Dopp in the Proceedings of the Tenth International Congress of Philosophy in Amsterdam in 1948, pp. 735–39.[18] Gilbert Ryle wrote an article that was considered very significant in the philosophical analytic movement, enti-

17 See below, p. 164.
18 J. Dopp, 'La notion d'existence dans la logique moderne,' in *Proceedings of the Tenth International Congress of Philosophy*, ed. E.W. Beth, H.J. Pos, and J.H.A. Hollak, vol. 1, fasc. 2 (Amsterdam: North-Holland Publishing Co., 1949) 735–39.

tled 'Systematically Misleading Expressions,' and one might say, though he certainly would not put it this way, that the thesis of the article was, 'Existential propositions do not exist.'[19] What he claimed was that you can always get along without an existential proposition, you can always put it in another way, such that the term 'existence' does not occur.

With regard to that question, the answer depends on what one means by 'existence,' and not everyone has the same notion of 'existence.' We will be finding that out next week when talking about the existentialists. What do I mean by 'existence'? I mean what you know when you answer yes to the question, Is it? Is there a horse? Is there a man? Are there men? If you answer yes, you know existence, and you know the existence through the yes. The *verum* is the *medium in quo cognoscitur ens*, and the yes, the 'it is,' is the act that is posited with respect to a compound of matter and form. Through the yes, you know existence just as through the affirmation of a definition, the definition as affirmed; the definition provides knowledge of essence, provided your definition is essential.[20]

If you take existence in that sense, it is very easy to handle Ryle or any other difficulties raised about existence by mathematical logic. Existence in that sense is beyond the horizon, generally, of mathematical logic, of the existentialists, of logical positivists, and so on. If they talk about existence, they do not mean that. For example, Ryle avoids existential propositions by saying, 'It is true that.' Well, 'It is true that' is precisely what I mean by existence. If it is true that there are men, men are, men exist. They all mean the same thing on the definition of existence that I have given. Therefore, Ryle implicitly admits all I mean by existence. However, if you do not have a precise idea of existence, or if your precise idea of existence is something else, then you have to find a different answer to the question. In other words, the answer to the question is not complete without having a determinate notion of existence. And I can only answer it insofar as I operate with one notion of existence. Others will have to find their own solution.

---

19 Gilbert Ryle, 'Systematically Misleading Expressions,' in A.G.N. Flew, *Logic and Language*, first series (Oxford: Blackwell, 1952), 11–36. Lonergan's reference may be to the discussion around p. 16: 'To put it roughly, "*x* exists" and "*x* does not exist" do not assert or deny that a given subject of attributes *x* has the attribute of existing, but assert or deny the attribute of being *x*-ish or being an *x* of something not named in the statement.' Ryle's paper dates from *The Proceedings of the Aristotelian Society*, new series, vol. 32, 1931–32 (London: Harrison & Sons, 1932) 139–70.
20 The perspective of p. 322 below is relevant here.

If anyone claims that a mathematical-logical system has any reference to anything beyond itself, he is affirming an existence in the sense of 'it is true that,' what you know when you affirm, 'Yes.' If he affirms that there is a difference between a myth or alchemy or astrology and, on the other hand, mathematical logic, then he is appealing to existence in the mathematical logic in some sense. What is wrong with the myth, the astrology, the alchemy, is that they do not refer to anything beyond themselves, except their fabricators.

4   Mathematical Logic and Substance

Does mathematical logic eliminate substance? Quite clearly, the notion of substance is not a fundamental notion in mathematical logic as it is in Aristotelian logic. In Aristotle, as we showed yesterday, when you ask *Quid?* with regard to a simple term, you ask for the form that goes into this matter and is the *causa essendi,* the reason why you have a thing, the reason why you have a substance. The notion of substance is all through Aristotelian logic as it is all through his metaphysics. Aristotle did not work out a logic that was incompatible with his metaphysics, his psychology, and so on. In mathematical logic, the notion of substance is bypassed. The propositional calculus deals with sentences, and the sentences can be about anything, and what they are about is not bothered with: P, Q, R. You go on to functional calculus $f(x)$; x is an argument, what you are talking about; f is a predicate, what you say about it. You are not committed to any doctrine of substance over *that.* Or you can have a multiple argument: John the book to Philip gives. You have a set of arguments and one predicate, the predicate of giving: John gives the book to Philip. You can have a function of several arguments, and you can have that complex, several arguments and a function, without any implication of substance whatever. There is no implication of substance in class relations. Alpha is a member of class B: no implication of substance. And there is no implication of substance in a theory of relations: there is some relation R of X to Y. Those are the fundamental things with which they operate. You can go through it without worrying your head about substance.

We have to acknowledge the fact that in Aristotle's day science barely existed. Consequently, if one's notion of science depends exclusively upon Aristotle, one does not appreciate the significance of relational structures. Modern science in its whole explanatory import is essentially a set of relational structures. It is explanatory insofar as it relates things to one another. Consequently, there is in Aristotelian logic an insufficient stress on relational structures. They are not unknown. I showed yesterday how you have the relational structure between phases and the sphericity of the moon, between being a sphere and going through these phases. There is a relational structure there that Aristotle analyzes very carefully. That relation exhibited by Aristotle as the real meaning of his middle term, when his syllogisms were of a scientific explanatory significance, comes out in the difference between a history of a science and a systematic exposition of a science. A systematic exposition of a science is from the *priora quoad se* – the periodic table, the theory of quantum mechanics – to the set of conclusions you deduce from that basis. That is a relational structure. Aristotle has the *causa essendi*, the *prius quoad se*, and this grounds the relational structure of quantum mechanics or relativity or chemistry. On the other hand, the start from the sensible data appears in the history of the science.

Further, while there is an insufficient stress in the Aristotelian notion of science on the significance of relational structures, and while mathematical logic is developed without any use, at least any explicit use, of the notion of substance, still de facto, as I have argued in *Insight*, you cannot have a science that is explanatory without invoking a notion of substance.[21] The reason for that is that in science, the relational structures are explanatory structures that have to be related to concrete matters of fact, the sensible data. To effect the relation between this relational structure and the descriptive predicates of concrete matters of fact, you need to be able to say that there are things of the same type that are described in one way and explained in another way. Without that relation, there is no possibility of discovering the explanatory structure or verifying it or changing it when it is not found to be verified. In other words, you need the notion of *thing* to operate scientifically, to go from descriptions to explanations, to bring explanations into closer harmony with the things described, to correct descriptions as science develops, and so on. There is a necessity of the notion of substance in an adequate formulation of a science that deals with the real. That is quite conspicuous in chemistry. In physics, of course, it does not appear as long as you are just talking about what we would say are properties like light, sound, heat, and so on; it is when you get to subatomic entities, the protons and mesons and so on, that you start talking about something like things, where you have unities with sets of properties.[22]

21 Lonergan, *Insight*, chapter 8, §§1 and 4.
22 Present physics strains, not always luminously, towards the identification of centers of predication and varieties of conjugates. See, for example, the

Finally, note that 'substance' is an extremely ambiguous term. The ambiguity of the term is connected with an ambivalence of subjects. When one uses the word 'substance' in the English language, one is most likely to mislead people. Historically the word 'substance' is usually interpreted as what lies underneath the phenomena, and of course if you shave off the top you only see more phenomena. By going further underneath, you get nothing more but more phenomena. That idea is pretty well fixed in people's minds, and it is not merely because of Locke's tradition and the views of false philosophies; it is also because there is an ambiguity in us. We can very easily take 'substance' to mean the term of the spontaneous extroverted animal that I am, as I respond to my environment. People can mean by that term of my extroverted spontaneity, 'That is the real,' and 'As long as I am in contact with that, I have solved all metaphysical and epistemological problems.' One has to be careful that one really is not thinking in that fashion in some concealed way. One has to be critical of one's thinking to make sure that one is not thinking in that fashion. What do we mean by 'substance'? We mean by 'substance' the concrete unity-identity-whole that is grasped by intelligence in a spatiotemporal multiplicity of phenomena. In other words, substance is *sensibile per accidens*. It is not what you see or feel, it is what you understand in what you see and feel. What you understand in what you see and feel is a concrete unity. There is a concrete principle of unity-identity-wholeness that makes this set of phenomena one thing: the *causa essendi* of this being what it is, a man, or a horse, or a cat, or so on. Grasping that intelligible unity, not finding simply a term with respect to which my spontaneous extroverted consciousness operates satisfactorily, is what is meant by substance.[23]

Here again, we have an example of what is meant when I said previously that the problem in philosophy is a problem of a development in the subject, in the student of philosophy, in one's raising himself up from whatever level he may happen to be on to the level of a Plato, and then an Aristotle, and then an Augustine, and then an Aquinas. You have to do an

---

synonymous use of the terms 'particle,' 'state,' and 'resonance' in David C. Cheng and Gerald K. O'Neill, *Elementary Particle Physics: An Introduction* (Reading, MA: Addison-Wesley, 1979) 268. The book gives a useful indication of the complexification of subatomic studies in the decades after Lonergan's lectures.

23 This challenge of chapter 8 of *Insight* is placed immediately here in the fuller precise challenge expressed in *Insight*, chapter 14, § 1: the challenge to take a stand on 'the basic position.'

awful lot of stretching to get up that ladder. You are not there already by the mere fact that you were baptized!

5   Conclusion

Finally, I put two questions with regard to logical atomism and logical positivism[24] but probably it would be more profitable first of all to answer any questions that arise from what we have said today or any time during the week. In other words, this is our last slap at this problem, and people may have questions of one kind or another that they want to raise.

24 See below, pp. 165–66.

PART TWO

Lonergan's Lecture Outlines

6

The Lecture Notes on Mathematical Logic

1   The General Character of Mathematical Logic

A Descriptive Approach
B Analytic Approach

A Descriptive Approach

1 Traditionally logic has been concerned with terms, propositions, inferences; the investigation of deductive systems has been considered too complex to be attempted.

The original and valuable contribution of mathematical logic (henceforth ML) has been the study of deductive systems as wholes possessing determinate properties.

Thus, given any apparently suitable set of definitions, postulates, rules of derivation, one may ask what will result; in particular, one may consider:

The problem of coherence: can both sides of a contradiction be deduced?

The problem of completeness: is there some relevant proposition that can be neither proved nor disproved?

The decision problem: is there some automatic procedure for solving all problems that arise within the proposed system?

2 This investigation has been closely linked with mathematics.

Advantage of concrete and easily controlled question, How much mathematics can be deduced from what basis?

Again, since mathematics is deductive, the question of possible deductive systems includes the question of possible mathematical systems; hence, ML becomes mathematics at the most general level.

Finally, at the beginning of this century mathematicians were concerned with legitimacy of an axiom of choice in set theory and with paradoxes involved in the notion of the set of all sets; the Bourbaki group illustrate the waning of this concern which, however, is only one aspect of the link between ML and mathematics.

3 The investigation has been symbolic and technical.

It has been symbolic because, inasmuch as a symbolism possesses determinate characteristics, it becomes possible to envisage the totality of possible deductive chains resting on a given basis.

It has been technical in the sense that the only permissible symbols, operations, changes of order, omissions, substitutions, additions, occur in accord with explicitly stated rules.

The significance of this combination of the symbolic and the technical is an objectification of mind and a consequent independence of any particular mind.

Illustrate by Euclid's lack of rigor and of necessity.
Illustrate by process of taking square root.
Recall great strides of nineteenth-century mathematics through elimination of geometrical intuition in calculus, through Dedekind's definition of real number, Cantor's synthesis in theory of abstract sets.

Value: unless all insights explicitly formulated in initial axioms and rules, then

you will be unable to exploit all their virtualities,
you will operate with concealed presuppositions,
you may be engaged in self-contradictory enterprise.

Note, however, that when ML is named 'formal' it is this symbolic and technical character that is meant; hence *Umdeutung* of 'formal' in name 'Formal Logic.'

Again, note pre-established harmony between this approach to logical issues and an empiricist, pragmatic mentality (recall B. Malinowski.)

4   The symbolic-technical character of ML gives rise to problems unknown in Scholastic (classical) tradition.

When a Scholastic comes upon a difficulty in his position, he draws a distinction.

A symbolic technique cannot draw distinctions; distinctions have to be included in initial definitions; they are equivalent to the introduction of a casual insight, to the transition from one logical formalization (henceforth LF) to another.

Unless distinctions excluded once deductive process has started, one is dealing not with a single well-defined LF but with an undefined series of LFS.

Scholastics accept the principle of excluded middle, but their acceptance does not mean that they waive the right to refuse disjunctions, to draw distinctions, to reformulate issues.
  ML either accepts or rejects excluded middle, and on that basis distinguishes different logical systems.
    Note principle of excluded middle takes on quite different meaning according as casual distinctions are admitted or excluded.

Besides distinctions of content, Scholastics have a retinue of structural distinctions, e.g., *re, ratione, exercite, signate*; first and second intention; etc.

Structural distinctions appear in ML as stratified series of languages:
  1 object language
  2 meta-language for syntax, for semantics
  3 meta-meta-languages ...

Scholastics employ analogous terms.
In ML there emerge indefinitely large stratifications; e.g.
Church's series of meanings for implication and quantification,
Curry's for canonicity,
Wang's for set.

5 Paradoxes (Cretan, reflective relation) have played a basic role in development of ML.

I think it important not to view the paradoxes as mere fallacies that a few apposite distinctions would clear away.

Aristotelian logic was a defense of mind against pretence of sophistry, distinction of proof from persuasion.

The techniques of Scholastic tradition correspond to natural dynamism of human mind.

But it is well to distinguish sharply between this natural dynamism and the exact idea of rigorously deductive system; else one will operate in accord with spontaneous dynamism and simultaneously entertain the illusion that one is rigorously deductive.

Again, it is impossible to investigate the properties of rigorously deductive system and admit distinctions to be introduced as need arises, for then one really is investigating series of LFs with no clear idea of the series.

B Analytic Approach

1 General Character of Technique

Pretechnical: grasp how to do it (practical insight) and do it yourself (sequence of movements).

Movement towards technical: increasing accumulation of insights (applied science, engineering, technicians, skilled workers) and increasingly detailed instructions for execution, both organizational (workers, foremen, supervisors, superintendents, ...) and mechanical (tools, machines, power-driven machines, automation).

Fundamental characteristic of technique:

A highly complex process controlled throughout by practical intelligence,
but only fragmentary intelligence of total process in any of the individuals engaged, hence division of intellectual labor.

2 Symbolism as Technique

Pretechnical: mental arithmetic.

Technical: plan of operations guided by rules.

One can learn and execute rules without understanding why they work; hence take square root of 1764 but not of MDCCLXIV. Hence,

Economy of intelligence: one can do it equally well though one does not understand; and even if one understands, one can do it without any effort of intelligence as mere effortless routine.

Economy of reason: one does it not only without understanding what one is doing but also without raising any theoretical question about correctness of procedure; all that is needed is practical checking (have rules been observed? does it work?).

Objectification of mind: liberation from individual variations and aberrations; complete irrelevance of 'I see it this way, I am inclined to think, etc.' The technique works independently of any particular mind; as such, it provides completely transparent communication.

3 Symbolism as Model

N.B. We are not using name 'model' in sense employed by mathematicians or in ML.

See A. Church, *Introduction to Mathematical Logic* 325, and note 451; or J. Ladrière, *Les limitations internes des formalismes*, introductory chapter; also P. Suppes, *Introduction to Logic* 253.[1]

Mathematician not only produces symbols but also perceives symbols he produces; at once writer and reader, speaker and hearer.

Moreover, while anyone can look at symbols, mathematician looks intelligently; he is intelligent, percipient, reader, hearer; cf. Plato, *Meno*, anamnesis; Aristotle, *De anima*, forms grasped by mind in images.

Symbolism as model is symbolism as potential object of intelligence.

Distinguish: model and object.
Model: sensible mark, sequence of such marks.
Object: what is conceived as result of insight into model.

---

1 For references to these works, see above, p. 19, note 19; p. 25, note 24; and p. 49, note 22.

Compare Euclidean diagram and defined points, lines.
In analytic geometry, double model: 1 diagram

2 algebraic symbols; single object, e.g., conic.

In physics, double model: 1 diagram

2 symbols; object, e.g. free fall.

Explicit and virtual elements in model.

Explicit: what is represented in diagram, expressed overtly in symbols.
Virtual: what is added to explicit model from suggestiveness of symbols,
  from familiarity with technique.

4 Symbolism as Both Technique and Model

The mathematician as agent and percipient, as speaker and hearer, as
writer and reader, effects transformation of symbols.

Translates problem into explicit model; virtualities of model suggest
possible sequence of changes; end result is solution of problem. E.g.,
prove De Moivre's theorem.

Note difference between this process (in which many premises sup-
pressed) and what traditionally has been meant by logical analysis.

5 Isomorphism
Consider: analogy of proportion; prolonged analogy of proportion; note
that it consists in similarity of relations and involves independence of
what are related.

Consider that the same symbolic technique can provide model for objects in series of isomorphic fields.

E.g., for algebraic, geometrical, physical relations.

Now algebra, geometry, physics include deductive processes; therefore,
the same symbolic technique will provide a model for these logical
relations.

Generally, there can be constructed symbolic techniques that serve as
models for total range of logical relations.

6 Mathematical logic (symbolic logic, logistics) is the investigation of the
field of logical relations through the development of suitable symbolic
techniques.

a Symbols: Russell-Whitehead, Hilbert, Polish

b Systems: see tables of thirteen systems in appendix to A.N. Prior, *Formal Logic*, Clarendon Press, 1955.
Add combinatory logic, outlined by R. Feys, 'La technique de la logique combinatoire,' *RPL* 44 (1946) 74–103; 237–70.[2]

c Note plurality of systems (very summarily)

Classical Propositional Calculus worked out differently by Russell, Hilbert, Łukasiewicz, et alii.
Lewis modal systems: add strict implication
Intuitionistic: omit excluded middle
Three-valued: admit third alternative to false and true.

d In general: one can have any type of logic one wishes, and one can construct any symbolism one pleases to attain more readily a specific objective, e.g., all rules of substitution of form $P_i a \to a Q_i$

e The obvious question.
'To sum all this up. The choice of languages is itself a problem that cannot be solved linguistically but only by some non-linguistic method.' L.G. Kattsoff, 'Ontology and the Choice of Languages,' *Proceedings of the XIth International Congress of Philosophy*, Brussels 1953, vol. 14: *Additional Volume and Contributions to the Symposium on Logic* (Amsterdam: North-Holland Publishing Co., 1953) 32.

2   The Development of Mathematical Logic

1 The development of ML has been the pursuit of an ideal, viz., a rigorous hypothetico-deductive system that from minimal suppositions would embrace the whole of known mathematics.

2 The ideal has been formulated as an axiomatic system or logical formalization.
It distinguishes terms and propositions, divides both into derived and not-derived, conceives not-derived as relative to system, and names them primitive.
Derived terms are defined by primitive.
Derived propositions are deduced from primitive.

---

2 See above, p. 35, note 35.

Rules of derivation must be stated explicitly, and no derivations are admitted except in accord with stated rules.

Let *P* and *Q* denote two LFs, and let *p* be any proposition that can be constructed in *P*.

Then *P* and *Q* are equivalent, if primitive of *P* are derived in *Q* and primitive of *Q* are derived in *P*.

P is complete, if one can derive either *p* or *Np*.

P is coherent, if one cannot derive both *p* and *Np*.

The primitive propositions of *P* are independent if no one can be derived from the others.

The primitive propositions of *P* are elegant if they offer the simplest basis for deriving in the simplest manner all the propositions of *P*.

3 Principal Lines of Endeavor

a Axiomatic set theory: Zermelo – Fraenkel – von Neumann

b Whitehead-Russell, *Principia mathematica*; aims to base whole of mathematics on logical axioms; a magnificent unitary view that remains one of the principal directions.

However, in both first and second editions there is a non-logical axiom of infinity.

In first edition there is also a 'theory of types' (to avoid paradox of class of classes that do not contain themselves) and an axiom of reducibility (to make possible Dedekind's definition of real number, excluded by theory of types).

In second edition the axiom of reducibility is eliminated, and there is employed a weakened theory of types that eliminates syntactical but not semantical paradoxes.

c Hilbert proposed a two-level approach.

First, a formalized deduction of the whole of mathematics from mathematical axioms; on this level there were to be admitted infinities of objects and of operations.

Secondly, a metamathematics that on a strictly finite basis would investigate logical properties (especially consistency) of the first mathematical level.

Results: short-term, geometry worked out with axioms verified intuitively in model that supposes validity of counting numbers; arithmetic could be shown to be consistent only if some axioms were omitted or all weakened. However, as will appear, this has proved most fruitful line of inquiry.

d Intuitionistic school: Brouwer, Heyting

Insists that LF is only tool, that mathematics is essentially constructive, that excluded middle cannot be invoked indiscriminately. Program involves lopping off more of classical mathematics than mathematicians are ready to sacrifice.

e Gonseth; review *Dialectica*

Tends to conceive axiomatic ideal just an outdated Euclidean avatar; insists on development, interaction between maths and cultural movements; relativist in tone.

f Bourbaki group: Hilbert's first level; metamathematics is a separate department of no particular interest to mathematician; weak point that rigid axiomatic structure neither accounts for past development of maths nor opens way to developments of future.

4 Gödelian Limitations

Jean Ladrière, *Les limitations internes des formalismes*, Louvain (Nauwelaerts) and Paris (Gauthier-Villars) 1957. 702 pp.

There have been demonstrated a series of theorems setting limitations to the possibility of reaching the ideal of the rigorously deductive mathematical system. The general form of the argument in such cases is approximately as follows.

a An LF is a symbolic technique capable of representing a manifold of deductive sequences.

Consider an LFL and an LFM, which symbolically are identical or sufficiently parallel, but differ inasmuch as LFL is interpreted logically while LFM is interpreted mathematically.

b Now in mathematics there exist non-enumerable sets, i.e., aggregates that do not admit a one-to-one correspondence with the positive integers, and so cannot be enumerated (counted).

Hence, to suppose that such a set is enumerable (e.g., the set of infinite decimals) results in a contradiction, and this contradiction can be demonstrated.

c With sufficient ingenuity it is possible to make the LFL sufficiently parallel to the LFM so that the proof of non-enumerability in LFM is matched by a proof of logical impossibility in LFL.

In other words, the proof that the proposition 'k has been enumerated' is contradictory is paralleled by a proof that the proposition 'k is a theorem,' or 'k is a soluble problem,' or 'k is true,' or 'k is definable' is contradictory.

d Such theorems are extremely complex; they have been worked out in a variety of manners; they arise when the LFL is sufficiently powerful to represent the theory of division, resolution into prime factors, and the unicity of such resolution.

e Their proximate significance was the refutation of Hilbert's proposal to settle the logical validity of arithmetic on a finitist basis.

Gödel's demonstration was followed by a demonstration by Gentzen that arithmetic was non-contradictory, where however the LFL had to employ transfinite induction.

f The ultimate significance, however, of such Gödelian limitations seems to be the same as of inverse insight; cf. irrationals, transcendental numbers, Galois on fifth-degree equations, Newton's first law.

5 The Transcendence of Gödelian Limitations

a Mere avoidance: J.R. Myhill (*JSL* 15 [1950] 185–96)[3] avoids such consequences by employing a logic without quantification and without negation.

b Use of indefinitely large stratifications (analogy)
Church: 'implication' and 'quantification' take on different meanings on different strata
Curry: similar procedure re his basic notion of canonicity.

c Skolem paradox shows that by different modes of stating one-to-one correspondence, 'enumerable' takes on different meanings.

d L. Henkin's study of relations between LFL and models showing that LFL lacks absolutely definite meaning.

e Hao Wang: indefinite series of sub-systems; at each level new resources of construction and new meaning for enumerable; the consistency and the theory of truth for any level, m, demonstrable at level $(m + 2)$.

3 See above, p. 63, note 47.

f Significance: ideal of ML moving from static and closed to analogous and open.

3   The Truth of an ML System

1 The truth of what?

a not the truth of what is seen or written that may be or may not be wf, but only wf expressions can be under consideration, since not-wf is just jumble of symbols

moreover, what is seen or written is usually only part of the expressions that are possible or necessary in the system; the larger range of what might be seen or written is also relevant and may be decisive

finally, strictly expressions are not true or false but merely adequate or inadequate; what is true or false is what is meant, intended.

b the truth of an MLS is the truth of what is conceived, considered, meant, intended by one who understands the MLS, who not only can judge whether any given expression is a wff in the MLS, but also can manipulate the MLS and so bring out all its virtualities of expression

c hence the truth of an MLS is the truth of a virtual totality of propositions

where the same proposition can be expressed in any of several languages, and the same expression in any given language can be uttered any number of times

where the virtual totality consists in all the propositions that can be formed within the system and can be derived in accord with explicitly stated rules from the axioms.

2 What is the general character of such truth?

a Aristotle, *Met.* M 10 1087a 15–20, distinguished science (*epistêmê*) in potency and science in act, affirmed that science in potency is itself indeterminate and of the universal and indeterminate, while science in act is determinate and of the particular and determinate.

b Clearly, then, the truth of an MLS is the truth of universal and indeterminate and determinable knowledge of the universal, indeterminate and determinable.

c Again, Aquinas distinguished abstraction of universal from particular, and abstraction of form from matter.

The type of abstraction of an MLS is of form from matter, of a *forma artificialis*, of a network of relations linking unspecified propositions, arguments, predicates, classes, relations.

That MLS is such a form appears from fact that an MLS is a logical interpretation of a range of symbolic expressions that admit other isomorphic interpretations.

3 What is meant by 'truth'?

a Distinguish definition and criterion of truth.

Truth is defined as *adaequatio intellectus ad rem*.

However, the criterion of truth is the precise element in knowledge by which one knows that such an *adaequatio* has been attained; commonly, this criterion is conceived by Scholastics as *perspicientia evidentiae sufficientis qua sufficientis*.

b The criterion of truth is analyzed more exactly in *Insight*, chapter 10, as a grasp of the virtually unconditioned, i.e., as a grasp of (1) a conditioned, (2) a link between the conditioned and its conditions, and (3) the fulfilment of the conditions.

c One of the modes of the virtually unconditioned is that of the analytic *proposition*, where

(1) the conditioned is the proposition in question

(2) the fulfilment of the conditions is the set of definitions of the terms contained in the proposition

(3) the link between conditions and conditioned are the syntactical structures in accord with which single terms in their defined sense coalesce to form a proposition.

d Another of the modes of the virtually unconditioned is the analytic *principle*, where

An analytic principle is an analytic proposition whose terms, in their defined sense, occur in true judgments of fact.

In other words, an analytic principle adds an existential reference to an analytic proposition, where existence is defined by its connection with factual truth.

On factual truth and on the three main modes of analytic principles, see *Insight*, chapter 10.

4 An MLS is by postulation a virtually unconditioned.

For an MLS is a virtual totality of propositions that result through explicitly stated rules of derivation from primitive terms and propositions.

Hence the MLS is a conditioned: it results from something inadequately distinct from itself.

The MLS has conditions, namely, the primitive terms and propositions.

These conditions are fulfilled by postulation, just as the definitions of the analytic proposition are fulfilled by postulation.

The MLS is a conditioned linked to its conditions, for the rules of derivation determine what is the totality of propositions that pertain to the MLS; and the rules of derivation are posited by postulation, just as the syntax of the analytic proposition is posited by postulation.

Accordingly, every MLS (satisfying the definition of an MLS) has the verbal type of truth that pertains to the analytic proposition.

In other words, if you were to speak or think in certain defined manners, you would *eo ipso* be committed to accepting such and such an MLS.

5 Various types of MLS contain fragments of factual truth.

The present assertion stands to the preceding as the analytic principle to the analytic proposition.

However, analytic principles may be absolute (as in metaphysics), provisional (as commonly is the case in empirical science), or serial (as in mathematics).

In our next lecture, we shall raise the question of the foundations of logic; this will be equivalent to the problem of putting logic on foundations of the same absolute type [as those] of metaphysics. However, the authors of the various MLS entertain no such intention.

Again, while MLS is closely related to mathematics and may be conceived as a generalization of mathematics, still there is no universal agreement among mathematicians upon the exact nature of mathematics, and so there is no possibility at the present time of a universal agreement upon the generalization of mathematics.

Cf. P. Bernays (Zürich), 'Zur Beurteilung der Situation in der beweistheoretischen Forschung,' *Revue internationale de philosophie* 8 (Brussels 1954) 7–13, considered that while the choice of method and of deductive framework for investigation of mathematical foundations may soon be settled, the determination of the notion of mathematics remains very remote.

On the basis, then, of our triple division of types of analytic principle (absolute, provisional, serial), it would seem that the various MLS possess factual truth of no more than the provisional type. In other words, the various MLS are just a series of hypotheses on the nature of deductive system.

This view is confirmed by our preceding lectures on the general character and on the development of ML; in other words, the history of ML is the history of the gradual and as yet incomplete discovery of what an MLS really is; a broad and a priori ideal has been gradually trimmed down into accord with the facts of logical possibility.

This view is also confirmed by the truth of the assertion that heads this section (5). For a very brief analysis serves to show that while many of the various MLS are incompatible, still the basic assertions contained in each are true as far as they go. Thus,

a There is no doubt about the existence, the factual truth of the occurrence, of propositions, negative propositions, compound propositions in the truth-functional sense. There follows the factual truth of the various elaborations of the classical propositional calculus, *in sensu aienti.*

b There is no doubt about the factual truth of strict implication. Thus, there is a strict implication of an MLS in its axioms and rules of derivation.[4]

There follows, *in sensu aienti,* the factual truth of some system of strict implication; but it is to be noted that Lewis's systems, s 4–5, are incompatible with s 6–8.

Further, since CPC [classical propositional calculus] makes no provision for strict implication, CPC can be factually true only *in sensu aienti.*

c There is no doubt about the factual truth that contingent futures are neither true nor false for minds to which they are future.

Hence, *in sensu aienti,* a three-valued logic (true, false, neither) is factually true; and this reveals the fragmentary character of two-valued logics.

---

4 In a set of these notes that Lonergan kept and that is now in the Lonergan archives in Toronto, he wrote in the margin here, 'as global judgment.' And at the bottom of the same page, 'Merely formal implication may be the limit of an operational pure [?] reason – (e.g. in perfect system, there is an instance of strict implication: If *A* then *B* without also If *B*, then *A*).'

d There seems to be no doubt that, as long as knowledge is still in genesis, *in potentia*, the principle of excluded middle cannot be built into a logical system and so applied indiscriminately.

Hence, 'intuitionistic' logic possesses a measure of factual truth; and this measure involves a restriction on the value of logics that accept excluded middle as an automatically operative principle.

See F.B. Fitch, *Symbolic Logic* (New York: Ronald Press, 1952), who weakens excluded middle and thereby frees himself from Russell's theory of types, which he argues to be self-referentially incoherent.

4   The Foundations of Logic

1 Originally logic was a clear-headed assertion of human rationality against sophistry (abuse of language to deceive mind) and apart from rhetoric (concerned with legitimate persuasion with regard to contingent).

However, it was not a pure assertion of rationality but an assertion implemented in technique of figures and moods of syllogism, etc. Nonetheless, this technical aspect was considered of minor importance, and no one felt it a matter of any moment what one thought, say, of the fourth figure.

Moreover, interest in epistemological issues led to a distinction between major and minor logic, with major logic a field for enormous differences of opinion, while minor logic remained the unquestioned object of acceptance for all sane people. One may perhaps add that such unquestioning acceptance played no small part in moving mathematicians to solve all their ultimate difficulties by basing mathematics on logic.

2 The development of MLS has changed the situation.

First, it should be conceded that now there exist techniques that are far superior in precision, in capacity of very complex refinements, in ability to deal with such enormously involved issues as the properties of axiomatic systems.

Secondly, it has to be recognized that the name 'logic,' as employed in a broad and steady stream of articles and books, denotes not any concern with the immutable laws of mind but rather familiarity with some or all of a set of symbolic techniques.

Thirdly, it would be, I think, a strategic blunder to be concerned over minor divergences between ML and Aristotelian logic. There are types of MLS that involve differences from AL; but it has been shown possible to construct an LF that coincides with the assumptions and implications of Aristotelian technique.

What, finally, is essentially new in the present situation is the invasion of the field of logic by the philosophic differences that formerly were confined to major logic. Accordingly, it no longer is possible to treat logic *adequately* without going into philosophic issues that previously could be neglected (by common consent) within logic. An instance of the interdependence of the whole of knowledge that should not be unwelcome to the real philosopher.

3   The Ambivalence of the Technical Achievement

There exist logical techniques that virtually are independent of any particular mind (equivalent to computers), that can handle problems too complex to be considered previously, and that thereby raise the question whether they are simply a tool for mind or rather a substitute for mind.

This issue has been suggested from the first lecture, and now it may be put explicitly as follows.

If one pleases, one may for one's personal satisfaction work out a theory of truth and come to the conclusion that an MLS by definition has the truth of an analytic proposition and that the various MLS represent a series of fragments of factual truth as well. On such grounds one becomes rationally committed to the acceptance of MLS either hypothetically or, under certain restrictions that vary with various systems, absolutely.

But one may, and many do, find it more pleasing to consider the matter of rational commitment to truth as a private and minor matter. What alone counts is the external fact that you employ some language and that any language presumably can be reduced to a logical calculus plus a vocabulary. What really counts is not anyone's private and internal dedication to truth (which is a grievously abstruse matter), but the public and external fact that he talks or ventures to write.

The ambivalence of the technical achievement of ML is that it seems to offer the alternatives of *either* a rational commitment to truth *or* a nonrational pragmatic acquiescence in the fact of talk.

4 Some Symptoms

H. Behmann (*Proceedings, Second Intern. Cong. of the Intern. Union for the Phil. of Science*, held Zürich 1954, published Neuchatel 1955, vol. II, pp. 97–108)[5] considered that strict implication was natural only in the sense that the notion of flat space is natural.

At the Colloque de Logique held in connection with 1953 Brussels Intern. cong. Phil., reported by R. Feys, *RPL* 1953, published *Rev. Intern. de Phil.* 1954,[6] F. Gonseth and A. Tarski met head on, when Gonseth put forth his highly nuanced views on the nature of proof, and Tarski stubbornly maintained that he could find nothing rational in such proposals. It would seem that by 'rational' Tarski meant 'symbolic and technical.'

In ML the CPC holds a dominant position, not, I should say, because of any merits of a logical character, but solely because it brings about a maximum assimilation of logic to the modes of mathematical technique. The philosophical movements of Logical Atomism (Russell) and of Logical Positivism (Wittgenstein)

(1) presuppose that a language is equivalent to a logical calculus plus a vocabulary

(2) presuppose that no questions are to be asked about the logical calculus itself

(3) and by their inability to carry out their own programs have brought about their own demise, at least at Oxford.

See J.O. Urmson, *Philosophic Analysis: Its Development between the Two World Wars.* Oxford, Clarendon, 1953.

Also G.J. Warnock, Fellow of Magdalen College, 'I am not, nor is any philosopher of my acquaintance, a Logical Positivist.' In *The Revolution in Philosophy* by A.J. Ayer et al., London, Macmillan, 1956, p. 124.

5 The Question of Foundations

a There are many MLS and, while some are equivalent, many are not. This fact posits as a problem the theory of choice of any given system either absolutely or relatively to a particular task.

b When the field to be formalized is as complex as arithmetic, the MLS becomes extremely complicated. Transfinite induction needed for logi-

5 See above, p. 99, note 19.
6 See above, p. 99, note 20.

cal theory of ordinary mathematical induction. It follows that logic is not a source of greater evidence but rather of bigger problems. What is the basis from which these larger problems are attacked?

c While it is the problem of the infinite that makes an LF of arithmetic so complicated, there seem to me to be parallel problems in the finite domain of ordinary matters of fact. For,

c′ ordinary concepts are not the simple smooth regular homogeneous nuggets needed to conform to an MLS, but they are open heuristic structures subject to enormous differentiation and variation; see *Insight* on Common Sense and on Classical, Statistical, Genetic, Dialectical Heuristic Structures. Also notion of 'person' in my *Divin. Pers. Conc. Anal.*[7]

c″ Hellmut Stoffer (Bonn), 'Die moderne Ansätze zu einer Logik der Denkformen,' *Zeit. f. Phil Forschung* 10 (1956) 442–66, 601–21, in a very fully documented pair of articles, argues for six types of logic needed to classify and deal with the expressed forms of thinking: (1) Plane, (2) Dialectical, (3) Existential, (4) Magical, (5) Mystical, (6) Hermeneutical.

All MLS would fall under the first category; discussion of the first five would fall under the sixth.

This direction of thought would be confirmed by the English experiment as described by Urmson.

d It would seem that (1) developing intelligence and (2) perfectly transparent expression and inference form a dialectical couple; the pursuit of either involves some sacrifice of the other.

6 Samples of Foundations of Logic

a Middle term: 'understand' means 'know cause'
   Causa essendi: phases of moon because of sphericity
   Causa cognoscendi: sphericity of moon because of phases
   Systematic expansion: textbook of a science vs. history of the same science.

b Subject term: Aristotle, Met., Z, 17, 'quid' means 'propter quid'

7 See below, p. 332, note 3, and the text on pp. 332–33.

c Predication (basic case): same data understood as individual (notion of thing) and as of a kind (notion of property: descriptive [heavy, hot]; explanatory [mass, temperature]).

d Judgment: virtually unconditioned; if A, then B; but A; therefore B, where A and B are propositions, sets of propositions; where A is simply experience; where 'if A then B' is merely implicit insight, invariant of thought, invariant of expression.

Significance of syllogism is not Kantian regress to 2^n premises but manifestation of conclusion as virtually unconditioned.

Even necessary object (God, analytic principle) known by us contingently.

Fallacy of waiting to be necessitated: inevitably heads toward skepticism (fourteenth century; rationalism).

Necessity of personal commitment, a personal responsibility; where commitment is intrinsically rational; where commitment is to absolute though in us it occurs contingently.

e Three levels to Aristotelian *syllogismos epistêmonikos*.

(1) The level of the words, symbols, sensible data to which words or symbols refer.

(2) The level of understanding: grasping subject (b above); grasping predicate and predication (c above); grasping ground of predicate's being in subject or being known as pertaining to subject (a above).

(3) The level of judgment and personal commitment (submission) to immanent rational necessity (d above).

The three levels effectively distinguished only insofar as one moves behind terms, propositions, inferences to their ground in the experiencing, intelligent, rational subject.

Again, the subject ek-sists philosophically only insofar as he distinguishes the three levels effectively.

To acknowledge explicitly only the first involves one in materialism, sensism, phenomenalism, positivism, pragmatism.

To acknowledge explicitly only first and second involves one in an idealism, relativism, essentialism, immanentism, Kantian criticism.

Only when all three acknowledged is Thomist realism reached.

Note that what is significant (effective distinction) lies not in subject's formulation of the three levels (for that may be superseded by more accurate formulations) but in the subject's immediate grasp in himself of his preconceptual, pre-judicial inability to get around fact of three levels. Cf. *Insight* chapter 11.

The subject in this self-knowledge is the foundation of logic; it is a foundation in the reality of the subject himself, and in every experiencing, intelligent, reasonable subject; it is a foundation in a reality, and so it is beyond the relativism of successively more nuanced and more accurate formulations (statements) of philosophic positions.

f This foundation of logic is also a foundation of metaphysics and of the general form of ethics. *Insight* chapters 14 to 18.

g This foundation is dynamic.

ML moved from naive ideal of grand deduction from simple set of axioms to necessity of set of deductive levels with each level far richer in resources than preceding (Wang)
because it started out with supposition of human knowledge, not as a process of knowing-coming-to-be, but as an aggregate of ready-made univocal terms and ready-made true propositions.

The above exhibits the coming-to-be of the subject-term, the coming-to-be of predication, the coming-to-be of the middle term, the coming-to-be of judgment.

The omission of strict implication in CPC and its derivatives is the omission of the coming-to-be of judgment through link between conditions and conditioned.
CPC can express only compound not complex sentence.

h This foundation grounds grand-scale analogy.

Experience : understanding : judgment :: potency : form : act

For every stage in the development of understanding, of science

i This foundation grounds sequences of developing concepts of same object.

Experience: the phenomena of fire.

Heuristic structure: What is fire?

Sequence: (a) fire is *X* manifested by these phenomena; (b) fire is one of four elements; (c) fire is manifestation of 'phlogiston'; (d) fire is a chemical activity of a certain specified type.

Because of (a), it is possible for (b), (c), (d) to be statements about the same object and so to be incompatible.

j   This foundation is sufficient not only for traditional and mathematical logic, but also for consideration of dialectical, existential, magical, mystical, hermeneutical logics.

Dialectical: start from what seems obvious; let it have its head; absurd conclusions bring to light the limitations of initial 'obvious'; e.g., development of ML; cf. repeated use of this technique in Hegel's *Phenomenology of Spirit*; Toynbee's 'Transfiguration' (problem of transportation becomes problem in ethics of motoring, traffic)

Existential: cf. e above.

Magical: cf. section on 'Metaphysics, Mystery, Myth' in *Insight* chapter 17.

Mystical: cf. H. Leisegang on St Paul (Denkformen, 1951 [2nd ed.])[8]

Hermeneutical: cf. Truth of Interpretation, *Insight* chapter 17.

5   Mathematical Logic and Scholasticism

1 A New Factor in Problem of Method

While ML is strictly indifferent philosophically and this philosophic indifference is maintained by the better and more intelligent writers, still it very easily is given an empiricist and pragmatic twist and so amounts to an invasion of logic by the enormous problem of philosophic differences.

Perhaps the *practical* procedure would be to continue to treat logic as philosophically indifferent, to use it as an introduction to philosophy, to add perhaps that it gives rise to questions and differences that at a later stage in the course will be treated more adequately.

From a *theoretical* viewpoint all seems to depend on the solution that one adopts of the general problem of philosophic difference.

8 See above, p. 93, note 11.

On a view associated with the great name of E. Gilson one should begin philosophy with metaphysics.

Now I have no doubt that the ultimate and decisive factor is *sapientia*, and that in St Thomas *sapientia* is (1) a gift of the Holy Ghost that is connected with mystical experience (*patiens divina*) and (2) within the natural order, Aristotle's *Metaphysics*.

Accordingly, I am quite ready to grant that this view has a solid foundation in tradition.

However, in Aquinas, while there is a distinction between natural and supernatural, between reason and faith, there is no separation. The Thomist distinction was followed four centuries later by the Cartesian separation. It is within the context of that separation, within the context of subsequent condemnations of fideism and of traditionalism, within the context of our own Anglo-Saxon cultural traditions, that we have to operate.

For such and many other reasons, I have endeavored in *Insight* to work out a genetic account of *sapientia*, an 'ascensio mentis per intelligibile et verum ad ens.'

Hence, in the fourth lecture, I sought the foundations of logic in the subject's personal appropriation of his own empirical, intellectual, and rational consciousness; on this view the foundations of logic are by identity the solution of the epistemological problem, the foundations of metaphysics, the foundations of ethics, and the foundations of natural theology.

2 Is Scholastic thought to be cast in the form of an axiomatic system?

There certainly exists a conception of Scholasticism as a deductive system.

Philosophy is the deduction of necessary conclusions from self-evident principles.

Theology is the deduction of further conclusions from the Word of God with (or without) the help of the self-evident principles of natural reason.

However, this view is more easily substantiated by appealing to Scotus than to Aquinas, and more by appealing to Aquinas's statements (which one interprets in the light of one's own deductivist horizon) than from Aquinas's practice.

I think it useful to distinguish (1) non-empirical (2) empirical and (3) comprehensive types of inquiry.

(1) In the non-empirical type there is little appeal to concrete matters of fact. Such is mathematics, and it is markedly a deductive science.

Note, however, that this deductive aspect is coupled with a constructive aspect. Mathematics starts from elements (geometrical entities, number, etc.) and out of these simpler and prior objects constructs ever more complex objects. And it is precisely this constructive aspect that makes the deductive aspect possible. Because x, y, z can each be constructed by beginning from a, b, c, it is possible to deduce the relations between x, y, z.

(2) In the empirical type of inquiry, there is a twofold movement.

First, there is the movement from the *priora quoad nos* to the *priora quoad se*: this is represented by any history of the origins and development of physics, chemistry, biology; all along the line the decisive factor is the sensible matter of fact.

Secondly, however, there is the opposite movement from the *priora quoad se* to the *priora quoad nos*: this is represented by the textbook of physics or chemistry, in which one begins from laws and systems and proceeds deductively towards the concrete and complex.

Note that neither the non-empirical deduction nor the empirical deduction occurs within a single plane.

ML seems to have shown that a mathematical LF has to be, not a single deduction from a single set of premises, but rather a series of levels each with more comprehensive premises and its own deductive expansion; where, observe, the series of levels is open, so that there is no top level.

Again, the history of physics, chemistry, biology has been the history of a succession of higher viewpoints, where on each later viewpoint the premises of the deduction have been different. Einstein includes Newton as a limiting case, but Einstein's premises were beyond Newton's ken.

(3) Philosophy and theology are inquiries of the comprehensive type: they are concerned, each in its own way, with everything.

Now if a philosophy is to include a philosophy of science, while the sciences are open, developing, changing, then the possibility of its being fixed is that it have the fixity, not of a monolith, but of a form that admits variable contents.

In other words, such a philosophy has to be an invariant structure in which the structured elements are free to change.

Hence, in *Insight* the structure of the knowing subject (a concrete unity of experiencing, understanding, judging) is shown to be invariant (not subject to revision) and to imply a corresponding invariant structure (potency, form, act) in the proportionate object of our knowledge. Where the philosopher knows that there are forms but the various departments of science investigate what the forms are.

Is philosophy so conceived to be cast in axiomatic form? I do not think it impossible. I do not think it to be very useful, because philosophy is not constructive after the fashion of mathematics, and because the real issues in philosophy do not lie in drawing conclusions but in the subjective intellectual development that is the 'ascensio mentis per intelligibile et verum in ens.'

Traditionally, Scholasticism is not a deduction from a single limited, well-defined set of principles: it is a sequence of theses that are based on deductive arguments with the premises of the arguments coming from all over the map.

Examine any manual of philosophy and you will find this to be so.

Make a logical analysis of the treatment of 'soul' in Aquinas's *Contra Gentiles*, II, about chap. 48 to 94.

3 Does ML eliminate 'existence'?

See for nuanced discussion of 'quantifiers,' J. Dopp, 'La notion d'existence dans la logique moderne,' *Proceedings*, Amsterdam 1948, 735–39.[9]

For a defense of what amounts to the proposition 'Existential propositions do not exist,' see G. Ryle's paper 'Systematically Misleading Expressions,' originally published in *Mind* about 1938–39, and reprinted in A.G.N. Flew, *Logic and Language*, First Series, Oxford, Blackwell, 1952.[10]

The answer to this question is dependent on what one happens to mean by 'existence.'

9 See above, p. 133, note 18.
10 See above p. 134, note 19.

I mean what is known through grasping the virtually unconditioned in a concrete judgment of fact; it is what is known by the 'yes,' the 'is,' in a concrete judgment.

On the whole, 'existence' in this sense is beyond the horizon not only of mathematical logicians, logical positivists, but also of so-called existentialists.

On the other hand, 'existence' in this sense can be shown to be admitted implicitly by anyone that claims an ML to possess a reference to anything that happens to be.

4 Does an ML eliminate substance?

An ML makes it possible to do an enormous amount of logic without raising any question of substance, and so for this reason it is welcomed by many.

Again, in Aristotle's day, explanatory science barely existed, and so in Aristotle and in those dependent on him for their notion of science, there is insufficient stress on the significance of relational structures in science.

Thirdly, one can do mathematics without a notion of substance.

Fourthly, one cannot dispense with the notion of substance in any science that is not only descriptive but also explanatory, as is shown in *Insight*.

Fifthly, 'substance' is an ambivalent term: it can mean the reality that is an intelligible unity-identity-whole (in spatiotemporal difference); it can also be used to denote the 'already out there now real' of spontaneous extroverted animal consciousness. This basic ambiguity rises from differences in the attitude, orientation, degree of ek-sistence, of the subject; and so it is an extremely difficult problem in philosophy.

5 What is relation of ML to (1) Logical Atomism, and (2) Logical Positivism?
See J.O. Urmson, *Philosophical Analysis: Its Development between the Two World Wars* (Oxford: The Clarendon Press, 1956). A brief, exact, illuminating, and cogent work.

Logical Atomism was the hope that a complete and satisfactory philosophy could be constructed by proceeding from the MLS of *Principia mathematica*, substituting ordinary words for the variables in the MLS, and showing that apart from the connectives supplied by the MLS nothing was needed but atomic experiences of the type 'red here now.'

Logical Positivism proceeds from a division of possible propositions based upon MLS.

A proposition may be true independently of the truth or falsity of its variables, e.g., *EANpqCpq*, and it is named a tautology; it is simply a circular combination of functors.[11]

A proposition may be false independently of the truth or falsity of its variables, e.g., *ENpp*; it is contradictory.

A proposition may be true or false according to the truth or falsity of one or more of its variables, e.g., *Kpq* is true if both *p* and *q* are true, *Apq* is true provided not both *p* and *q* are false, etc. Such a proposition needs some extralogical means of verification.

Hence, propositions are either tautologies (mere circular combinations of words) or they are empirically verifiable. Only in the latter case have they much in the way of meaning. The crucial problem of LP was to find a verifiable verification principle.

11 Lonergan is using here the notation of the Polish logician Łukasiewicz, where *N, K, A, C, E* symbolize respectively negation, conjunction, disjunction, conditional, biconditional, propositional symbols are lower case, and the operators are placed to the left of the connected propositions. On the variety of symbols, see Kneale and Kneale, *The Development of Logic* (see above, p. 42, note 10) 513–24; there is a comparative table on p. 521.

7

The Lecture Notes on Existentialism I: Orientation and Authors

1 General Orientation

1 By 'existentialism' we shall understand the types of method and doctrine exemplified by K. Jaspers, M. Heidegger, J.-P. Sartre, Gabriel Marcel.

The name is admitted by Jaspers and Sartre; it was admitted for a while by Marcel who after 'Humani generis'[1] and, perhaps, to disassociate himself from Sartre, rejected it; Heidegger says he is concerned with *Eksistenz.*

Jaspers is Kantian and Lutheran; Heidegger an apostate and agnostic; Sartre an atheist; Marcel a convert to Catholicism.

2 They are concerned with what it is to be a man, not in the sense of having a birth certificate, but in the sense employed by President Eisenhower last fall when, asked whether it was not risky to send the fleet into the Mediterranean during the Egyptian crisis, he answered, 'We have to be men.'

'Being a man' in that sense results from a decision, is consequent to the use of one's freedom, makes one the sort of man one really is, involves risk (in the present instance, the risk of nuclear warfare and all that implies).

---

1 See below, page 219, note 2. Paragraphing in this and the next chapter, including the spacing between paragraphs, follows Lonergan's autograph of these lecture notes.

3   It is anti-positivist: 'being a man' is not any set of outer data to be observed, any set of properties to be inferred from the outer data, any course of action that can be predicted from the properties; it springs from an inner and 'free' determination that is not scientifically observable.

It is anti-idealist: the various transcendental egos are neither Greek nor barbarian, bond nor free, male nor female; they don't suffer and they don't die; we do.

Positivism and idealism have been major determinants in producing the contemporary world; in the measure that the contemporary world is found unsatisfactory or, frankly, disastrous, existentialism has a profound resonance.

Sein und Zeit[2] quickly ran through five editions; Jaspers's *Geistige Situation der Zeit*[3] was through five editions in about a year and has been translated into six languages including Japanese; Sartre was a cafe hero in Paris.

This contemporary resonance fits in with existentialist concern for time and for history.

Since 'being a man' is not a fixed essence with which we are endowed from birth but the result of the use of our freedom, and further since 'being a man' is not a property that necessarily remains with us but is maintained by us precariously in the continuous use of freedom, 'time' is an intrinsic and necessary component in 'being a man.' Hence, Heidegger's *Sein und Zeit*, Marcel's *Homo Viator*.[4] However, concern with history on the grand scale appears only in Jaspers, e.g., *Vom Ursprung und Ziel der Geschichte*.[5]

4   It is unconcerned with propositional truth and with man's per se capacities for truth or anything else.

This unconcern arises in Heidegger, Sartre, from phenomenological concentration on the sources, grounds, whence spring concepts and judgments.

It arises in Jaspers from Kant who is believed to have shown that any objective statement deals only with appearance.

2   See below, p. 370, number 4.
3   See below, p. 371, number 5.
4   See below, p. 224, note 23.
5   See below, p. 372, number 15.

It arises in Marcel from his concern with being a good man as opposed to mere existence as a man, and the commonsense attitude (buttressed by dissatisfaction with idealism) that technically correct propositions have little or nothing to do with what you really are.

In all it arises from a turning away from the universal, necessary, abstract, per se, to the unique individual, the contingent, the concrete, the de facto.

Jaspers repeatedly insists that freedom is not definable; Sartre establishes the fact of freedom by asking whether you have been in the torture chamber with the Nazis and made the experiment of freedom by not giving your comrades away; none of them would dream of discussing 'man' as what is common to mewling infants, people sound asleep, and the mature man facing a crisis in his life.

Gabriel Marcel: 'Plus il s'agit de ce que je suis et non de ce que j'ai, plus questions et réponses perdent toute signification. Quand on me demande, ou quand je me demande, en quoi je crois, je ne puis me contenter d'énumérer un certain nombre de propositions auxquelles je souscris; ces formules, de toute évidence, traduisent une réalité plus profonde, plus intime: le fait d'être en circuit ouvert par rapport à la Réalité transcendante reconnue comme un Tu.' Quoted by R. Troisfontaines, *De l'existence à l'être*, ii, 352.[6]

5 This unconcern with propositional truth and this distaste for the per se is de facto connected with an incapacity to provide foundations for either propositional truth or the per se.

It is my firm conviction that, while there is much in existentialism on which we should practice the patristic maxim of despoiling the Egyptians, still we cannot simply take existentialism (even Marcel's) and incorporate it within Scholasticism.[7]

---

6 See below, p. 228, note 29.
7 The following is crossed out in the autograph of these notes (but see below, p. 229): 'Marcel (in a private conversation with F. Copleston) intimated that he had little or no idea how one could go about proving the existence of God; the statement quoted above, taken in its concrete context, virtually eliminates the possibility of any evolution of dogma or any dogmatic theology in the traditional Catholic sense; in other words, while it fits in with the requirements for a Biblical theology, it has no room for the Councils of Nicea and Chalcedon, to say nothing of Trent and the Vatican.'

6 Existentialism is concerned with the human subject qua conscious, emotionally involved, the ground of his own possibilities, the free realization of those possibilities, the radical orientation within which they emerge into consciousness and are selected, his relationship with civilization, other persons, history, God.

7 G. Marcel is not a systematic thinker; in his preface to R. Troisfontaines, *De l'existence à l'être*, he congratulates the author on having done for him what he could not do for himself.

G. Marcel is a penetrating thinker and an extremely effective writer: he can put a concrete idea, orientation, criticism of life, across with extraordinary brevity and skill.

He reviews his intellectual history in 'Régard en arrière,' a paper added to the collection *Existentialisme chrétien: Gabriel Marcel*, introd. by E. Gilson; contributors include Delhomme, Troisfontaines, et al.[8] See Bochenski.[9]

His *Journal métaphysique*, I, was published in 1927,[10] the date of *Sein und Zeit*. His background is idealism (including Bradley) and Bergson; Kierkegaard is acknowledged to have influenced him indirectly.

8 K. Jaspers began with abnormal psychology, of which he became professor and wrote various technical articles; he has a profound respect for science and is a mordant critic of scientists; forty years ago he was ridiculing the mythology of the brain and the mythology of the unconscious in the psychologies of his time.

He is a Kantian with the *Critique of Practical Reason* brought to life by Kierkegaard and Nietzsche.

He is the most broadly cultivated of the existentialists and with the widest range of interests; he writes very intelligibly, explains exactly what he means, strikes one as very balanced and sane.

In his *Philosophie* (1932) he explains that *Existenz* and *Transzendenz* correspond roughly to what are named by mythical consciousness the soul and God.[11]

8　See below, p. 230, note 32.
9　See above, p. 51, note 26.
10　See below, p. 230, note 31
11　See below, p. 371, number 6. In the autograph, the word 'Cyphers' is handwritten at the end of this sentence.

Since then he has developed the notion of *das Umgreifende* (which corresponds roughly to the notion of being in *Insight*) and has come to place a great deal more emphasis and reliance on reason (more perhaps to disassociate himself from Sartre and similar tendencies than from assignable grounds) and to speak openly of God (as a necessary philosophic postulate).

9 M. Heidegger is perhaps the most original and profound of the lot; his immediate source is Husserl; from Heidegger by way of a strong dose of French clarity and logic comes Sartre, who figures as the *reductio ad absurdum* of the movement.

2   On Being Oneself

1 Subject is subject of ...; a relative term; meaning varies with correlative.

> Grammatical: function in sentence.
> Logical: function in proposition.
> Metaphysical: recipient: matter, form; potency, act, etc.
> Psychological: subject of stream of consciousness.

2 Consciousness streams in many patterns: dream, biological, aesthetic, intellectual, dramatic, practical, mystical.

Contrast: subject of stream as orientated on knowing, and subject of stream as orientated on choosing.

Of old: speculative and practical reason; now, concrete flow orientated on knowing and orientated on choosing.

3 Intellectual pattern is intellectual by its detachment, by non-intervention of alien 'subjective' concerns,[12] by concentration of attention, effort, on observing, understanding, judging.

Subject is involved, but as involved he is subordinated to dictates of method, to immanent concretion within himself of principles of logic, of scientific aspiration, of absolute criteria; commitment is to submission to norms.[13]

---

12 The word 'subjective,' in inverted commas, is added by hand in the margin of the autograph.
13 The last phrase, from 'commitment,' is added by hand in the autograph.

Subject is headed towards object, universe; he himself enters into picture only within objective field, as a particular case in a broader totality; the data of his consciousness may be a source of information, but they are relevant not qua his.

Subject has a responsibility: his judgment is his, and 'personne se plaint de son jugement'; still, it is a limited responsibility, for he can frame his conclusions as positive or negative, certain or probable, etc.; in brief he is bound to say what he knows and no more than he knows, re object and re mode, but he is not committed to reaching definite results.

4 The practical pattern of experience demands the intervention of the subject.

He may choose *A* or *B*, *A* or Not-*A*; or he may consent to drift, permit himself to be other-directed, where however the consenting and permitting are equivalent to choosing, though an inauthentic equivalent.

The choice, decision, drift are determined neither externally, biologically, psychically, nor intellectually.

Even when one knows everything about everything, an *operabile* cannot be demonstrated; it admits no more than rhetorical syllogisms. But in fact I do not know everything about everything; I do not know everything that ultimately is relevant to the choices I have to make; and nonetheless I already am alive, thinking, acting, under a perpetual necessity of drifting or choosing, choosing *A* or Not-*A*, *B* or Not-*B*, ...

Hence, choosing is within an atmosphere of incertitude, and so it involves an acceptance of risk.

Choosing not only settles ends and objects; it gives rise to dispositions and habits; it makes me what I am to be; it makes it possible to estimate what I probably would do; it gives me a second nature, an essence that is mine in virtue of my choosing; still, it does not give me an immutable essence, achievement is always precarious, radical new beginning possible.

In choosing I become *myself*; what settles the issue is not external constraint nor inner determinism nor knowledge but *ut quo* my will and *ut quod* myself; in the last analysis the ultimate reason for my choice being what it is, is myself;

if left to mere balancing of motives, impulses, then I consent to drift; I consent to being other-directed; I implicitly choose as *myself* the *On*, *Man* – inauthenticity;

if not left to mere balancing of motives, impulses, then I intervene, I knowingly assume risk and responsibility;

in either case what ultimately is operative is purely individual, unique;

in the drifter what results is another instance of the average man in a given milieu;

in the decisive person what results is what he chooses to be;

in the drifter, individuality is blurred; his individuality is his consenting to be like everybody else;

in the decisive person there comes to light both his individuality and the total-otherness of other individuals; my choice is what it is because that's what I choose; yours is because that's what you choose; even when what is chosen is the same, still the sources are simply different.

Finally, there are limiting situations: the drifter can no longer just drift; and the decisive person is powerless to change things by deciding:

general: historical period, social milieu of birth, opportunities, male or female, old or young;[14]

particular: death, suffering, struggle, guilt;

confronted with limiting situations, the drifter may try to forget, but ultimately he cannot succeed; he is totally involved, all of him is involved, and he is totally unprepared;

on the other hand, the decisive person can be as decisive as he pleases, but the limiting situation is not thereby removed.

5 'Oneself' is the irreducibly individual element whence spring the choices of the decisive person and the drifting, forgetting, of the indecisive.

What springs from that source is free; for it, one is responsible.

What results from that source is not only the sequence of activities but also the character of the man, the second nature, quasi essence, by which precariously one is what one is.

Nor does choosing wait upon learning, the acquisition of as much knowledge as might be relevant; it involves risk and incertitude.

Finally, in choosing is involved everything that concerns me.

6 Being oneself is being the subject of free acts. It is existential. In the limit, ex-sistence implies transcendent, absolute.[15]

Within a satisfactory synthesis, there is possible an alternation, a Withdrawal and Return, a mutual complementarity.

14 Written by hand in the margin of the autograph: 'in them emerges exigence for absolute value; absolute value needed for it to be worth while to be myself in any significant sense.'

15 This paragraph is handwritten in the autograph, at the bottom of the page, with an arrow in the left margin pointing to it. Crossed out in the autograph

In the intellectual pattern of experience I am choosing because I choose to submit entirely to the exigences of knowing in order to know; and without that knowing there would be, not merely a residual incertitude and risk to choosing, but a total blindness that makes choice indistinguishable from mere force, instinct, passion.

In the practical pattern of experience, there is an ultimate moment of 'being myself,' of incertitude and risk, and nonetheless total commitment; but it is a known ultimate moment, and it is within a context of knowing and with respect to a largely known.

3   On Being Oneself: Philosophic Significance of the Theme

1 It provides a ready rationalization for those who do not wish to endure the restraints of knowing. Let's drop philosophy, speculative theology, science.

Love of neighbor, zeal for souls, dialogue, *disponibilité*, prayer.

2 Breaks through positivist science of man.

It denies that there is any ready-made essence or nature with predictable properties.

L'homme se définit par une exigence.

Eisenhower: 'We have to be men.' It implies that we might be less than men, that there is an exigence for us to be men, that the exigence is to be met by a decision.

3 Breaks through pragmatist science of man.

One learns from experience about things, about one's own potentialities.

---

right above this handwritten paragraph was the following typewritten start on this §6:
'In brief, the intellectual and the practical patterns of experience are incompatible.
'In one, free decisive intervention despite risk and with total commitment is essential; in the other it is barred from the movement towards truth to be attained, though total commitment is demanded from the subject by truth as attained.'
The following paragraph ('Within ...') begins a new page, is also numbered 6, and is really concerned with the relation of the intellectual and practical patterns treated in the crossed-out paragraphs. In the lecture outlines that Lonergan distributed to his audience, his handwritten paragraph was given (with 'free' misread by the typist as 'fine'). The 'Within' paragraph followed.

But[16] the issue is not one of knowing, whether a priori or a posteriori; given all the knowledge possible, all the human experiments desirable, there still remains the whole issue of deciding, which even then would involve incertitude and risk.

And meanwhile one already is living, and one has only one life. The decision to risk nuclear warfare is not justifiable pragmatically.

4 Breaks through the idealist view of man.

The idealist's absolute or transcendental ego is neither Greek nor barbarian, neither male nor female; it neither dies nor suffers nor struggles nor acknowledges guilt.

The idealist's world is a world that is pure intelligibility, rational throughout; it is not a world of free choices springing from unique individuals that are totally concerned in the once-for-all of the momentous moment.

5 Sets problems for contemporary Scholasticism.

(a) What meaning is possible for the fact that I become myself?
Ambiguity that comes to light in metaphysical theory of person, subsistence.

Rests on issue: is metaphysics knowledge of things through their causes or through the *decem genera entis*?

Is the thing just its substance or is the thing a whole that includes both substance and accidents?

(b) *Verum et falsum sunt in mente; bonum et malum sunt in rebus.* But in the concrete, there are no abstractions, and so there is no abstract good.

If no abstract good, abstract moral precepts do not suffice to reach the good; they can be no more than pointers to the direction, location, in which the good lies, or limits indicating where the good does not lie.

But there remains for each one to work out concretely what the good really is.

There remains an order of the universe, but it is not an order deducible from abstract essences and schematic hierarchies; it is a concrete

16 The following is crossed out in the autograph: 'But the process of experiencing oneself is already one's becoming oneself; one has only one life; and the problem is not one of sacrificing oneself to the determination of pragmatically validated knowledge of man, since the real issue is not one of knowledge but of deciding.'

unfolding in concrete situations; and the concrete situations are proximately the product of individual decisions about the concrete good.

There remains the natural law (situations do not change moral precepts) but there arises the significance of *kairos*, of my situation, my opportunity, my duty; and while these can be illuminated by moralists, by spiritual directors, the ultimate issue is whether or not I am to take a risk and assume a total responsibility and rise to the occasion as I alone see it.

There is to the order of the universe the emergence of good from evil, the heightening of evil to a maximum that sets the alternative of conversion or destruction,

where the evil is to be met not by being included as intelligibility within the order but as surd violating the order, as a demand not for justice but for self-sacrifice and charity.

The order of the universe is not a mechanistic plan flowing from essences; it may descend to that through sin; but it rises from it inasmuch as the order is a matrix, network, of personal relations.

Situation, surd, *kairos*, charity.

(c) The need of an *ancilla* that will supply theology with the categories necessary to assimilate the doctrine of the Bible.

The possibility of such an *ancilla*: can existential questions be handled by the Catholic philosopher? Do they not suppose knowledge of theology by their very nature?

(d) Withdrawal and return: not simply the mutual dependence of willing to know and knowing to will.

There is the problem of conversion (reorientation, reorganization of mind and life).

Kierkegaard's spheres: aesthetic, ethical, religious *A* and *B*.

Upward change is not in virtue of knowledge on lower plane; it is not in virtue of will following knowledge on lower plane.

There has to be the apparent irruption of a latent power, the possibility of a radical discovery where the discovered has been present all along, the fact of an obnubilation that prevented prior discovery.

This sets the radical question in all philosophizing.

It is relevant for Scholastics with their innumerable disputed questions, and no method of solution, not only not in sight but not even desired, sought, seriously believed in.

In various measures it is the concern of the thinkers named existentialists.

Proposal: to face our existential question and through it move towards some understanding of this question for others.

4   Husserl: Later Period[17]

1 Enormous literary remains, mostly in shorthand, preserved at Louvain and being classified and edited under H.L. van Breda, O.F.M.; there is some parallel institute at Cologne.

Die Krisis der europäischen Wissenschaften und die transzendentale Phäno-.menologie, ed. W. Biemel. Published, Haag, M. Nijhoff, 1954.[18]

Husserl's last work; about the first third published in his lifetime; the rest put together from his remains; probably owes something to the stimulus of his most brilliant (and disowned) student, Heidegger.

A general idea of this work provides a good introduction to Heidegger and offers the advantage of not involving us in the complexities of the development of Husserl's ideas on Phenomenology, *Reduktion, Epoché.*

2 It might seem paradoxical to speak of a crisis in modern science: its achievements are unmistakable; its labors in endless fields continue apace; and what unsolved problems there are will be solved either by the methods of the past or by the discovery of new methods to complement and perfect those of the past.

Still, the need of new methods can be discovered only by a critical survey; and if the need exists at present, then the survey will not only discover the existence of the need but also provide a signpost to point the way towards a solution.

Such a survey demands a criterion, and the criterion that can hardly be rejected is an act of recall in which we reenact within ourselves the original intentions of the scientific enterprise.

These intentions had two principal manifestations: fourth-century Athens; and the Renaissance.

17 In the autograph, this section is headed 'The Late Husserl.'
18 For further details, see below, p. 247, note 2.

3 The formulation of the aim of science in fourth-century Athens consisted in an *Umdeutung* (shift in meaning) of popular notions of *sophia, alêtheia, epistêmê*; this shift took place through the Platonic contrast of *epistêmê* and *doxa*, of *dialektikê* and *eristikê*; it consisted in setting up an ideal of knowledge and truth that involved (1) a sustained effort, (2) a methodical procedure, (3) a rigor, (4) an attainment of evidence, (5) a solid immovable basis in certainty, that simply were not contained in the previous customary connotation of such terms as *alêtheia, epistêmê*; finally it unfolded in the works of Aristotle, Euclid, Archimedes, the historians, and the medical doctors.

4 The Renaissance brought forth a far more grandiose proposal: it discovered in the ancients

    (1) an ideal of knowledge and truth vs. merely traditional opinion

    (2) as a principle of transforming society vs. merely traditional power.

In the measure that that ideal and that principle are valid, Western man is the exemplar of mankind, the realization of the meaning of what it is to be a man.

In the measure that that ideal and that principle are not valid, [Western] man is just another anthropological classification; he is of concern to us, not because of any intrinsic value or significance, but merely because he is the type or species to which we belong.

5 Hence, if we are to judge modern science by the criterion of its original intentions, we must ask what hope modern science offers

    (1) of the attainment of knowledge and truth

    (2) of a principle that frees man from merely traditional opinion and power and enables him rationally and responsibly to place human society on a basis of truth and reason, freedom and responsibility.

6 Judged by this criterion, modern science can be criticized

    (a) for its tendency to splinter into specialties: any university catalogue; congresses; *Deus scientiarum Dominus*;[19]

(b) for the autonomy of the splinters: what counts effectively within each of the departments, sections, subsections, is what is recognized as 'good' in that department, section, subsection;

19 See below, p. 252, note 9.

discussions of knowledge, science, truth, are just so many other specialties, and their relevance to other fields is a mere matter of opinion;

(c) for the drift to the criterion of technical competence;
upon a background of traditional norms that are not questioned (*Selbstverständlichkeiten*), the effective principle of change is technique: what counts ultimately is 'getting results,' and what counts proximately is the approved technique, how one goes about it, all the wrinkles of observation, experimentation, all the apparatus of bibliography and footnotes;

(d) for the position of the human sciences;
scientific medicine is based on scientific anatomy, physiology, pharmacy, chemistry, physics; folk medicine (recipes, cures) for the individual patient has disappeared; but for human society the only medicine remains folk medicine; endless nostrums are proposed and, scientifically, they are of no value; de facto, techniques are unified by totalitarian state and by mass democracy: unifications of power not reason;

(e) for the impossibility of a reorientation on the present basis;
a reorientation demands a general view, and no general view is possible; only a shifting set of best available opinions in more or less unrelated fields;
a general view is the work of a mind, and no mind can master all the techniques and so no mind can present a scientifically respectable general view. *Bodenlösigkeit!*

7 If we have found that modern science does not fulfil its original inspiration, intention, aim, we can go further and ask if there has been some radical defect or oversight in its program.

H's diagnosis of the malady is that scientific clarity floats on popular obscurity, scientific evidence on popular *Selbstverständlichkeit* (Marcel: *tout naturel*), in brief the real basis of science has not been explored, examined, evaluated.

(a) For there exist two truths and two worlds.

There is popular truth in the sense of telling the truth in the home, in business, in law courts, in newspapers and periodicals, in autobiography.
There also is scientific truth in the sense of a validated set of propositions: logic, maths, physics, chemistry, etc.

These two reflect the original duality and bifurcation of *doxa* and *epistêmê*, of setting up a scientific ideal within a context of popular notions (one might compare the Hebraic ideal of 'man before God' within the unity of Hebraic tradition).

There is the popular world of poets and men of common sense, of everyday assumption, opinion, activity.

There is the quite different world of the scientist and philosopher: mass instead of weight, temperature instead of heat, dimensions instead of size, elements instead of bodies.

(b) There have occurred a series of *Unterschiebungen*.

The scientific or philosophic world is shoved under the popular world as the underlying reality, as what really is 'out there.' Popular notions are considered mere ignorance or naivete.

(c) But the fundamental truth and the really basic world is not the scientific or philosophic but the popular.

One has only to take any scientific procedure or conclusion and with a little probing it will come to light that the ultimate evidence lies in the popular world, the *Lebenswelt* with its *Selbstverständlichkeiten*.

Science claims to rest on experience, but what is experienced is not the scientist's 'real world' but the 'popular world.'

Science rests on the testimony of observers, experimenters, etc., and they are operating (1) in the *Lebenswelt* and (2) after the fashion of the *Lebenswelt*. E.g., there is no investigation of the psychophysical parallelism (or whatever you please) that has to be postulated to proceed from the results observed by Michelson and Morley to the conclusions announced by Michelson and Morley. Indeed, scientists may find this objection a mere oddity, but it is an oddity, not from any scientifically established viewpoint, but merely from the viewpoint of the *Selbstverständlichkeiten* of common sense.

8   If a malady and a diagnosis, then also a remedy, cure.

(1) The priority of the subject: the subject is the source of both truths and both worlds. There is a *natürliche Einstellung* that yields popular truth and the popular world. There is a cultivated (Athens, Renaissance, *Aufklärung*) *Einstellung* that yields the conceptual worlds of scientists and philosophers.

(2) What the subject is the source of is intentional, namely, what he means, symbolizes, represents, intends, ...

Cf. Cassirer, *Essay on Man*, Man is the symbolic animal.[20]

Cf. Köhler's apes, incapable of free images;[21] man's capacity for free images is also man's capacity for envisaging a world, in fact, many incompatible worlds.

(3) What is needed is a return to Descartes's *Cogito*.

Let the subject realize that all he thinks, believes, is certain of, whether on popular, scientific, philosophic grounds, is just intentional.

Let him ask how much he can primarily, irreducibly, immutably hold: e.g., 'I doubt,' 'I think thoughts,' ...

Let him refuse to leap from Cartesian acceptance of *Cogito* to Galileo's mathematized world of real bodies.

Similarly let him refuse to leap from the intending 'I' to Descartes's metaphysical substance, the soul.

For both these leaps are erroneous: they postulate an objective reality that is more than and other than the range of the intentional products of the constructing subject.

And both of these transitions are disastrous; for while everything comes from the subject, still science has a 'real world' of protons, electrons, etc., and an utter incapacity for *Geisteswissenchaft*, and scientific psychology is an absurd attempt to study the subjects (from which everything proceeds) in terms of the outer observable objects.

(4) The solution is the identity of *Transzendentale Phänomenologie, T. Psychologie, T. Philosophie.*

Epoché: the immediately evident is the intentional (withdraw from interest in, concern with, commitment to the 'really real,' the way a man forgets his business to live in the intimacy of his family, or vice versa).

Transcendental Reduction: not the mechanist or behaviorist reduction of the intentional to the 'real' but of all intended terms to the intending subject.

20 Ernst Cassirer, *An Essay on Man: An Introduction to a Philosophy of Human Culture* (New Haven: Yale University Press, 1944; issued as a Yale paperbound, 1962).
21 See below, p. 264, note 13.

Secure science and philosophy an immovable ground: not some flimsy ideal construction within an obscure context of *Selbstverständlichkeiten*; not the dubious products of some historical cultural process; but seek in the *Lebenswelt* what is primarily given, really primitive.

5   Critique of Husserl's *Krisis*

(a) There is a real problem set by science and especially human science; and its only solution lies in a philosophy.

Natural science can get along somehow (with a bias towards practical and neglect of basic research) by relying on pragmatic criterion of success; but human science, since the scientist is one of its objects, is involved in philosophic, indeed theological, issues (cf. problem of synthesis today and in MA [Middle Ages]).

(b) Husserl pursued philosophy 'als strenge Wissenschaft,' as grounded in necessity and yielding absolute certitude.

This ideal with its Greek and Cartesian antecedents is in need of distinction.

All human judgments rest on virtually unconditioned; they are true as a matter of fact; the pursuit of absolute necessity and absolute certitude is doomed to failure because it seeks more than there is to be had.

(c) The correlations of *Abschattung-Horizont* and *Einstellung-Welt* are valuable contributions to cognitional analysis.

Still, the alleged two worlds are but one set of beings considered from two standpoints: as relevant to human living; as constituted by inner relations of things to one another; 'being' is the unifying notion.

Again, the alleged two truths are simply the result of applying the different criteria relevant and appropriate to the different standpoints.

(d) Science does not rest de facto on evidence and procedures of *Lebenswelt*.

There has been a failure to attempt the phenomenology of the scientist and phenomenologist: Thales, Archimedes, Newton, Einstein are just odd and strange from commonsense viewpoint.

This failure has been buttressed by subsequent exclusive concern with engaged as opposed to contemplative consciousness.

One must not expect scientist to be able to detail what he really does. Einstein's advice to epistemologists: Don't listen to what scientists say; watch what they do.

(e) Greek, Renaissance, subsequent normative accounts of truth, science, method

are not just artificial ideals floating on popular obscurity, though their non-philosophic or inadequate philosophic statement may be such;

they are expressions and clarifications and objectifications of the immanent normativeness of human intellect itself, which is *participatio creata lucis increatae.*

This fact coming to light in Heidegger's *Erschlossenheit.*

(f) There is a real priority of the subject in knowledge.

Human sensitive psyche is not animal: free images; development of imagination.

Participatio creata ground of questions, intellectual activity.

But this priority is not to be interpreted in Greek and Cartesian fashion with exaggeration of absolute necessity and absolute certitude. Moreover, *epoché* is involved in confusion of 'animal faith' and 'rational judgment'; and transcendental reduction properly is to 'being' and not to 'intending,' which also is.[22]

6   Phenomenology: Nature, Significance, Limitations

1 Nature

Phenomenology is an account, description, presentation of the data structured by insight.

(a) Of data, what is given, what is manifest, what appears.

Not just external data, phenomena, but also inner; hence, opposition to mechanism, behaviorism.

Not exclusively inner data: the inner intentional act terminates at the outer datum; and the outer datum is just the term of an inner intentional act.

No exclusions: not primitive as opposed to derived, natural as opposed to cultural, sensitive as opposed to intellectual, cognitional as opposed to emotional, conative.

(b) Data structured by insight (my way of putting it)

22 Lonergan's autograph of the lecture outlines seems to proceed directly to the notes on Heidegger (see below, §7). The notes on phenomenology in §6 were, it seems, written separately.

Selective: not exhaustive description of all and any data significant: seeks basic universal structures; 'Eidetic,' '*Wesensschau*,' Aristotle's parts of form in *Metaphys. Z.*

Takes effort, time: not first bright idea, calls for scrutiny, penetration, contrasts, tests; may have to overcome spontaneous tendentiousness, systematic oversight, common over-simplification, preconceptions arising from 'scientific,' 'philosophic,' or other sources.

(c) Not insight as such

Extremely elusive.

Would lead immediately to unity (viewpoints, higher viewpoints, theory of judgment).

There is no such tendency towards unity in Husserl (forever discovering new fields to be explored)

and similarly there is no such tendency in his successors.

(d) The data as structured by insight and not the subsequent conceptualization, definition, theoretic statement of the data in their essential features.

Perpetual appeal to prepredicative manifestation.

Basic distinction between what is given, manifest, appears, and thematic treatment of the given by the phenomenologist (*Phainomena legein*).

2 Significance

(a) It provides a technique for the exploration and presentation of whole realms of matters of fact that are significant and have been neglected or treated superficially.

Bias in favor of outer data, in favor of measurable, countable: 'Scientific' psychology; comparable in this respect to the opening of new vistas and fields effected by Freud.

Traditional psychology: either rough and ready statements of what was presumed to be obvious or, when effort for precision attempted, bogging down in account of 'indefinable something.'

Husserl on perception: *Abschattung* and *Horizont*

F.J.J. Buytendijk. *Phénoménologie de la rencontre* (Desclée, 1952), *La Femme* (Desclée, 1952 or earlier), perhaps *Wesen und Sinn des Spiels* (Berlin 1933).[23]

23 See below, p. 270, notes 5–7.

S. Strasser, *Das Gemüt*. Freiburg i. B.: Herder, 1956. *Le problème de l'âme. Etudes sur l'object respectif de la psychologie métaphysique et de la psychologie empirique*, French trans. by P. Wurtz. Desclée.[24]

M. Merleau-Ponty, *La structure du comportement*, 1942. *La Phénoménologie de la perception*, Paris, Gallimard, 1945.[25] (Brilliant on significance of one's own body in one's perceiving; sentient and sensible [spatio-temporal]; neither purely *pour soi* nor purely *en soi*; not ghost in machine but incarnate subject; neither subject nor body intelligible without the other.)

(b) It provides philosophical psychology and philosophy with a powerful instrument.

Husserl's Quest: *Logische Untersuchungen; Ideen zu einer reinen Phäno-menologie, Formale und transzendentale Logik; Erfahrung and Urteil.*[26]

Strasser; Merleau-Ponty (*Une philosophie de l'ambiguité*).[27]

Heidegger: A man's understanding of himself as implicit in his projects is the intelligibility of that man, the de facto *Sein* of that *Seindes*; just as phenomenology has to get beyond obvious and superficial, so must each man; hence inauthentic and authentic living, and priority of inauthentic.

L. Binswanger (*Traum und Existenz*) dreams of night (somatic determinants), dreams of morning (the human subject begins the projection of a world; interpretation of dream in terms of itself vs. interpretation as fragmented waking, conceptualization of dream symbols).[28]

R. Bultmann: *Pistis* is *christliche Seinsverständnis*; the rest is myth (objective is science or myth, and Xtianity is not science).

H.W. Bartsch. *Kerygma und Mythos.* I, II, III, IV, V, and *Beiheft* to I–II. Hamburg 1948–55.[29]

R. Marlé. *Bultmann et l'interprétation du Nouveau Testament.* Paris: Aubier, 1956. Théologie 33.[30]

3 Limitations

As phenomenology is essentially prepredicative so also essentially it is preconceptual and prerational.

24 See below, p. 271, notes 8–9.
25 See below, p. 271, notes 10–11.
26 See below, p. 249, notes 5–6.
27 See below, p. 272, note 12.
28 See below, p. 273, note 14.
29 See below, p. 274, note 16.
30 See below, p. 274, note 17

It provides the evidence in which the phenomenologist and his reader can grasp the virtually unconditioned; but as far as I know it has not penetrated to the analysis of that reflective rationality; and so it fails to give it due weight in psychology and in the consequent philosophy.

Hence, its criterion of true is the manifest, the evident; what becomes manifest, evident, when one lets the phenomena appear, does not brush them aside, is not living the life of an escapist.

Per contra, as affirmation based on manifestness of what is, so negation based on manifestness of what is not, of nothing. In Heidegger and Sartre, the basic role given to the anxiety crisis as the manifestness of Nothing.

Hence, possibility of Husserl's *epoché*: withdraw from interest in, concern with the 'really real'; concentrate on intending and intended

Radical difference between direction and redirection of attention, and the *als ob* of suspension of judgment; possibility of *epoché* connected with this ambiguity.

Hence, impossibility of returning from *epoché*.

If by intentional acts I regard the given as just what appears (and I can do so), then by what sleight of hand can another intentional act of affirming or anything else restore the 'really real?' H.J. Pos.[31]

Cf. *Problèmes actuels de la phénoménologie. Colloque internationale de phénoménologie.* Bruxelles, 1951. Desclée, 1952. H.L. Van Breda.[32]

Real difference between
(a) *natürliche Einstellung,* Santayana's 'animal faith'
(b) reaching absolute 'is' through grasp of virtually unconditioned.

Hence, incapacity of phenomenology for dealing with issues of speculative thought. E. Fink. loc. cit.[33]

'... das Seiende ist Phänomen und weiter nichts. Eine Prüfung dieser Unterscheidung liegt gar nicht im Bereich der phänomenologischen Methode, weil sie alle und jede Prüfung grundsätzlich als Ausweisung der selbstgegeben Phänomen versteht' (p. 72).

'Dass das Ausweisbare allein ist ... kann nicht wiederum durch Ausweisung dargetan werden. Das Ercheinen des Seiendes ist nicht etwas, was selbst ercheint' (p.70).

31 See below, p. 277, note 21.
32 Ibid.
33 Ibid.

Hence, Heidegger bogged down in remote criteria of truth and untruth: 'being in the truth' 'being in the untruth.'

A. de Waelhens. *Une philosophie de l'ambiguité: L'existentialisme de M. Merleau-Ponty*. Louvain, 1951.[34]

M. M-P preparing a book *L'origine de la vérité*.[35] A. de Waelhens. *Phénoménologie et vérité*, Paris: PUF, 1953.[36]

Das Seiendes: brute existence. *Sein*: its intelligibility which is in man and from man. Heidegger confined to art.

Lotz. H's method excludes the possibility of his proving the existence of God.[37]

7   M. Heidegger

(a) What he has to tell us about man.
(b) What he thinks about being.
(c) What he thinks of history of philosophy.[38]

1 Phenomenology as Method

Phainomena: whatever is manifested, appears;
not appearance vs. underlying reality
not sense vs. art culture sentiment
not outer public vs. inner private
not immediate but also what takes time attention scrutiny

Legein: Read off, let appear, discover, un-veil.

Truth: Based on evidence of letting phenomenon appear; what is true is what is manifest, un-covered, un-veiled, re-vealed.

2 Transcendental Phenomenology

Eidetic; concern with ego as transcendental, as constituted by the characters necessary for any possible 'intending'; what has no presuppositions, must be presupposed by every other knowledge (since every knowing is an intending), provides rock on which all philosophy science can be securely founded.

34 See below, p. 272, note 12.
35 See below, p. 278, note 23.
36 See below, p. 279, note 24.
37 See below, p. 278, note 22.
38 These three lines are written by hand in the autograph, right next to the heading 'M. Heidegger.'

3 Heidegger: phenomenology of conscious living, of stream of consciousness

Let stream appear, come-to-light, reveal itself.

Since no inquiry knowledge can occur except within stream, phenomenology of stream is basic, first, presupposed by all other knowledge.

Since eidetic is universal, necessary, abstract, cannot but omit individual, existential, concrete.

Hence phenomenology of conscious living is a fundamental ontology, the sole basis from which one can tackle question, What is being?

4 Stream basic: as basis of horizon; but also from viewpoint of a phenomenology

For stream of consciousness is itself a manifesting, a coming-to-light; it is not just living but conscious living; it is the coming-to-light of a consciousness-in-its-world.

If the stream is only a partial-coming-to-light, then phenomenology will dis-cover what remains to come to light.

It will distinguish authentic and inauthentic conscious living.

The truth of phenomenology will be a dis-covering what it is to be in the truth, and what it is to be in untruth.

It would seem that only by being in the truth can one hope to have a stream of consciousness in which one truly can come to answer the question, What is being?

5 Now if there is a stream of consciousness, the streaming, flowing, direction, postulates a finality, a basic drive, and this as conscious, as the root of consciousness is *Sorge, Besorgen, Fürsorge*, concern, preoccupation, care for.

Because the stream is an organizing of contents, it is an *in-der-Welt-sein*
insofar as the organizing rests on *Besorgen*, the organized consists of tools, the referential system of tools linked to one another for the stream is *Zuhandenheit*, and the total complex of tools constitutes the *Umwelt*
insofar as the organizing is *Fürsorge*, there is the *Mitwelt* of persons that also use the tools.

Because the stream is self-organizing, there are
Verstehen: a grasp of concrete possibilities of the stream; preconceptual
Entwurf: project of what is to be done

Rede: articulation of *Entwurf*, seriation of its elements
Sprache: concretization of articulation.

Because when conscious one already is concerned, preoccupied, caring (condition of stream as stream), there is
Befindlichkeit: le sentiment abrupt de se trouver-là
Geworfenheit: sentiment of being tossed into world, abandoned.

Because the being of a stream is its flowing, it is essentially temporal: *Sein und Zeit, Homo Viator.*
Because the being of a stream of consciousness is a flow of presentation to one present, it is *Da-sein*, where the *Da* is pregnant; 'there' not the way a stone is present to a stone, not the way things are present to us, but the way we have to be present for things to be present to us.

Later emphasis upon *Erschlossenheit* opens way to '*ens* ≡ *omnia*,' intellect *lumen naturale.*
De Waelhens seems a bit 'taken in' – others much more.[39]

6 Inauthentic Dasein

Dependence on world: any possibilities I can realize involve me in a network of conditions; plenty of alternative possibilities, but none without involvement in network.

As Jaspers would put it: technical society
(a) creates the possibility of the masses 10^9 increment in 150 years, and thereby ensures its own necessity
(b) it defines the set of jobs to be done: there is some optimum use of tools machines etc. in total process of extraction transformation distribution; the actual is best approximation possible to this optimum (else obsolescence elimination); and man's work is residual
(c) it defines the product, creates man's world: what is produced is what can be produced and through advertising techniques sold to the masses, to the average of desire taste
(d) standards ideals values basic criticism irrelevant: the one question is to keep things going; if that is not your norm standard rule, then uncooperative, trouble maker, unwanted
conformist comes to top where his freedom is hazard of making misjudgment of objective possibilities on grand-scale significance.

39 The last two paragraphs were handwritten in the autograph.

(e) personal worth: skill, experience, character tend to vanishing
jobs are standardized, depts of standards; person has to meet average
standards as a replaceable interchangeable part.

(f) field of freedom contracts: carrying out ideas rising from my creative imagination, not as mere eccentricity, but as significant contribution.

Flight into world: Inauthentic *Dasein* wants things that way; wants to be a
realization of '*On*' '*Man*' 'One.'

Wants release from being one's own self, freely and responsibly discovering and realizing one's own potentialities with all the risk involved.

Finds security, assurance, peace of soul in being like everyone else.

Why? *Selbstverständlich, évidence journalière,* 'obvious' in the sense that it
does not seem helpful to call it in question, that commonly it is taken for
granted, that obviously there are so many other ways of occupying oneself.

Gerede, Bavardage quotidien, Talk.

Cuts articulation of *Verstehen* from real; means becomes an end; *Mitsein*
becomes talking to one another, being preoccupied with the talking
authoritative: things are so because said to be so
all-embracing: only from and against talk can one reach genuine
evident and certain: doubt excites deep indignation, resentment, because Talk hides inauthenticity
curiosity: concerned with new because new; not wanting to understand
anything but to be distracted, to escape
ambiguous: talk about everything but really understand nothing; doing all sorts of things yet nothing that is *my* doing.

Verfallenheit: all this without any effort, without taking thought, a spontaneous accomplishment in which we become estranged uprooted from
ourselves, the selves that really are ours
a permanent aspect of human existence; new civilization would involve
only superficial change; there are only two basic alternatives; this is one,
and the other is intolerable
permanence of instability: changes have to keep coming; no device of
escapist is effective for any length of time.

7 Authentic *Dasein*

The Critical Experience: *Angst,* Anxiety crisis, the collapse of stream of
consciousness as organized.

Dis-covery of aggregate of brute existents, of existents as stripped of all the meaning significance conferred upon them in stream of consciousness.

Dis-covery of *Sorge* (root of stream, reality of *Dasein*)
être anticipant déjà jeté dans un monde dans lequel il s'est perdu
summation of anticipations, projects: the ultimate project is dying, quitting the world.

Selbst, Selbstheit (opp. *Man-selbst*) the tension through time of the option between authentic and inauthentic mode of *Dasein.*

Unauthentic re Death: all the ways of hiding it; slip it into generalities; everyone dies.
Authentic: *Durchsichtigkeit,* face it: *Erwarten,* I am expecting; *Freiheit zum Tode,* Detachment.
Not a matter of stopping living, projecting, doing; but of continuing without being a dupe.

Earlier: tragic attitude. Later: emphasis more and more on art, poetry, and finally nature mysticism as conferring an intelligibility on the existent.

8   Heidegger's Claims

Explicit claim to have made only a beginning, a *Fundamentalontologie.*
Explicit rejection as misunderstanding of practically all interpretation.
However, *Dasein* is fundamental fact; stream of consciousness in its basic formation influences all subsequent philosophic efforts at creating a horizon.
Tools solidify into things: *Dasein* interprets itself as thing: *Deus se habet ad naturalia sicut artifex ad artificiata.*
Proclaims philosophy to have taken a wrong turn with the Greeks; have to go back to early nature philosophers.

Existent: what's there in anxiety crisis.
Sein: intelligibility conferred on existent and on self by *Dasein;* rather Being in truth of authentic, than being in untruth of inauthentic *Dasein;* yet negation of value judgment.

9   Heidegger's Position in History of Phil.

Descartes: Rational *Cogito;* Absolute object in Spinoza; switch through Kant to absolute Subject in Fichte, Hegel, Schelling.
Material substance as extension; mechanism; empiricist philosophy

informing scientific thinking; elimination of man as man in drift of modern civilization.

Late Schelling: from Indifference of Subject-Object to Philosophy of Mythology and Revelation.

Post-idealists: Kierkegaard, Feuerbach, Marx, Nietzsche, Dilthey.

Heidegger's *Dasein* is an indifference or rather simultaneity of subject-object in a concrete living; abstract indifference in Husserl's 'intention.'

10   Critique of Heidegger

(a) H.S. Sullivan: psychic development occurs along lines of minimum anxiety.

(b) Psychic development in man is liberated above flow of animal consciousness; understanding and free image go hand in hand; basic feature of stream of human consciousness.[40]

(c) The stream of consciousness defines a horizon; and horizon is a philosophic concept of fundamental importance; nor can the constructed horizons of the philosophers ignore the fundamental horizon of *Dasein*. *Insight*: Self-Appropriation ≡ *Fundamentalontologie*.[41]

(d) Much of human living is infra-rational
tribal consciousness, group feeling, group decision
pragmatic tendency in science and logic (cf. Trobriand Islanders)
modern civilization is drift determined mainly by technical possibilities of production, and organizing human living by social engineering (advertising, press, escape literature, state education).

(a′) *Sorge* – Pure desire to know. Limit effective in scientific endeavor.[42]

(b′) Truth as 'letting appear' – Truth as Unconditioned.[43]

(c′) Being as simply intelligible, God, *ens per essentiam*;
material being; simply intelligible as form; differently intelligible in other, as potency, as act.

(d′) This is intellectualist: but intellectual pattern of experience sole

40 Handwritten in margin of autograph: 'humanistic education.'
41 Material beginning at '*Insight* ...' added by hand in autograph.
42 Handwritten in margin of autograph: 'common-to-many.'
43 Handwritten in margin of autograph: 'proximate' 'remote is being in truth.'

absolute; it knows and judges others; to do so, it has to differentiate itself and, once it has done so, then bring action and feeling into line.

8   Single Page on 'Horizon'[44]

Discussions of 'Horizon' because topic conceived as 'What about existentialism?' rather than 'What is existentialism?'

Existentialism is an attempt, carried out in a variety of manners, to do justice to the fact of human living (freedom, responsibility, commitment, interpersonal relations, communication, death, God)
without breaking through the frontiers of knowledge set by Kant, namely, that
sense alone is not constitutive of human knowing and that
true judgment can be the *medium in quo* the real is known only if the real is already known prior to true judgment.

Heidegger, preliminaries to a solution that in thirty years has not been reached
Sartre, a premature ontology that is sheer negation though its coherence and penetration light up the insufficiencies of existentialist thinkers
Jaspers, a full and brilliantly technical exploitation of the resources at his disposal
Marcel, detached from theoretical issues, reaches true concrete conclusions about being through the 'good'

can't do justice to details of these efforts in time at our disposal

no great point in attempting to do so, since the brilliance of the efforts is matched by the failure to break out of the closed circle

On the other hand, there is a notable point in attending to significance of existentialism for Scholasticism

Scholasticism is a philosophy of being, but it suffers from a multiplicity of schools, it rests upon a bog of disputed questions, it is not marked by any conspicuous desire and labor to eliminate QQ DD,[45]

---

44 The material from here to the end of this chapter appears on one page in the autograph, separate from all the other pages, which were stapled into separate sheaves. It introduces the material of the next chapter.

45 A Latin abbreviation for *quaestiones disputatae* (disputed questions).

because of half-hearted acceptance that truth is *medium in quo* real is known – not denied – but very commonly it is not really believed

with result of enormously weakened capacity to influence ground unify the sciences and to be useful to theology.

Existentialism invites Scholasticism

to move from per se (subject, principles) to actual order

to move from being a philosophy among philosophies to being a philosophy of philosophies, from non-historical to historical.

8

The Lecture Notes on
Existentialism II: Subject and Horizon

1   The Dilemma of the Subject

1 The major premise of the dilemma is that either the real-for-me is defined as the immediately given or else it is the object known through the true *tamquam per medium in quo.*

The minor premise is the psychological fact that, without introspection, the subject is never the object and, even in introspection, the difficulty is not eliminated but merely displaced, since the subject as subject is never the subject as object.

Hence, if the real-for-me is the immediately given, then there follows the existentialist opposition between objective science and, on the other hand, real knowledge of the subject which is non-objective. Hence, exclusion of metaphysics in any traditional sense; invention of new types of metaphysics, for dealing with all that concerns man.

On the other hand, if the real-for-me is what is known through what is true, then I am confined to a universe of objects; the subject as subject is inaccessible to me; and because the subject is inaccessible, I remain the victim of unscrutinized horizons, incapable of taking a place on the contemporary level of philosophic discussion, capable of complete openness of horizon only *per accidens* and not philosophically.

2 Subject: many meanings in different contexts.

Grammatical s.: a word or phrase fulfilling a specified function in a sentence.

Logical s.: whatever admits a predicate, has one; red is a color
Scientific s.: subject : habit :: object : act.
Psychological s.: the human conscious subject.

3 Conscious: predicated of subjects, acts, processes.

Subjects: he was knocked unconscious; dreamless sleep; dreaming waking
Acts: growth of beard, metabolism of cells, vs. seeing, suffering
Processes: circulation of blood, digestion of food (in no malfunctioning), vs. inquiring to understand, reflecting to judge, deliberating to decide, deciding to enter course of action.

4 Object: the motive, product, end of conscious act.

Motive: color moves sight, illuminated phantasm moves intelligence
Product: imagining produces image; understanding produces concept
End: ens, verum, bonum; biological ends.

What conscious act centers on, brings about, heads for.

5 The ambiguity of awareness, presence.

I see colors, but I do not see seeing, I do not see myself seeing.
In seeing colors, the colors are present (presented) to me, but they are presented not to me as absent but as present.
Inasmuch as colors are presented to someone also present, there is consciousness in the direct act of seeing; I do not see unconsciously, though I may see indeliberately, inadvertently, without noticing what nonetheless I see.

Consciousness is not a matter of reflex activity, of introspection; it is the possibility of reflex activity having something to turn back on, of introspecting having something to introspect.
Consciousness is a property, quality, of acts of a given kind: sensitive and intellectual, cognitive and appetitive.
Consciousness always accompanies waking and even dreaming states. The direction of attention to the conscious component in such states is a secondary phenomenon that would be meaningless aimless were there not the primary phenomenon.
Consciousness of Christ.

Presence, awareness, ambiguous.

Presence of objects to the subject and, concomitantly in a quite different sense, presence of the subject to whom objects are presented.

Object is presented as intended.

Subject is presented as intending.

Object is what one is aware of, what one sees, hears, desires, fears, investigates, understands, conceives.

Subject is one who is aware, and one cannot be aware and be unconscious, just as one cannot see and be unconscious, etc.

But 'being aware' is quite different from 'being what one is aware of.'

Hence in primary stream of consciousness (a) the subject is never without an object and (b) the subject is never the object. Between subject and object there is a cleavage, a radical opposition.

In infinite act, subject, act, primary object coincide; in finite act, act and object differ, for act is limited by something, by what it is about; in human act, subject, act, object differ, for not only is act finite, but also subject does not know himself by his own essence.

6 Introspection does not eliminate but displaces cleavage.

In an incomplete and elusive fashion the subject can shift his attention from object to act and subject.

On this basis he can proceed to classify, describe, relate, explain from hypotheses, theories, systems, devise tests, verify, judge: subject, capacities, habits, acts, objects.

Apart from its basis in shift of attention, this process is essentially the same as in all human knowledge.

Experience :: Understanding conception :: Reflection judgment.

Moreover, just as in knowledge of other objects there are known, known unknown, and unknown unknown, so also in knowledge of the subject.

The phenomenon of the horizon remains, only here the horizon is more difficult to tackle because of the difficulty of the basic shift of attention.

Throughout this process the cleavage remains.

The human subject does not know himself by his essence; he begins from objects, defines acts by objects, habits by ranges of acts, potencies by ranges of habits, essence of soul by sets of potencies.

In shift of attention: what is attended to, who attends; what is attended to is subject as object; who attends is subject as subject, so that subject still remains inaccessible except as peculiarly present.

What is classified, described, understood, is not the subject classifying, describing, understanding.

Hence, Hume (a) knowledge he describes (b) knowledge he uses.

7 The Dilemma

If real is known through true, then only subject as object known; if only subject as object known, the whole inquiry is conducted within horizon, prejudged by horizon, and no possibility of philosophic attack on radical problem of horizon.

If real is the immediately given in its immediate intelligibility (phenomenology à la Heidegger), if immediate truth is this uncovering, revealing, if judgment is just the articulation of what is revealed

Again, if no idea of unconditioned, true, *ens* (Jaspers, Marcel)

then either new type of metaphysics concerned with the reality of the subject as subject or at least *Existenzerhellung* or truth as *Unverborgenheit a-lêthê*.

2   Subject and Horizon

I The Notion of Horizon

1 Human knowledge is in process.

Intellect: *quo est omnia facere et fieri*; but though unlimited in range, it begins from *tabula rasa*.

Process is raising and answering questions: *Quid, Propter quid, An, Utrum*. Or manifestation of process is ...

2 Hence at any stage of development, a threefold division.

Known: the range of questions I can raise and answer.

Known unknown: the range of questions I can raise, find significant, worthwhile, know how they might be solved, but de facto cannot answer and know I cannot answer. *Docta ignorantia.*

Unknown unknown: *Indocta ignorantia*; the range of questions that I do not raise; if raised I would not understand nor find significant nor judge worthwhile nor know how to go about solving.

3 The *horizon* is the limit, the boundary between *docta* and *indocta ignorantia*.
What is beyond my horizon consists not merely of answers but also and
principally of questions that are beyond me, meaningless-to-me, insignifi-
cant-to-me, not worthwhile-to-me, insoluble-to-me. 'I haven't got a clue.'

As defined, the horizon is a relative term: what is meaningless-to-me
may or may not be meaningless absolutely.

By way of contrast, we shall also speak of the *field*: what is beyond the
field is meaningless absolutely, insignificant absolutely, insoluble absolutely.

The field is *the* universe, but my horizon defines *my* universe.
Both are relevant to metaphysics: for metaphysics deals with *ens*, with
omnia, with the universe.
The field regards metaphysics as such, but the horizon regards metaphysics as possible-to-me, relevant-to-me.

4 The existence of the horizon comes to light not directly but indirectly.
Not directly: it can be sharply defined only by going beyond it, by
reaching a wider horizon in which the old appears as a part. From within
any given horizon, its limits are not clear and sharp in focus, but obscure,
hazy, distant: for what is beyond the horizon is what we pay no attention
to, and what is at the horizon is what we pay little attention to.
Indirectly: for we can study instances in which the recession or contrac-
tion of the horizon occurs.

II The Horizon in Science (or Mathematics)

1 The scientific (or mathematical) horizon recedes if there occur

(a) a crisis: existing theories, methods, modes of thought cannot handle the facts, results, satisfactorily;

(b) a fundamental revision of concepts, postulates, axioms, methods;

(c) the development of a radically new scientific structure; e.g., non-
Euclidean geometry, calculus, Galois, Einstein, Quanta, Copernicus, Dar-
win, Freud.

2 Recession of the horizon meets with resistance.

Max Planck on what makes a scientific theory accepted: not clarity of
observation, exactness of measurement, coherence of hypothesis, rigor

of deduction, decisiveness of verification, but retirement of present generation of professors.

3 Eventually the resistance is overcome.

Universally: scientific results are equally accessible to all; at any time, roughly, contemporary scientists are abreast.

Permanently: the new theory covers all the old facts, and many more; there is no tendency to revert to earlier positions, to revive old views.

4 Hence, science is characterized by such universality and permanence, by the contrasting absence of permanent division into opposed schools of thought, of the survival and revival of what to others seems to be definitely superseded.

Resistance to scientific advance is a subjective phenomenon; it is eliminated by a new generation of professors.

The old have intellectual habits without the suppleness needed to develop new habits; they have invested their intellectual capital in a point of view, and they are not prepared to declare themselves bankrupt.[1]

III The Horizon in Human Science, Philosophy, Theology

1 In these fields there occur recessions of the horizon in the same fashion as in natural science or mathematics.

I.e., crisis, radically new viewpoint, radically new structure.

Plato: *aisthêta noêta*; the *noêta* are *ontôs onta*.

---

1 The following appears handwritten in the autograph, but with a marginal note 'Omit':
  '5 Still the universality and permanence of scientific achievement has in fact no more than a pragmatic foundation.
  'Bronislaw Malinowski: no myth no magic re planting crops, tending them, reaping harvest
  'Science does not conquer mythic consciousness; it finds techniques to broaden the field in which myth and magic and gnosticism never gained any foothold.
  'Hence, scientists cannot give an account of foundations of science; their science is or is not ultimately a superstition by the amount of philosophy the individual may have; human science omits *man*; scientists can exhibit utter stupidity or silliness once outside their specialty.'

Aristotle: the *noêton* is the *aition tou einai* immanent in the material object; extrapolation to immovable movers.

Augustine: real is body; real is true.

Aquinas: a transformation of existing theology (Gilson; Scotus was the traditionalist, augustinisme avicennisant).

Descartes: philosophy as an independent and separate subject; not merely distinction but separation from theology.

2 In these fields the recession of the horizon does not result in a straightforward universal and permanent difference.

Not universal: per se, an original philosopher founds a new school; he changes philosophy only *secundum quid*; he gives rise to new topics, new approaches, new techniques, but the basic differences remain – there is a family resemblance between different realizations of the materialist, idealist, realist tendencies respectively, from fourth-century Athens to today.

Not permanent: the original thinker founds a new school, but the school splinters
   further, there occur periods of decadence, loss of vigor, of influence
   as there occur insensible changes with changing times, in which the original message can be lost; devaluation of meanings
   so also there occur revivals, second spring, recoveries of vigor and influence.

3 The difference between the phenomena of the horizon in maths and natural science, on the one hand, and in human science, phil., theol., on the other is not too difficult to account for.

In the latter case the new horizon on the object involves a new horizon on the subject; for the subject is one of the objects.

And a new horizon on the subject involves not merely new concepts, postulates, axioms, methods, techniques, but also a conversion of the subject, a reorientation, a reorganization.

A new concept of oneself, new principles to guide one's thinking, judging, evaluating, all that concerns oneself, is a conversion.

Without the conversion, the new ideas not only are inoperative in one's own living, but also they are insignificant, without real meaning, without any vital expansiveness, in the domain of objects.

The original thinker founds only a school, because he cannot effect the conversion of subjects, he can only promote conversion in, for the more ready.[2]

His school splinters, is subject to periods of decadence and revival, because even his followers can succeed in subjective conversion only up to a point.

IV The Existential Gap

1 The existence of philosophical and theological schools, the possibility of decadence and revival within any given school (the words of the master are repeated but his meaning is lost), the fact that human science to be science systematically tends to omit what is human,

reveal the fundamental significance and importance of horizon in studies concerned with man, directly or indirectly.

This significance is: The reality of the subject can be beyond the horizon of the subject.

The subject can suffer from an *indocta ignorantia* with regard to himself.

This *indocta ignorantia* is not a matter of what the subject might very well be excused from knowing: depth psychology, social conditioning, history, biology, biochemistry.

It is a matter of the subject's own intelligence, his own reasonableness, his own freedom and responsibility.

On the one hand, he is intelligent, reasonable, free, responsible; he manifests these characteristics in many fashions; he would be insulted if told he was stupid, unreasonable, irresponsible, a victim of catchwords.

Yet at the same time in a very true sense his own intelligence, his own reasonableness, his own freedom and responsibility stand beyond his horizon.[3]

2 The words from 'he can only' are added by hand in the autograph.

3 At this point in the autograph, the following is handwritten in the margin: 'Many Scholastics elim[inate] psychology from psychology – mind is a metaphysical machine.'

2 The existential gap is the difference, greater or less, between one's horizon on oneself and what really one is.

> Again, the existential gap is the gap between
> what is overt in what one is and
> what is covert in what one is
>
> what Hume asserted human knowledge to be
> the knowledge Hume manifestly employed in stating proving his assertion.

3 The existential gap is not eliminated by affirming the propositions that are true and denying the propositions that are false.

> The decadent school repeats the propositions of the master, but it has collapsed the master's meaning into something less that will fit into a contracted horizon.
>
> The problem of the existential gap is the problem of a conversion that is proportionate to the objective development:
>
> it is not the problem of agreeing with Augustine that the real is the true; it is the problem of meaning as much as did Augustine when he spoke of *veritas*.[4]

4 Hence study of the existential gap is concerned with
> *immediacy*: not a matter of true or false props, but of conversion genuine authentic
> *obnubilation*: discovery movement from covert to overt
> *norms*: there is something normative; conversion should occur
> *freedom, responsibility*; else norms really meaningless
> *Transcendent*: the norms involve an absolute value; the subject takes his stand by them even against the world, against himself, finds in them a symbol, an indication of God
> *Existenz*: the subject becomes himself in his relation to Transcendent.

4 The following appears handwritten in the margin, but prefaced by 'Omit':
   'Existential gap is a gap in existing: man is not just an animal in a
   habitat; he is a responsible self-orientation in a universe.
   'Human existing is just becoming, advancing into the gap, closing it
   not completely and not securely definitively.'

3   Horizon and Dread

1 The horizon is grounded in the subject: it is the boundary at which begins his *indocta ignorantia.*

Still, this is merely an objective aspect of the horizon: it is defined in terms of what the subject not only does not know but also considers meaningless insignificant insoluble.

We have to inquire into the subjective phenomena of horizon: how is it constituted; how is it maintained.

2 To consider single acts involves violent abstraction.

Sensitive acts are involved in a multiple correlation:

see: approach, look, focus.

Intellectual acts suppose sensitive, operative with respect to sensitive stimulus and manipulation of sensitive flow.

Hence, study of consciousness is study, not of isolated acts, but of flow, stream, direction, orientation, interest, concern.

3 Study of such streams of consciousness, at a first approximation, is erection of ideal constructs.

Cf. motion of mass in central field of force, Carnot cycle.

Hence, patterns of experience:

(1) biological: beast of prey and quarry

(2) aesthetic: release from biological interests: free creation

(3) dramatic: primary aesthetic creation is in oneself and with regard to others; extravert if successful; else introversion

one is the hero in one's dreamland – one has to make, constitute, oneself[5]

(4) intellectual: Thales and milkmaid, Newton working on gravitation

(5) practical: getting things done

4 Limit to patterns of experience: underlying biological manifold has to have higher sensitive integration.

A stream of consciousness that runs too freely has the nemesis of compulsions, invasions, neurotic phenomena, anxiety crises.

Anxiety crisis: breakdown of stream, pattern; objects there but meaningless, i.e., no dynamic significant integration.

5 The words after the dash are handwritten in the autograph.

Anxiety: minor phenomena; development of a type of consciousness takes place along lines of minimum anxiety. Sullivan.[6]

Abnormality: development has had to avoid anxiety by extreme measures.

5 A world: what lies within a horizon; a totality of potential objects. Not some particular object, but a possibility of some types of objects and not of others.

World, horizon, corresponds to the concrete synthesis that is my conscious living, and that concrete synthesis does not admit change without experience of anxiety, dread;

it is not the reality of my world that is the anchor, the conservative principle; it is the dread I experience and spontaneously I ward off whenever my world is menaced.

My concrete synthesis in conscious living is (a) integration of underlying neural manifold (b) set of modes of dealing with *Mitwelt* of persons and *Umwelt* of tools; or any other combination.

To change it, to be converted to new world, to let my horizon recede is to invite experience of dread and to release a spontaneous, resourceful, manifold, plausible resistance.

This dread and release not a function of objective evidence for my world; it is a function of my mode of life, my solution to total range of problems arising in my concrete living.

6 Hence, a series of corollaries: (1) Conversion a leap.

To convert someone, to be converted oneself, is not exclusively a matter of proofs arguments evidence; there is a pivot on which one turns, a heritage to which one clings; it is pearl of great price – all else goes.[7]

There is for everyone a problem of integrated conscious living.

In childhood: illness, fever, easily moves to delirium, if I may quote my own experience.

The problem is solved only more or less satisfactorily: whole range of

---

6 'Sullivan' is handwritten in the autograph. The reference is undoubtedly to H.S. Sullivan, whose book *The Interpersonal Theory of Psychiatry* (New York: W.W. Norton, 1953) is mentioned in *Insight*. All three comments on Sullivan in *Insight* (on pp. 227, 279, and 556) are relevant in the present context.

7 The words from 'there is' are handwritten in the autograph.

types of unsatisfactory solutions, from psychoses to neurotic phenomena of minor type.

The problem exists *because* man is capable of free images: Köhler's apes; literature to develop imagination; to provide intelligence with a tool that will make possible the movement of intellect to *ens, omnia; and because* 'free images' is not an unlimited, unconditioned freedom.

Conversion, moving to new horizon, entering into new world, is tampering with a hitherto successful solution to the problem of conscious living.

If I can get by the initial anxiety, I shall be better off; just as analysand if he can stand anxiety involved in cure will be cured.

But not merely a problem of standing the anxiety; it is also a problem of dealing with the resistance.

The would-be convert appeals to his *Selbstverständlichkeiten*;[8] he indignantly appeals to what is obvious to everyone with an ounce of common sense; he moves round in a circle within his established horizon; and as long as it remains, his brand of logic and his set of premises will be unshakable-to-him.

Moving to a new horizon, conversion, involves a leap: a leap from *Selbstverständlichkeiten*, which are mostly misunderstanding what in some sense is true, but also are props to present position, to another concrete solution to problem of conscious living.

To experience such dread, seriously suppose that some philosophy (that is not your own) were true.

Real distinction: not a problem of distinction but a problem of reality, of what really is, of horizon, of horizon buttressed by dread, and avoidance of dread rationalized by *Selbstverständlichkeiten*.

Again, *neglect of sense* ≡ transformation of mode of concrete living.

7 Corollary (2) The self-constituting subject.

Freedom of will: rational alternatives and free choice.

Prior freedom: the solution that has been the concrete synthesis in my living.

Cooperation of subconscious, imagination, intelligence yielding projects

8 Handwritten in margin of autograph: '"tout naturel" Marcel.'

within aesthetic (play), dramatic, practical, intellectual patterns of experience.

The drama we do not think out and then execute; the drama that spontaneously arises already charged with image emotion appetite.

It is a freedom not had by animals.

It is an 'ontological' freedom by which the conscious subject is this conscious subject, develops this solution to the problem of concrete living.

It is that by which we become what we are before we are able to think out alternative courses of action and choose between them.

It sets the horizon within which occurs our thinking and choosing, so that

while any particular project can be vetoed, yet the veto has to have its grounds within my world, my horizon

and no project can arise unless it is such as to fall within the world that is mine.

Still, if we have made ourselves without any awareness of what we were up to, so we later can remake ourselves in the light of better knowledge and with a full responsibility.

Nor is the refusal to remake ourselves any escape, for that is just assuming responsibility for whatever we happen inadvertently to have made ourselves in the past.[9]

8   Corollary (3) The basic function of philosophy.

Philosophy is the attempt to illuminate the effort of intelligent, reasonable, free, fully responsible self-constitution.

Hence, philosophy is concerned with *good*: what is freely and responsibly chosen and effected; what is concrete (*verum et falsum in intellectu; bonum et malum in rebus*).

Point of comparison with Scholasticism is with Schol account of good: *bonum particulare*: corresponds to particular appetite

---

9   The autograph contains handwritten comments on Kierkegaard's spheres of subjectivity and Sorokin's division of cultures, along with mention of Heidegger's *Ek-sistenz* and a reference to Marcel and *De l'existence à l'être*. The section is marked 'omit.' The points on Kierkegaard and Sorokin correspond to Lonergan, *Topics in Education* 42.

bonum ordinis: series of particular goods, series of coordinated activities, habits of apprehension appetition, interpersonal relations (communication in good, congruent with coordination of activities, rising from habits)

bonum per essentiam: the absolute norm; possibility of individual willing good against world, others, self; transcendent.

Concern with *good*.

(a) concerned with improving my operating solution, functioning synthesis in concrete living; with transition from freedom of images to freedom of enlightened responsible choice; conversion

(b) concerned with improvement as mine; not truths but the truth I live by, that is involved in my free self-constitution; not notional but real apprehension and assent

(c) concerned with a solution of living: not abstract living but living in a world, with others, in a technical civilization; study critique of personal relations, of technical society

(d) concerned with concrete possibility of that living at its highest point; ultimate self-affirmation [self]-constitution in relation with transcendent, as person, Thou (Marcel), with my *Existenz* as awareness of self as gift given to self (Jaspers)

(e) concerned with history; as everyone, philosopher responds to problems of age; his specific character is to respond to these common problems at deepest level, at point of maximum consequence for human welfare or human disaster

Jaspers: primitive cultures; organized civil.; *Achsenzeit*; present as momentous as discovery of fire tools speech; old ways relentlessly being dissolved; masses; one world history

(f) the philosopher is *open*: by definition, going beyond horizon based on dread; philosophers his educators qua obscure, for such obscurity is revelation of my blind spots, my horizon

(g) the philosopher has to be *genuine*; not talking beyond his own horizon, devaluating the currency, collapsing the great into a narrow horizon world

(h) philosophy has to be *relevant*; not analytic propositions; not analytic principles with a per se relevance that is per se only because fact of horizon overlooked; not relevant to man in general, but to me in my age and those with me

(i) philosophy can only *illuminate*; it looks not to a theoretically compelled assent, but to a free conversion; one cannot be another, do his thinking, judging, deciding, living for him.

4   Horizon and History

1 An Enlargement of the Significance of the Existential Gap
Not merely a matter of a difficult and doubtful technique in the study of the totality of philosophies,
but a critical issue within the historical process.
Existential gap is not merely a call to authenticity of subject in his private existence; it is a call to authenticity in all subjects, an invitation to understanding at a critical moment in human history, a summons to decisiveness, an exploration of the techniques of communication. (Existentialists write novels, plays.)
History, as a total field of human operations in this life.

2 Begin from Notion of Dialectic

Familiar: dialectic of an idea; e.g., dialectic of rigor: exclude casual insights; axiomatization; paradoxes; new basis.
Unfamiliar: dialectic of a reality, of man, of history.

Still, if dialectic of an idea, there is some dialectic of man, of history
Not of man as what recurs by reproduction; no transmission of acquired characters.
But of man as technical, social, cultural; for in these respects, what man is results from man's ideas on man.

Man as technical, as using tools: not merely satisfies animal necessities, but creates human environment, the city, the state, as a totality of material products facilities
Man as social, as organizing and organized; institutions such as family, education, economic system, political system, systems of alliances and enmities
Man as cultural (culture in anthropological sense): the current effective totality of:

immanently produced and symbolically communicated contents of imagination, emotion, sentiment; of inquiry, insight, conception; of reflection, judgment, valuation, of decision, implementation.

In these respects man (a) presupposes nature but (b) makes himself by taking thought.

Man as technical social cultural is difference between aggregate of babies born and abandoned in jungle and the aggregate of human beings operating in a civilization.

3 The Objective Functioning of the Dialectic

(a) There is a circuit, a mutual causation, in man's making of man as technical social cultural.

The objective situation (technical social cultural) is at once a *product* of and an *occasion* for

imagination, sentiment, emotion; inquiry, insight, conception; reflection, judgment, evaluation; decision, policy, implementation.

(b) As product the objective situation *objectifies, reveals*, what man has been feeling, thinking, deciding about man.

As occasion, the objective situation *suggests* and *motivates* changes in what man has been feeling, thinking, deciding.

(c) Insofar as there is an effective existential gap, an operative limited horizon,

the situation as product will objectify and reveal the existential gap in overemphases and oversights

but the situation as occasion will be powerless to suggest and motivate the correct solutions, remedies, as long as existential gap remains;

hence, situation progressively deteriorates; more and more liberal use of useless solutions, remedies

in the limit, either existential gap is closed, or else the civilization liquidates itself.

4 Resolute and Effective Intervention in the Dialectic

(a) Everyone participates: everyone contributes to the production of human situations; everyone has to respond to the human situations in which he finds himself.

Still, such participation may be mere drifting: one does not understand what is going on; one has no clue as to what is wrong; one has no idea what one could effectively do about it.

Man as historical, man as making man, is beyond man's horizon, is in a dreamland.

(b) Resolute and effective intervention presupposes subjects in which the existential gap has been, is being, closed;[10]
else they will merely increase the confusion and accelerate the doom.
Resolute and effective intervention means that these subjects do not remain within an ivory tower admiring their own deeper profundity, the incomprehension of the mass of men.

(c) Resolute and effective intervention heightens the operation of the dialectic.

The situation objectifies the existential gap; intervention *crystallizes the objectification*; it is there; it is obscurely evident to everyone; but it is not articulate, it is unexpressed, it is not effectively noticed.

The situation suggests and motivates the necessary changes in the subject; intervention clarifies the suggestion and drives home the motivation; *clarifies*, by linking old errors with present evils; *drives home*, to retain errors is to perpetuate evils.

Intervention constitutes the correction by communication;
(1) what man felt, thought, decided made things as they are
(2) different feelings, thoughts, decisions will make them different
(3) communication results in different feelings, thoughts, decisions.

5 The Essence of the Dialectic

(a) lies in a conflict between what man is, is to be, and what man feels, thinks he is, is to be.
Objectification is of what man thinks; but the objectification is also a revelation of overemphases and oversights, insofar as there is a conflict between man's plans for himself and what man really is.
Revelation is motivation for change, insofar as what man has made of himself is in conflict with what man really is.

10 The words 'is being' are added by hand in the autograph.

(b) The dialectic, then, does not operate within the field of concepts and judgments, terms and propositions; it is not based on a conflict between opposing philosophies.

It is based on a conflict between any defective philosophy (implicit or explicit) and what man really is, is to be.

(c) The verdict of the dialectic is not a label of approval on a philosophy;

it lies in the facts of the situation, in its tension, in its basic hopefulness, its ultimate desperateness, its stimulus to affirmation or its imposition of nihilism

(nihilism: don't care; what happens to me, to man, could not mean less than it does to me).

Still, the facts are significant only to those whose horizon does not preclude knowledge of what it is to be a man.

If the facts are not significant, then they are destructive of societies; because the effective horizon continually forces a misinterpretation of the facts.

6 In first lecture, concluded that there exists a valid and important field of inquiry concerned with subject in his immediacy, obnubilation, capacity for change, authenticity, freedom, responsibility.

Now must further conclude that such a field is also relevant to man as technical social cultural.

History is concerned to bring to light man as he really is; hence to study this generalized existential field is to get to the heart of historical process.

Again, study of horizons eliminates the horizon that keeps man as historical beyond one's field of vision.[11]

---

11   Handwritten in autograph, another '6' prefaced by 'Omit':
        'Situation: vs Order determined by fixed essences in hierarchy
        'Kairos: what it is up to me alone to do alone
            vs limit on possible action set by abstract ethics
        'Charity: love neighbor; not prefer higher values in order
            'Order ≡ personal relations   order ≡ mechanical,
                                                    metaphysical
                                                    legal structure
                            ↑tension↑

5   Horizon as the Problem of Philosophy

1   De facto there exist many horizons; this is also de iure since man makes man (generation, technique, society, culture) and within these limits man makes himself if he chooses, or drifts into what he happens to be if he fails to choose.

2   This multiplicity may be considered
>   as a mere matter of fact: history of culture, thought, opinion
>   as a problem to be explained: *Psychologie der Weltanschauungen*
>   as an issue calling for judgment, decision: philosophic issue.

3   The multiplicity of horizons as philosophic issue arises when we ask
>   (a) Is some horizon the field, or is there no field?
>   (b) If some horizon is the field, how can it be determined?

To deny that there is a field is to deny that philosophy has a positive content; still, that denial is itself philosophic though perhaps unconsciously so.

>   positivism: let's do science
>   pragmatism: let's experiment, see what happens
>   skepticism: let's inquire some more
>   relativism: there are no definitive answers, just points of view

To affirm that there is a field involves one in the second question, which is at once ontological and epistemological.

It is ontological in its consequent: beyond such and such a limit there is nothing to be known and so no *indocta ignorantia*; it settles where reality ends, where meaninglessness begins.

It is epistemological in its antecedent: to define the field raises the question of the truth of the definition; and the definition is true in virtue of known evidence; what then is the evidence? This evidence is of some reality; hence ontological also in antecedent.

4   The simultaneity of E and O is intrinsic to the positive answer.

Simultaneity: E as antecedent; O as included in antecedent, though not O as formulated in consequent. It is the antecedent ontological evidence or 'ontic' evidence that in existentialism gives rise to metaphysics. In detail:

(a) Any determination, justification, evidence for a horizon, arises within a stream of consciousness and so arises within what already is constituted as a horizon.

(b) The justification of the horizon cannot rest on the consequent ontology, on the realities known within the horizon, for then every horizon would automatically be self-justifying; and that is the negative solution.

(c) It cannot rest on the norms, invariants, principles that de facto characterize, determine, constitute any given horizon; for again on that showing, every horizon would be self-justifying.

(d) It has to involve a discovery of the evidence, norms, invariants, principles
  that naturally, ontically, possess a cogency, inevitability, necessity, normativeness
  that thereby constitute a self-justifying horizon, stream of consciousness, and so field
  that nonetheless admit the possibility of other horizons, through the whole gamut of human differences
  that account for the actual existence of these differences at least in principle
  that account for them in such a manner that at the same time they discredit them, reveal them to be, not self-justifying, but self-destructive
  that discredit them in such a manner that nonetheless their actual occurrence remains possible, plausible, convincing.

(e) The prior reality that both grounds horizons and the critique of horizons and the determination of the field
  is the reality of the subject as subject.

It is not any object known objectively, and it is not the subject known objectively, for all objects are known within some stream of consciousness and so within a horizon; and it has been contended that such objects cannot justify any horizon without thereby justifying all horizons.

It is the reality of the subject as subject; for the subject as subject is both reality and conscious the subject as subject is reality in the sense that we live and die, love and hate, rejoice and suffer, desire and fear, wonder and dread, inquire and doubt

it is Descartes's *Cogito* transposed to concrete living

it is the subject present to himself, not as presented to himself in any theory or affirmation of consciousness, but as the prior (non-absence) prerequisite to any presentation, as a priori condition to any stream of consciousness (including dreams).

The argument is: the prior reality is not object as object or subject as object; there only remains subject as subject; and this s as s is both reality and discoverable through consciousness.

The argument does not prove that in the s as s we shall find the evidence norms invariants principles for a critique of horizons; it proves that unless we find it there, we shall not find it at all.

PART THREE

Lectures on Existentialism

9

General Orientation[1]

The subject of existentialism is difficult to line up because there are as many different shades of opinion on every topic as there are writers, and the differences of opinion go right back through the entire history of each thinker. On the other hand, there are certain broad tendencies that are common, and we will concentrate mainly on those. In particular, the question this week, as last week, will be, What about? and not just, What is? We cannot hope to go through the various authors on existentialism in a week. One could spend a week on any particular one and even then not do an entirely thorough job.

1   The Term 'Existentialism'

By 'existentialism,' then, we shall understand the types of method and of doctrine exemplified by Karl Jaspers, Martin Heidegger, Jean-Paul Sartre, and Gabriel Marcel. The term refers to types of method and of doctrine. The name 'existentialism' was admitted by Jaspers without any difficulty, and of course, Sartre has claimed to be *the* existentialist. On the other hand, Marcel admitted it for a while, but after *Humani generis*,[2] and perhaps more

---

1 The first part of the sixth lecture, Monday, 15 July 1957.
2 The 1950 encyclical of Pope Pius XII, *Humani generis*, speaks of an 'existentialism' that has a low estimation of unchanging essences and devotes its attention rather to the 'existence' of singulars ('... immutabilibus rerum essentiis posthabitis, de singulorum "exsistentia" tantum sollicita sit'). See *DS* 3878.

particularly to disassociate himself from Sartre, he decided to refuse to call himself an existentialist. And Heidegger from the beginning claimed that he was concerned with *Ek-sistenz.* (He is very fond of re-writing words with hyphens in them.)[3]

The four that I named are the principal people in the field – the most notable – and they differ remarkably. Jaspers is a Kantian and a Lutheran. Heidegger is an apostate; he was a Jesuit novice for a while, and then he went to a seminary for the secular priesthood, where he lost his faith and went into philosophy. Sartre is a professional atheist. Marcel is a convert to Catholicism.

2   Bibliography

With regard to a bibliography on these people in general, check Bochenski's *Bibliographische Einführung in das Studium der Philosophie.*[4] In one of the fascicles, Jolivet, the dean of the faculty of philosophy in the Catholic Institute of Lyons, has a bibliography of some twenty pages up to 1948 on French existentialism,[5] where you will find everything of relevance to Marcel and Sartre and all the writing about them up to 1948. Since then the movement has quieted down. For Jaspers and Heidegger, there will be provided a mimeographed list of their works.[6] I cannot guarantee that I have included absolutely everything. The earlier writings of Jaspers, when he was professor of pathological psychology, have been omitted, but starting with his *Allgemeine Psychopathologie* in 1913[7] the list includes pretty well everything.

I will mention some few points to note particularly regarding Jaspers's works. There is the *Philosophie* of 1932 in three volumes: they come together to about a thousand pages.[8] A very famous book of his is *Die geistige Situation*

---

3 William Richardson says that the spelling 'ek-sistence' translates a later expression of Heidegger's to emphasize 'the essentially ec-static character of There-being.' See William J. Richardson, s.j., *Heidegger: Through Phenomenology to Thought* (The Hague: Martinus Nijhoff, 2nd ed., 1967) 39.

　　On the various authors' acceptance of the term 'existentialist,' contrast Lonergan's comment in an aside, in *Topics in Education* 81 note 9: '... Jaspers is the only one who accepts the title, as far as I know; perhaps also Berdyaev.'

4 See above, p. 50, note 25.

5 See above, p. 51, note 26.

6 See below, pp. 369–72.

7 See below, p. 371, number 1.

8 See below, p. 371, number 6.

der Zeit.[9] It ran through five editions in German in about a year,[10] and it has been translated into six languages including Japanese. *Von der Wahrheit* of 1947 was over a thousand pages.[11] His work *The Origin and Goal of History* is a remarkable book of about 350 pages.[12] Other works of his are more popular. Some of his historical works, for instance on Nietzsche[13] and Descartes,[14] are of a different kind, and some of his titles represent conferences that he gave here and there.

Heidegger's key books, *Sein und Zeit*[15] and *Kant und das Problem der Metaphysik,*[16] along with *Was ist Metaphysik?*[17] and *Vom Wesen des Grundes* (on the nature of a foundation or a basis),[18] belong to the period 1927–1929. There is also *Vom Wesen der Wahrheit,*[19] a later work. But apart from his earlier publications of around 1927, he writes rather deep oracular pronouncements.

3   'Being a Man'[20]

A fundamental feature found among all these writers is that they are concerned with what it is to be a man, not in the sense of having a birth certificate but in a simple sense that is obvious to everyone and that has no technical backing of any kind. Last fall, when questioned during the Egyptian crisis, President Eisenhower – I think it was at a press conference – was asked by someone if it wasn't risky sending a fleet or making some other move in the Mediterranean, and he answered very briefly, 'We have to be men.'[21] 'Being a man' in that sense provides a very brief clue to what the existentialists are concerned with. 'Being a man' in that sense results from a decision. It is consequent to the use of one's freedom. It is something we

 9 See below, p. 371, number 5.
10 Later (see below, p. 223) Lonergan speaks of 'six editions in German in a few years.'
11 See below, p. 372, number 13.
12 See below, p. 372, number 15.
13 See below, p. 372, number 9.
14 See below, p. 372, number 10.
15 See below, p. 370, number 4.
16 See below, p. 370, number 5.
17 See below, p. 370, number 6.
18 See below, p. 370, number 7.
19 See below, p. 370, number 11.
20 On this theme, see also *Topics in Education* 79–81.
21 For details see *Understanding and Being* 190, note 16, and *Topics in Education* 80, note 6.

have to be. It makes one the sort of a man one *is*, and it involves risks. (In the present instance, the risk of nuclear warfare was in the back of the mind of the journalist asking the question.)

Now that simple notion of being a man as something that one has to be and that one is not necessarily, that comes about freely and makes one what one is, is perhaps an extremely complicated notion when one gets to analyzing one's ethical foundations, but it is a notion that anyone can grasp, and it is common to existentialists generally. It is a laicization of Kierkegaard's 'being a Christian.' Kierkegaard wanted to know whether he was a Christian. And he proposed to himself the answer, 'Well, after all, I'm a Dane, and everybody in Denmark automatically is a member of the Lutheran Church of Denmark as by law established. So what is this nonsense of asking me whether I'm a Christian?' Then he showed that the question had a deeper meaning. A similar question can be asked about 'being a man.'

4   Relation to Positivism and Idealism

That question is, in the first place, anti-positivist. 'Being a man' is not any set of outer data to be observed. It is not any set of properties to be inferred from the outer data. It is not any course of action that can be predicted from the properties. With that simple negation, you cut off entirely any scientific definition within the contemporary secular notion of science. If the external data are not relevant, then the properties you can infer from the data are not relevant, and any prediction you can make on the basis of the outer data or the inferred properties is not relevant. 'Being a man' springs from an inner and free determination that is not scientifically observable. And, in general, the existentialists emphasize this point. They cut off from human science any positivistic interpretation and any secularistic interpretation of the term.

In the second place, the question is anti-idealist. The various transcendental egos are neither Greek nor barbarian, bond nor free, male nor female. They do not suffer and they do not die. We do. And again, the existentialists grasp the point. They cut themselves off entirely from the idealists. They are separated, then, on the one hand from the positivistic scientific attitude, and on the other hand from the idealist position that to a great extent has dominated the philosophic schools of Europe.

Now positivism and idealism have been major determinants in producing the contemporary world. From the Renaissance on, man has been deliberately making man. An ideal of man was born at the Renaissance. The Renaissance discovery of the ancients was really the formulation of a human ideal, and it was popularized and has spread over Western civilization from its source in the Renaissance into the cultured classes in the period of the Enlightenment, and in the present century throughout the whole of the population. It is connected with the breakdown of Christianity and the domination of a humanly inspired way of life. Insofar as positivism and idealism have been major determinants in producing the contemporary world, and in the measure that the contemporary world is found unsatisfactory or even disastrous – a common attitude on the continent of Europe after the last World War and the domination of the Nazis – existentialism has a profound resonance. It stands for something that is utterly different from the types of thinking that produced the mess we are in, and so it has a great appeal on that account.

That fact would seem to be connected with the success of the existentialist writers. *Sein und Zeit* was published in 1927 in Husserl's periodical, *Jahrbuch für Phänomenologie und phänomenologische Forschung*, an annual for philosophy and phenomenological research. Within four or five years it had run through five or six separate editions. I have already mentioned Jaspers's *Die geistige Situation der Zeit*, which ran through six editions in German in a few years and has been translated into various languages. Sartre became popular particularly after the war, but his *L'être et le néant* dates from 1942.[22] He was a cafe hero in Paris. Gabriel Marcel is the hero of European seminarians and of people interested in bringing Christianity back to the masses; he can put Christian doctrine across with amazing skill and effectiveness.

5   Time and History

This resonance with contemporary history fits in with existentialist concern with time and with history. 'Being a man' is not a result of your being born. It is something that results from your use of freedom, and what results from your use of freedom is not a property that necessarily remains with you. It is maintained by the continuous use of your freedom. It is precarious. One is ever being a man or becoming a man; one does not achieve it. You can play

22 Jean-Paul Sartre, *L'être et le néant: Essai d'ontologie phénoménologique* (Paris: Gallimard, 1942); in English, *Being and Nothingness: An Essay on Phenomenological Ontology*, trans. Hazel E. Barnes (New York: Philosophical Library, 1956).

the hero for a day and be just the opposite the next day. So this business of 'being a man' is intrinsically connected with the notion of time, and that intrinsic connection with the notion of time involves a connection with history in the sense of the development of human institutions, human cultures, and human ideas.

The intrinsic connection between 'being a man' and time is highlighted in Heidegger's title *Sein und Zeit* (*Being and Time*) and in Marcel's title *Homo Viator*,[23] man as a being on the way. However, concern with history on the grand scale appears only in Jaspers, as far as I know. His book *Vom Ursprung und Ziel der Geschichte* (*The Origin and Goal of History*) was published in Munich by Piper in 1949, and they released the eleven-thousandth copy in 1950. The other thinkers probably do not have the capacity to take such a broad view of history as Jaspers does in that book. Of course the center of history for a Christian is the birth of Christ, and there is 'before Christ' and 'after Christ.' But Jaspers starts out from the viewpoint of the history of culture, and for him the decisive period is what he calls the *Achsenzeit*, the axial period, which he locates in the period extending from about 800 B.C. to 200 B.C. There we find the rise of individualism and of philosophies of all types in Greece, the Hebrew Prophets, Zoroastrianism in Persia, Buddhism in India, Confucius and Lao-Tzu in China. Jaspers maintains that, while there were higher civilizations prior to that *Achsenzeit*, in Mesopotamia, in Egypt, in the Indus Valley, still those higher civilizations were simply enormous organizations of man. There was not man the individual, in the sense that the self-responsible individual thinking for himself was not significant in those civilizations. In the civilizations created during the *Achsenzeit*, in those five areas and within approximately a span of six hundred years that in Greece stems roughly from Homer to Archimedes, there emerged the individual. Any civilization that has come into contact with the Greeks since the *Achsenzeit* has either absorbed the Greek capacity for philosophy or else has vanished. People who do not come in contact with that development remain on the level of folk cultures or primitive cultures. His judgment on our own time is that it is not another *Achsenzeit*, but is rather something comparable to the initial discovery of fire, or of language, or of the fundamental art on which human living is based. He sees this in technical society

---

23 Gabriel Marcel, *Homo Viator: Prolégomènes à une métaphysique de l'espérance* (Paris: Aubier, 1944; Paris: Association Présence Gabriel Marcel, 1998, postface de Pierre Colin); in English, *Homo Viator: Introduction to a Metaphysic of Hope*, trans. Emma Craufurd (London: Gollancz, and Chicago: Regnery, 1951; New York: Harper & Row, 1962; Gloucester: Peter Smith, 1978).

and in the unification of the world. Jaspers is the most cultivated, the most broadly humanistic, and in a sense the most balanced and sane of any of these writers. He is a man of considerable talent, and he has none of the oddities that you find in, for example, Heidegger or Sartre.

So much for what is positive.[24] First, then, we are concerned with what it is to be a man, in the sense that 'being a man' results from the use of freedom and that 'being a man' is not a method that you acquire once for all. Second, 'being a man' remains something predicated precariously on the continuous use of freedom. It stands in radical opposition both to any positivistic approach to a consideration of man – to human science based upon the externals of man, on his behavior, on what you can observe or infer or predict on that basis – and again to idealism: the transcendental ego doesn't suffer or die, and we do.

6   Existentialism and Scholasticism

On the other hand – and this is a point of great importance to Scholastics – at first appearance the existentialists and ourselves seem to be all at one. We affirm reality, and we affirm morality and freedom. What more can you want? There is here a challenge to Scholasticism. Scholasticism has to differentiate itself very clearly from this movement. For existentialism by and large is unconcerned with propositional truth and with man's per se capacities for the truth of anything else. This unconcern arises, in Heidegger and Sartre, from phenomenological concentration on the sources, the grounds, whence spring concepts and judgments. The Scholastics deal in concepts and judgments, in definitions and truths, and in what they can infer from them. But in Husserl and still more pronouncedly in Heidegger, and similarly in Sartre, who is mainly giving a new twist to Heidegger's doctrine, the concentration of attention is on the *pre-predicative*, the preconceptual. They are concerned with the man who is the source of the concept, the man who is the source of the judgment. They are concerned with foundations, with the ground, with the origin, with the source. The same concern arises in Jaspers from his Kantianism. For Jaspers, Kant showed that what is called objective knowledge, knowledge of objects, is just knowledge of appearance. But insofar as Jaspers is dealing with the subject himself that is the source of those concepts and judgments by which we

---

24 See the first three points under 'General Orientation' in the distributed notes (above, pp. 167–68).

order experience and have a knowledge of appearance, he is getting close to a reality that, as such, is reality, though not as expressed, as conceptualized, as objectified.

We have in Thomas two types of *verbum interius,* the *verbum interius incomplexum,* the definition or concept, and the *verbum interius complexum,* the judgment. By the judgment we answer the question, *An sit?* By the concept we answer the question, *Quid sit?* Both of these spring from acts of understanding. It is insofar as you understand that you can define, and it is insofar as you grasp the evidence that you can judge. In the metaphysical analysis we have *species intelligibilis,* the *intellectus possibilis,* a phantasm in which the *intellectus possibilis* grasps the *species,* and the *intellectus agens* that moves the whole thing along.[25] But the existentialists in general do not emphasize concepts and judgments. Concepts and judgments are what you talk about, and that's all; they are not what you *are.* What you are is the ground. They have no metaphysical analysis to get you there, but that is what they are talking about: you, the subject, the concrete subject experiencing, understanding, judging. They do not use those terms, but that is what they are talking about. Husserl is looking for the foundations of the conceived objects represented by concepts and judgments. The same is true of Heidegger and Sartre: they too are looking for foundations.

In Jaspers the setup is somewhat different. He is not dependent on Husserl, but he is a very solid Kantian, and according to Kant judgments and concepts refer to appearance. A divine mind that creates its objects knows its objects inside out, and the object known as such is the *Ding-an-sich,* the *thing itself.* That is one interpretation of Kant. Another way to interpret Kant's *thing itself* is that it is what Galileo was puzzled about behind the appearances. But whichever way you take it, what we know are appearances as long as we are talking about objects, even when the object is ourselves. If we are talking about ourselves, then there is the self that talks and the self that is talked about. The self as talked about is the self as object, and as talked about it is what is known as appearance. However, in Jaspers, since we know the subject qua subject, the subject as not yet objectified, the subject as ground and source of concepts and judgments and free acts, then the *Ding-an-sich,* the *thing itself,* is performing; and consequently Jaspers has

---

25 Lonergan's blackboard work here makes the text patchy. His view of Thomas is presented, of course, in *Verbum.* See the index there under such words as Agent intellect, Possible intellect, Phantasm, etc. Lonergan here is sketching his usual diagram: see below, appendix A, p. 322.

an entry into reality. It is not an entry into reality that is objectified, or into anything but experience as appearance, but insofar as the subject is objectifying and not objectified, there is a contact with reality and something that for Jaspers is significant.

Finally, the same tendency exists in Marcel. He exemplifies, from one perspective, the commonsense attitude. Technically correct propositions often have little or nothing to do with anybody's living. What difference does it make to the kind of man you are whether you hold this or that proposition in metaphysics or ethics? This is a very commonsense attitude, and it is a little hard to overcome, as you know in your experience as teachers: what difference does it all make? Again, you can do your noviceship and learn to be a Jesuit without having done any philosophy, and you can trot out all numbers of saints who are pretty weak on abstract speculation. The business of 'being a man' does not involve that. In Marcel the fundamental idea is the good. He approaches being through the good. We will try to say more about that later.[26] But Marcel is no more concerned than the others with propositional truth. In all of them there is an unconcern with propositional truth and with the per se, because existentialism is a turning away from the universal, the necessary, and the abstract to the unique individual, the contingent, the concrete, the de facto.

Jaspers, though he talks a lot about freedom, insists that you cannot define freedom. Any definition of freedom is just a statement about appearances. But you can *be free*, and insofar as you are free you will have something better than merely knowing freedom. Perhaps that is a way of putting it. Again, Sartre establishes the fact of freedom by a very simple question, Have you made the experiment of liberty? By the experiment of liberty he means, Have you been in the torture chamber with the Nazis and refrained from giving your comrades away? If you go through that and do not give them away you will know what freedom means. It is not something that anybody is going to argue about. Everyone knows about the French Resistance movement. It was burned into the mind of France, and the people who betrayed their companions and their friends were outcasts. Trying to make out that the people who did betray could not have done otherwise was a very unpopular attitude in France, to say the least. Sartre could put all this across very concretely and very effectively. You do not have to have any abstract arguments with principles to put across such a doctrine of freedom.

26 See below, pp. 294–97.

Finally, none of them would dream of discussing man in the abstract, what is common to mewling infants, to people sound asleep, and to men facing a crisis in their lives. They are concerned with people who are awake and preferably confronted with a crisis. Consequently they are not dealing with what per se is so. They want to get their man in a concrete situation and talk things over with him, and they can put across such points in an extremely effective fashion.

Regarding this unconcern with propositional truth I have quoted a passage from Gabriel Marcel,[27] who is a Catholic convert and exerts considerable influence on the continent among Catholics and seminarians. (He is quite a problem for professors of dogmatic theology for that reason!) He says, 'The more the question is one of what I am and not of what I have, the more the whole business of questions and answers loses all meaning. When someone asks me or when I ask myself what I believe in, I cannot be content to enumerate a certain number of propositions to which I subscribe. Quite clearly, such formulae are the translation of a reality that is much more profound, much more intimate: the fact of being in open communication with a transcendent reality that I recognize as a Thou.'

There you have clearly stated an opposition, a disjunction, between subscribing to a list of formulae, of propositions – the decrees of the Council of Trent ('Si quis dixerit, anathema sit'[28]) and the Vatican Council and so on right down the line – and on the other hand the reality which these propositions are a translation of for theologians and people like that, but a reality that is much more profound and much more intimate: my spiritual life, the fact that I pray to God and that God is another person to whom I can speak as a Thou. (You will find that passage quoted by Roger Troisfontaines in his two-volume work on Marcel in volume 2, p. 352. The title of the work is *De l'existence à l'être, From Existence to Being.*[29]) Marcel's 'being a man' is a matter of moving from merely existing (we all do that) to being; and in the last analysis being is loving, being is having charity. Knowing a bit of Marcel's view makes Sartre's presentation from approxi-

---

27 See the French as given in Lonergan's notes, above, p. 169. Lonergan here translates the passage somewhat freely.

28 'Anathema sit,' 'let him be *anathema*,' was the standard expression of 'excommunication' in the councils of the Catholic Church. The usage is biblical ('an accursed thing') – see, for example, Deuteronomy 7.26 – rather than classical ('something offered up or dedicated').

29 Roger Troisfontaines, *De l'existence à l'être: La philosophie de Gabriel Marcel,* vol. 2 (Louvain: Nauwelaerts, and Paris: Vrin, 1953) 352.

mately the same basis so devastating, for Sartre goes to work in *L'être et le néant* on an analysis of love and finds two types: one which heads towards masochism, and the other towards sadism. This shows the weak point in existentialism: you can push it any way you want, and that is the weak point in unconcern with propositional truth.

This unconcern with propositional truth, this distaste for the per se, is de facto connected with a theoretical incapacity to provide foundations for propositional truth. At the present time, those concerned with existentialism, at least in Paris, I believe, among the professors at the Sorbonne, have come to the point where they see that the whole issue is, What is truth? And that is the point that you cannot answer, I would say, on the phenomenological basis. However, we will return to that later.[30]

While there is a great deal in existentialism on which we can and should practice the patristic maxim of despoiling the Egyptians, taking what is good in it and bringing it into our own work, we cannot just take it over wholesale without a critical appraisal and a revision in some fundamental points. It cannot simply be taken over whole, even the existentialism of a Marcel, and put right into Scholasticism. Very specious arguments of course have been made. Their concrete mode of approach is the concrete mode of approach of Christ in the gospels, with his parables. It fits in beautifully with scriptural theology, with biblical theology. You can build biblical theology on an existentialist basis. But if you have nothing but an existentialist basis you cannot go on to the councils of the church, to Nicea and Chalcedon and Trent and the Vatican and the rest of the councils. The councils are concerned with propositional truth. Involved here is the opposition between Protestant and Catholic on the nature of faith: faith as not simply confidence in God, *fides fiducialis*, but faith also as recognizing propositions to be true, faith as *assensus intellectus in verum*. The existentialists do not provide a basis for that.

The first point, then, was 'being a man' in opposition to positivism and idealism. It is concerned with time and liberty. The second point is unconcern with propositional truth. Finally, to sum up, existentialism is concerned with the human subject qua conscious, not with the unconscious subject, the subject as sound asleep or the infant, but with the human subject qua conscious, qua emotionally involved, with the aspect of emotion or sentiment. Existentialism is concerned with the ground of the subject's own possibilities. To be free you have to have some grasp of possibilities.

30 See below, pp. 275–79.

You are the ground of your own possibilities and of the free realization of those possibilities. There is the radical orientation within which those possibilities arise. The sort of thing that would be possible to Ignatius at Manresa is not a practical possibility to most of us in our ordinary living: we are not living on that level of sanctity. But for existentialism, what the subject makes himself is connected with what possibilities he is going to realize in his relationship with civilization, with other persons, to human history, and with God.

7   Marcel and Jaspers

To give an outline account of Heidegger or Sartre presupposes a knowledge of Husserl's phenomenology, so we will not touch on them yet. But we can give a few indications on Marcel and Jaspers.

Marcel is not a systematic thinker. In his preface to Troisfontaines's two-volume work, he congratulates the author on having done for him what he could not do for himself. Marcel's fundamental work is his *Journal métaphysique*,[31] which he started about 1910. He published the first part of it in 1927, about the same time as Heidegger's *Sein und Zeit* appeared. He can write a private journal of metaphysical reflections. He can write very effective dramas. But apart from that what he writes is very good, I think, in content. He can put a point across beautifully, a concrete idea, an orientation – a criticism of life, actually – with full force and extraordinary brevity and skill. He is a penetrating thinker and an extremely effective writer.

In a composite volume introduced by Gilson, under the title *Existentialisme chrétien: Gabriel Marcel*, Marcel has an essay 'Regard en arrière,' 'Looking Backwards,' in which he presents a history of his thought and his intellectual development.[32] His background is idealism, including not merely French idealism but also Bradley, and he is also influenced by Bergson. Kierkegaard is acknowledged to have influenced him indirectly. It is more or less impossible to summarize this piece because it moves right down to discussing concrete issues such as fidelity, personal relations, and concrete moral issues.

Karl Jaspers began with abnormal psychology, of which he became a professor and in which he wrote various technical articles. He has a pro-

31 Gabriel Marcel, *Journal métaphysique* (Paris: Gallimard, 1927); in English, *Metaphysical Journal*, trans. Bernard Wall (Chicago: Regnery, 1952).
32 Gabriel Marcel, 'Regard en arrière,' in *Existentialisme chrétien: Gabriel Marcel*, Présentation de Étienne Gilson (Paris: Plon, 1947) 291–309.

found respect for science and at the same time is a mordant critic of scientists. Forty years ago he was talking about the mythology of the brain against the positivistic approach to psychology, and the mythology of the unconscious against the Freudians. He is a Kantian, and what distinguishes him is the way in which Kant's *Critique of Practical Reason* comes to life through Kierkegaard and Nietzsche. Kierkegaard is on the religious side, an expression of religious subjectivity. Nietzsche is an expression of cultural subjectivity. You will find a very fine brief exposition of Nietzsche in Collins's work on the history of modern philosophy.[33] Fr Copleston also has a book on Nietzsche.[34] But the fundamental idea, very briefly, is that God is dead, in the sense that he is not living in the minds and hearts of people in the nineteenth and twentieth century: not as God of people of culture in the nineteenth century, and not as God of the general population today. Because God is dead, because he is not a force in human living, the whole morality that Western culture has inherited from Christianity has lost its foundation, and we have to think out a new morality. There has to be a transvaluation of values, and that is what Nietzsche works at. Both Nietzsche and Kierkegaard give new significance to Kant's *Critique of Practical Reason*. Most of the followers of Kant became just positivists; the *Critique of Practical Reason* tended to be looked upon as more or less the rearguard action of a man who wanted to retain the amount of Christian idealism that was not thrown overboard by the Enlightenment; but in Jaspers, the significance of Kant's *Critique of Practical Reason* comes to life through Kierkegaard and Nietzsche.[35]

Jaspers is a thoroughgoing Kantian. For example, in his *Philosophie* he explains his terms *Existenz* and *Transzendenz*.[36] *Existenz* and *Transzendenz* mean, in the categories that are possible in his philosophy, roughly what is expressed by mythical consciousness as the soul and God. Why? Because anything that is objectified is just an expression of appearance; there is no room for it within Kantian philosophy. But within personal experience, when the subject comes to the stage where he lives, where his being is, as it were, being a gift given to oneself, *that Existenz* – when you look upon

33 James Collins, *A History of Modern European Philosophy* (Milwaukee: Bruce, 1954) 774–808.
34 Frederick Copleston, *Frederick Nietzsche: Philosopher of Culture* (London: Burns, Oates and Washbourne, 1942).
35 The editing here, particularly around the expression 'comes to life,' is based on Lonergan's distributed notes. See above, p. 170.
36 The terms recur passim in the three volumes. For an orientation, see the first volume, *Philosophy* 43–95.

yourself as a gift given to yourself you have quite a profound idea – there is the equivalent in his philosophy of what in mythical consciousness, as he calls it, is called the soul. (He is using 'soul' here in the sense of saving one's soul and losing it rather than in the sense of the metaphysical category 'soul.') Similarly, in limiting situations, in situations such as suffering, struggle, death, the element of the transcendent is a necessary element in the proper exercise of freedom. In other words, he presents something like what we would call a moral argument for the existence of God. The world to Jaspers is a set of ciphers, a code, a symbol, and what we have to do is, as best we can, decipher these symbols. But we have no objective statement of what life is. His Kantian foundation excludes that.

Since his *Philosophie* of 1932, he has developed the notion of *das Umgreifende*, the encompassing. It does not occur in the *Philosophie*. It corresponds roughly to what I call the *notion of being* in chapter 12 of *Insight*. It is what embraces everything. He does not, however, identify it with the notion of being. But it is a fundamental advance. It was the basis for further advance in his philosophic position. Moreover, he has come to place a great deal more emphasis and reliance on reason, probably as a reaction against Sartre and to differentiate himself from Sartre. This appears in conferences he gave, as you will see for example in number 17 in the bibliography, *Reason and Anti-Reason in Our Times*.[37] His notion of *Vernunft* approximates very closely the Aristotelian and Thomist notion of wisdom. One has *Vernunft* in the measure that one considers all possible systems or opposites, faces every possible issue, and takes a stand on the total view.

Finally, Jaspers has come to speak openly of God in *Der philosophische Glaube*, which has been translated into French as *La foi philosophique* and into English under the title *Perennial Scope of Philosophy*.[38] Heinemann, in his book on existentialism, says that that translation into English is not satisfactory,[39] but what is not satisfactory about it I cannot say. He speaks openly of God as a necessary philosophic postulate. So there has been a development in Jaspers, but there has not really been any *fundamental* move, except perhaps for his concept of *Umgreifende*, which opened things up considerably.

37 See below, p. 372, number 17.
38 See below, p. 372, number 14.
39 F.H. Heinemann, *Existentialism and the Modern Predicament* (New York: Harper & Row, 1953; the reference in the second, revised edition published by Harper Torchbook in 1958 is on p. 65, in note 2).

So we have three outstanding men who, more or less at the same time, presented the anti-positivistic, anti-idealistic, concrete approach in philosophy. There is Jaspers with his Kantian background, Marcel with an idealist and Bergsonian background, and Heidegger with a phenomenological background in Husserl. We have not yet said much about Heidegger. It takes more time to get into him. And Sartre is a derivative from Heidegger.[40]

40 At a later point (see below, p. 359), Lonergan replied to a question regarding Kierkegaard, who of course is another key figure in what is called existentialism.

10

On Being Oneself[1]

We could get into the more technical side of existentialism rapidly. However, I propose to proceed somewhat differently, by treating first the topic entitled 'On Being Oneself.' We are not concerned simply with being a man, but with being the man that *I* am, not any general man.

My presentation will be more or less from a Scholastic background, because if we were to get into their terminology we would just get into more and more complications; we would have to go further and further back to take up Kant and others in a way that would consume our time. So the presentation will not be in their terms.

1   The Subject

The idea of being oneself comes up in them all. It is just a more precise statement of the normative tendency that exists in existentialism. Jaspers is the one who talks most of all about being oneself. He begins with the notion of subject. 'Subject' is used in a variety of ways. It is a relative term, where the subject is always subject *of* something. The meaning of 'subject' varies with what the subject is subject *of*. Grammatically, the subject denotes the function in the sentence. Logically, the subject denotes the function in the propositions. Metaphysically, the subject is the recipient: matter is the subject of form, potency of act. Psychologically (and that is the relevant

1 The second half of the sixth lecture, Monday, 15 July 1957.

sense for us) there is the subject in the stream of consciousness, the subject of the stream of consciousness.

2   Patterns of Consciousness

Consciousness streams in many patterns. In other words, if you talk about a sensation such as seeing, you are being extremely abstract. You do not have just seeing by itself; seeing is within a flow of consciousness, within an orientation, within a pattern modified by interest and by background. And so one can say, in a general sort of way that describes the concrete, that consciousness streams in many patterns. There is the stream of consciousness in a dream that is rather broken up and incoherent; you need an interpreter to tell you what you really were dreaming about. If you dream of lions that you are not afraid of, interpreters will remark that it is not really lions that you are dreaming about. There is the aesthetic pattern of experience: the child at play, the artist that creates something, the poet writing a poem. There is the intellectual pattern of experience. When Newton was writing his *Principia mathematica philosophiae naturalis*, he lived in his room day and night, had his meals brought to him, and could barely stand any interruption. He was just absorbed in the business of getting out the big idea, the idea that put modern science on the map. There is the dramatic pattern of experience, dealing with people. If I were all alone in this room I wouldn't be talking exactly the same way as I am. My consciousness would be flowing in a different pattern. There is the practical pattern of experience, getting things done. And mystics describe a pattern of consciousness all their own, in which not much happens or very enormous events happen.

We are familiar in Scholasticism with the distinction between practical and speculative intellect. Kant has critiques of speculative, pure reason and of practical reason. But the existentialist focus would be the subject in a concrete flow of consciousness, a flow of consciousness orientated on knowing, on trying to know, or orientated on choosing. The significance within existentialism of the flow of consciousness orientated on choosing is enormous. That is the flow of consciousness relevant to being a man. That is the existential flow. We noted the work done by the scientist, the positivist, the idealist, and the Scholastics, most of them. If you are just trying to know, well, you are not in contact with real life, you are not the real man, you are just sitting back as a spectator, you yourself are not involved, you are not living. We will try to characterize as well as we can the distinction between those two streams of consciousness.

3   The Intellectual Pattern

An intellectual pattern of consciousness is intellectual by its detachment, by the non-intervention of subjective concerns. A scientist cannot do his work properly if he is worried about something else. He has to be given totally to his job of planning and executing his experiments, making his observations; the whole man has to be just at that. Similarly in philosophy, one has to be engaged entirely by the problem. And that engagement is not a personal engagement; that is, it is not what suits me, what suits my temperament, what brings out my idiosyncrasies that will make me a good scientist or a good philosopher in the objective sense. The more impersonal, impartial, detached, pure, so to speak, purely intellectual the work is, the better it is from an intellectualist viewpoint. One can speak of a total engagement in intellectual work, but it is an engagement of submission. One is committed to a submission to norms: to the norms of logic, to the norms of scientific method, to the norms of philosophic procedure. One is headed to a universe, to the object. The subject has more or less to disappear. The subject enters into the picture only as another item in the objective universe, and one of no greater importance than any other. If Aristotle were writing simply about Aristotle or Kant simply about Kant, they would not be of interest to all mankind in the following centuries. It is precisely insofar as one is given over to an objective universe in which one is just one item along with endless others that one is doing good work.

Again, just as he is committed in his objective scientific or philosophic inquiry, so too the subject has a responsibility. But it is a limited responsibility. He can frame his conclusions either as positive or negative, as certain or probable, or he can say that this is a question that he cannot solve. He is committed to arrive at results of a certain kind. In fact his work will be regarded as strictly intellectual, strictly scientific, strictly philosophic, insofar as he is not obliged on external grounds to arrive at results of a certain kind. For that reason, Catholic philosophers are looked upon more or less as hired men to defend the Catholic Church. There is a question mark put against them as philosophers simply because they are Catholic; they have to arrive at those conclusions anyway. They are not simply this purely detached intellect working out of the intellectual pattern of experience, as represented by the example I gave of Newton or the famous example of Thales who, seeing the stars, could not see the well at his feet.

4   The Practical Pattern

On the other hand, the practical pattern of experience, the practical flow of consciousness, *demands* the intervention of the subject. The subject may choose *A* or *B* in difference of specification, *A* or not-*A* in difference of contradiction. He may be content to drift, to be one of the crowd. He may permit himself to be other-directed. But even that permitting himself to be other-directed, to drift with the crowd, to be like everybody else, is his decision. It may be an inauthentic decision just given by lack of resistance, lack of initiative. But at least *he* is making the choice, and he is responsible for it. The choice that he makes is not determined physically, chemically, biologically, sensitively, or intellectually. The lower orders do not determine what the will will do. Intellect does not determine what the will will do. See St Thomas in the *Pars prima*: he has not got on to his full theory of will yet,[2] but his view is based entirely on the fact that you cannot have a demonstrative syllogism with respect to a course of action, an *operabile*. I can't prove that I shall eat my dinner; I can find very good reasons for it, but they are all rhetorical syllogisms; they are not of the type of the arguments in Euclid. As Aristotle puts it, the doctor can either kill or cure. The same knowledge that enables him to cure you also enables him to kill you, very simply. Giving you a hypodermic needle with an air bubble in it is an effective way of doing it. He knows how to do it, he has the opportunity, and because science does not determine what he will do, Aristotle proves the existence of will as something that intervenes slowly, to move one way or the other. Even when you know everything about the issue you cannot demonstrate it. But in fact we do not know everything relevant to our choice. Our knowledge is incomplete. We do not know all the consequences of our choices. We do not know everything about the situation in which our choices are made. We do not know just what effects our choice of this type or that will have, what results it will have in this concrete situation. Connected with choice is an element of risk, an element of incertitude. Because there is uncertainty connected with choice, because knowledge is not complete, there is also an element of risk. A person within the practical

2 On Thomas's theory of the will, see Bernard Lonergan, *Grace and Freedom: Operative Grace in the Thought of St Thomas Aquinas*, volume 1 in Collected Works of Bernard Lonergan, ed. Frederick E. Crowe and Robert M. Doran (Toronto: University of Toronto Press, 2000), chapters 1-5 and 11-4.

flow of consciousness has to choose one thing or another, this or that. He may choose under his own steam or simply because he is going along with the crowd, but he has to choose, and that choice is not settled for him either by lower determinants or by his intellect.

Choosing not only settles ends and objects; it also gives rise to dispositions and habits. It makes me what I am to be. It makes it possible to estimate what probably I would do in a given situation. It gives me a second nature, and that second nature is what existentialists frequently refer to as the essence of man, when they talk about essence as consequent to existence. It is the essence you get insofar as you develop habits and dispositions by your choices. This essence that I get through my choices, that puts the stamp of a drifter or a resolute man upon me, is not an immutable essence. Achievement is always precarious. A radical new beginning is always possible.

Now in the light of these characteristics of choice one gets a meaning for the statement, the expression, 'Be yourself,' 'Be oneself.' 'What else could one be?' is the first question that arises. Our proposed topic is 'on being oneself,' but what else could you possibly be? But it is in choosing that I become *my*self. The self that results consists in habits and dispositions that emerge from my past choices. That self is mine. It is not determined by external constraints or by any inner mechanism, nor by my knowledge, but by my will *ut quo* and by myself *ut quod*. In the last analysis, the ultimate reason for my choosing to do this and not that is *I* choosing. The 'I' is the ultimate ground. Even if my choices are a mere balancing of impulses and motives, even if my choice is a consenting to drift, to being other-directed, implicitly I am choosing as *my*self the average man, the member of the crowd: that is what myself is going to be. I am *on*, in the French expression, and *Mann*, in the German expression – the indefinite individual. I can choose to be just the indefinite individual, 'someone,' 'one.' If my choices are not left to mere balancing of motives and impulses, the 'I' intervenes. I knowingly assume risk and responsibility. In either case what ultimately is operative is purely individual, purely unique. In the drifter, what results is another instance of the average man in a given milieu. In the decisive person what results is what he chooses to be. In the drifter, individuality is blurred. His individuality is consenting to be like everybody else, wanting to be like everybody else. In the decisive person there comes to light both his own individuality and the total otherness of other individuals. Just as *I* am *my*self, because *I* choose in the manner and mode in which I choose, so also is that true of the other person. And when you have two strongly marked individuals, that otherness that results from the individuality, the uniqueness, involved in choices becomes very manifest. But even in the case of the drifter, still *he* drifts, and the fact that he is a drifter is due to *himself*, or *his* self.

Finally Jaspers speaks of *limiting situations, Grenzsituationen.* There are general limiting situations, for example, a historical period, or the social milieu of my birth. If I had not been born at a certain time, in a certain family, with a certain home, a certain elementary schooling, and so on, would I have become a Jesuit? It is very improbable. There are several other ways in which I might have been born, and what I would have become would have been something entirely different. But it would not be anything that I could exert any control over. It is beyond the limit of what I can control. Again, gender is something one cannot control, and it makes a terrific difference to one's life. Again, age places definite limits upon the sort of things you can do and the sort of thing you can choose. There are things I could do if I were younger, and young people think about the things they could do if they were older.

But there are the limiting situations of a particular type, namely, the limiting situations that are totally mine and that involve me totally: death, suffering, struggle, guilt. Confronted with the limiting situations, the drifter may try to forget, but ultimately he cannot succeed. He is the one that is going to die. No one can die for you. There is the fact that one can sacrifice one's life for another, but even then the other fellow has to do his own dying later on. We all have to do our own dying. In suffering you cannot see your way through it, but you feel that if you can just hang on you will be doing very well. I read about the way the communists were getting confessions in a certain period in Poland by putting a man naked in a cell with a rat on his lap; and if he managed to kill the rat they would put in two. You do not know how you would come out in situations like that. What would my self amount to? And there are so many different types of suffering. Again, struggle is between you and another individual, or between one nation and another nation. And sitting about doesn't help. You are in the limiting situation. And finally there is guilt.

Those four notions come right out in the crucifixion, in the Christian notion of the redemption. There is Christ's suffering and death, and the struggle between Christ's preaching and those who opposed him, and the redemption of man, the liberation of man from guilt. There is a deep spiritual element in these notions. Just as Romanticism exploded in the nineteenth century in the literary world because religious emotions that had been, as it were, channeled by Christianity prior to the French Revolution had lost their normal modes of expression, so the spiritual starvation of modern man is an important element in existentialism. In fact they insist upon things like that. That is where the individual, my self, and the 'I' are forced right out into the open. The drifter may still try to drift, he will be totally unprepared for the situation, but he has to decide on being involved totally. On the other hand, in these situations the decisive person can be as decisive as he pleases, but it makes no difference.

5   'Oneself'

What is it, then, to *be oneself*? What is the *oneself*? The *oneself* is the irreducibly individual element whence spring the choices of the decisive person and the drifting or forgetting of the indecisive person. What springs from that source is free; for it one is responsible. What results from that source is not only the sequence of activities but also the character of the man, what he is, the second nature, the essence or quasi essence by which precariously one is what one is, and by which one will remain what one is if one is not tested too severely. In the words of St Paul, God makes issue with temptation.[3] That choosing does not wait upon learning, upon acquiring as much knowledge as might be needed or relevant to making the decision. Making decisions involves uncertainty. It involves taking risks. Finally, in choosing is involved everything that concerns me.

That is the existential subject. When you are talking of the subject in that sense you are talking of the subject that ex-sists, you are talking of the concrete individual, man to man, the *cor ad cor loquitur* of Newman.

Now in existentialism there is a marked tendency to want everything out that does not pertain to that concrete flow of consciousness orientated upon the individual's choice. What is apart from that is the sort of work done by idealists and positivists, by scientists and philosophers of the older type. For the existentialist philosopher thought and thinking is *engaged* thinking. He himself is committed. A scientific, objective approach is considered an unrealistic approach. So we find here another powerful emphasis upon considering simply what is de facto, concrete, contingent, unique, individual, and upon getting away from giving any significance to propositional truth and the per se.

3 See 1 Corinthians 10.13.

6   Withdrawal-and-Return

How does one unite the two? What is really the relation between the speculative pattern of experience, on the one hand, in which we do our Scholastic philosophy like the scientist, like the idealist philosophers, and, on the other hand, the concrete practical pattern that we cannot reject? We cannot reject it, for it is a component in Christianity and in Christian morality and in spirituality and sanctity. What are the relations between the two? Very briefly, I think we can find something in the way of a formula in one of those couplets coined by Toynbee in his *Study of History* – *withdrawal-and-return*.[4] You may remember that Toynbee gives a series of examples of men who have left their mark on history, who withdrew like St Paul into the desert or Ignatius at Manresa, who withdrew from practical life to a life of quiet and a lack of association with others, and who returned to the world and made a terrific difference. They made a transforming contribution to human society. We can also say that the point to the speculative, purely intellectual work is the point of the withdrawal, while the point of the practical is the point of the return. A man has to have both.

There is, in the intellectual pattern of experience, an element that is choosing. When I am in the intellectual pattern of experience, I am choosing, because I choose to submit entirely to the exigences of knowing and to meet completely the demands of the effort to know. Without that knowing there would be not merely a residual incertitude, a risk that I have to assume in my choices, but a total blindness that makes choice indistinguishable from mere force or instinct or passion or arbitrariness. In other words, we cannot get along without the proper development of intelligence. On the other hand, in the practical pattern of experience, while there does exist an ultimate moment of being myself, an element of uncertainty and risk and nonetheless of total commitment, still this ultimate moment is known, and it is within a context that is known and with respect to what largely is known. To give oneself over entirely to the practical is to become blind, whereas to give oneself over entirely to the speculative is to become ineffective. To be bold, open-eyed, and effective, one has to go in for both. It is a threat to Scholasticism, of course, if one were to admit entirely this

4 See the indices in Arnold J. Toynbee, *A Study of History*, vol. 3: *The Growth of Civilizations* (London: Oxford University Press, 1934); vol. 6: *The Disintegration of Civilizations, Part Two* (1939), and vol. 10: *The Inspiration of Historians* (1954).

existential insistence upon only the kind of thought that is not simply the thought of the detached spectator. They do not want the latter at all. They identify it with science and idealism, and that is anathema to them. But we cannot jettison intelligence and reason and truth.

7   Philosophic Significance of the Theme

There are, next, the few notes on the significance of existentialism for Scholastics,[5] which we have already established. Existentialism breaks through a positivistic science of man. It denies that there is any ready-made essence for a positivist to investigate: external data, properties, predictable results. That just does not exist. 'L'homme se définit par une exigence,' man is defined not by a formula but by an exigence: be a man, be yourself. Again, it breaks through a pragmatic attitude, a pragmatic science of man. It is true that one learns from experience; one learns about things and about one's own potentiality. But the issue is not simply knowing. One can know all about one's potentialities and all about the objective situation that it is possible for one to know, and still there remain uncertain elements, the element of risk, and the necessity of my choosing. And meanwhile, while one is living, one has to be making one's choices all the time anyway. Again, there is no pragmatic justification for the decision to risk nuclear warfare, but 'we have to be men.' Existentialism also breaks through the idealist's view of man, as already we have noted. The idealist's world is a world of pure intelligibility; it is rational throughout; it is not a world of free choices springing from unique individuals that are totally concerned in the once-for-all of the moment. The existentialist type of thinking also provides a ready rationalization for those who do not want to go to class and study: Let's forget all the books, all this nonsense, and get down to something practical. There are all kinds of practical things that can be most admirable: love of one's neighbor, zeal for souls, conversation with the individual, dialogue, *disponibilité* (a readiness to do whatever others really need), the life of prayer. The temptation is particularly noticeable in objections raised by students from Europe in Rome. Finally, existentialism sets more fundamental problems. In metaphysics our traditional disputes about the nature of subsistence and the person have to be cleared up if you are going to say exactly what you mean by becoming oneself and being oneself. I cannot go through the various opinions: Tiphanus, Cajetan, Capreolus, and so on.

5 See above, pp. 174–77.

But there are ways of conceiving the person in which this 'being oneself' just has no meaning. I treated the problem elsewhere in a little book that came out in 1956, *De constitutione Christi ontologica et psychologica.*[6] The ontological part goes into that.

Again there is a problem that arises in the theory of ethics. According to Aristotle and St Thomas, who constantly repeats it, *Verum et falsum sunt in mente; bonum et malum sunt in rebus. Bonum* is not some abstraction. It is concrete, and consequently there arises the difficulty that if there is no abstract good, then abstract moral precepts do not suffice to determine the good. They can be no more than pointers to the direction or location in which the good lies, or limits indicating where the good does not lie. There remains the problem for each one to work out concretely the good he can do by his decision in his concrete situation with his potentialities and possibilities.

There is an order of the universe, but the order of the universe is concrete. It is not a matter of a set of abstract essences having properties that are thrown together in a certain way, with the levels and possibilities all determined by these abstract essences. It is a concrete unfolding in concrete situations, and the concrete situations are the product of individual decisions about the concrete good. For St Thomas,[7] in the *Contra Gentiles* the tendency is to conceive the order of the universe simply as a plan that is the finest thing that God created and realized; the notion of the person is somewhat deferred; but in the *Summa theologiae,* the notion of the person comes to the fore. The order of the universe and interpersonal relations are two notions that slightly tend to cover one another when the order of the universe is conceived concretely. There remains the natural law. Situations do not change moral precepts. Still, there arises the significance – we hear about it particularly in Protestant theology – of *kairos,* of *my* situation, *my* opportunity, *my* duty. My situation and opportunity and duty are things that can be illuminated by moralists and by spiritual directors, but I cannot communicate entirely to another just what *my* potentialities are, what *I*

6 Bernard Lonergan, *De constitutione Christi ontologica et psychologica* (Rome: Gregorian University Press, 1956; reissued 1960, 1964); to be published in English, *The Ontological and Psychological Constitution of Christ,* as CWL 7.
7 Lonergan mentioned in an aside that 'Fr Wright of the Oregon Province finished his thesis a year ago on St Thomas's doctrine of the order of the universe.' See John H. Wright, 'The Order of the Universe in the Theology of St. Thomas Aquinas,' S.T.D. thesis (1957) directed by Lonergan at the Gregorian University, Rome.

think I can do. I cannot settle concrete issues entirely by considering simply general norms. If you want an example, I wrote a book called *Insight.* I could not explain what I wanted to do before I got it written. How do you arrange to get a thing like that done? The easiest way is to do it in your spare time. And even after it is done, there are not too many who know right away just what you are trying to do and what is the good of it. There is that notion, then, of *kairos.* The individual has to see what can be done and take upon himself the risk and the responsibility of getting it done, or, on the other hand, the responsibility of nothing being done.

There is in the order of the universe the emergence of good from evil. St Augustine, and after him St Thomas, remarked that God permits evil, on the one hand, because general principles are not to be upset ...[8] [Such a perspective is] without the help of faith: it depends upon the natural light of reason. There is a further domain that is handled in the light of faith and divine revelation. Now that is a neat separation, but it is not adequate when we are confronted with a situation that existentialism is talking about. The existentialist is not talking about human nature per se. He is talking about the concrete men that exist and that are acting. He is talking about me and about you. And men living in concrete situations in this world are men with the effects of original sin, with the offer of divine grace, with the acceptance or the rejection of divine grace. A full understanding of existential man introduces theological questions that cannot be handled by the Catholic philosopher in the traditional sense of the word 'philosophy' among Catholics, as a field of natural reason. So there is a problem there.

Again, we offered a general indication of the reconciliation between the speculative and the practical in terms of withdrawal-and-return. But the issue is much more complicated than that. There is a problem of conversion, of a reorientation, a reorganization of one's mind and living. That reorganization, reorientation, transformation of self (what Toynbee calls transfiguration) involves both at once. It involves truth and will, truth and resolution. On that point we shall have more to say later.[9] When a person has one's living organized on a lower level, the movement to a higher level involves something like the apparent eruption of a latent power, the possibility of a radical discovery, where the discovered has been present all along but where there has been a hiding of what has been discovered. These

---

8 There is an indeterminate gap in the recording here. The bracketed phrase following in the text is an editorial addition.
9 See below, pp. 302–10.

notions of obnubilation, dis-covery, uncovering what has been there all along, conversion, transformation of one's living, are all right in the center of existentialism, and they lead to the most fundamental questions that can be raised with regard to the philosophic enterprise.

Finally, with regard to this last point of a personal conversion, intellectual and moral, something of a leap is involved. Kierkegaard distinguishes three or perhaps four spheres: the aesthetic man, the ethical man, the religious man, where the religious man is subdivided into two categories, *A* and *B*, what we would call natural religion and supernatural religion. The movement from one sphere to another cannot be in the light of the principles on which the lower sphere is based. If the person is in the aesthetic sphere, he pays attention to what interests him for as long as it interests him; he does not raise fundamental questions from the nature of the case; he wants to be amused; he wants to have something to do; he wants to keep going; he does not want to be bored; and there is not in that way of living the means to move up to the other. You cannot lift yourself by your own bootstraps; you cannot stand in a tub and lift it too. It is only by the dis-covery, the unveiling, the revealing, of things that have been hidden, repressed, given a systematic oversight, that the change can take place. You may remember from what I said last week that the ultimate issues come back to the subject. It arises right in here.[10] And this issue is relevant to Scholastics with their divisions into schools, with their endless disputed questions, with the absence of any systematic method of solving these disputed questions, with the absence of anything much in the way of a desire, a serious effort, a conviction that some method, some procedure, has to be found to put an end to all that. While that sort of thing may not bother the philosophers too much, it strangles theology. You eliminate speculative theology from the minds of your pupils if all you can do is say, 'There is this opinion, and this opinion, and this opinion, and how do you choose between them? Well, we really don't know, we have no way of settling it.' The reaction of students of theology to that sort of thing is, 'Well, let's study the Fathers, let's study the scriptures,' and Scholastic theology goes by the board. It is an issue of survival for Scholasticism, particularly at the present time, to think of the critical problem not as a problem of Kant or of people outside the Church, but as *our* problem, as the elimination of unnecessary disputed questions, as finding a methodical solution to them. It is an issue of giving our Scholastic

---

10 'Here' may mean either a pointing to an element in the diagram (see above, p. 101, note 24 and p. 322 below) or a pointing to himself.

philosophy and particularly theology – and theology is what concerns me, that is where the shoe pinches me – a systematic, methodical solution so that, at least among ourselves, our philosophy will be scientific and not the sort of thing that, as soon as you start pressing here and pressing there, you run into disputed questions that go all over the map. That *is* a difficulty that I think exists.

What we are trying to do is to get a general orientation on the type of thinking that is called existential. We have seen its oppositions to positivism, idealism, pragmatism; its insistence on a fundamental normativeness that initially can be stated as 'being a man,' and that ultimately is a matter of being yourself, being oneself.[11]

11 Lonergan asked if there were any questions on what he had covered in the lecture. There were none.

11

The Later Husserl[1]

1   Husserl's Last Work

We start moving toward the technical elements, which are the really significant elements in existentialist thinkers, by considering the last work on which Edmund Husserl was occupied. About 1934 he gave a public conference that was a tremendous success. (He was about seventy-five years old at the time – perhaps not quite that old, but about that. It was the first time in his life that he had been anything in the way of a popular success, and he was considerably moved.) About a third of the work was published during his lifetime, and then it became unavailable. But from his literary remains, the other two thirds have been added on. It was published by Martinus Nijhoff at The Hague in 1954: *Die Krisis der europäischen Wissenschaften und die transzendentale Phänomenologie* (*The Crisis of European Sciences and Transcendental Phenemenology*).[2] It is edited by Walter Biemel, who belongs to an

---

1 The first half of the seventh lecture, Tuesday, 16 July 1957.

2 Edmund Husserl, *Die Krisis der europäischen Wissenschaften und die transzendentale Phänomenologie: Eine Einleitung in die phänomenologische Philosophie*, ed. Walter Biemel (The Hague: Martinus Nijhoff, 1954; 2nd printing, 1962); in English, *The Crisis of European Sciences and Transcendental Phenomenology: An Introduction to Phenomenological Philosophy*, trans. David Carr (Evanston: Northwestern University Press, 1970). Carr's 'Translator's Introduction' to the English text, p. xvii, indicates that the series of lectures that served as the basis of the work was delivered in Prague in November 1935: 'The Crisis of European Sciences and Psychology.' Earlier, in a letter of 30 August 1934 to the International Congress of Philosophy at Prague Husserl had

institute in Cologne. Husserl left enormous literary remains, many of them written in a peculiar shorthand. They were conserved and are being classified and edited at Louvain under the leadership of Fr van Breda, O.F.M. There is a somewhat parallel institute at Cologne, though just exactly what its functions are I do not know. Walter Biemel is at Cologne, and he is the man who edited this work. He has done a number of works in collaboration with the people at Louvain, where he did two translations of Heidegger in collaboration with de Waelhens and a translation of a book of Jaspers, his *Die geistige Situation der Zeit*, in conjunction with Ladrière, the man who also did the work on Gödel's theorem.

This work on the crisis of European science was probably under the influence of Husserl's most brilliant and disloyal student, Heidegger. It provides a good introduction to the type of idea and problem that was of concern to Heidegger when he wrote his *Sein und Zeit*.[3] Considering simply

---

responded to a request to comment on 'the mission of philosophy in our time,' and in May 1935 he had lectured in Vienna on 'Philosophy in the Crisis of European Mankind.' Appendix I of Carr's translation presents the Vienna lecture.

3 Carr's 'Translator's Introduction' states (pp. xxv–xxvi): '... Husserl's phenomenology was being eclipsed in both academic circles and the public mind by the increasingly popular *Existenzphilosophie* of Jaspers and Heidegger. Husserl had no kind words for this philosophy itself; but the *Crisis* and the Vienna lecture make clear his growing awareness of the profound malaise among the educated to which the existentialists were directly speaking. Husserl's bitterness, especially against his former protégé Heidegger (who, like many others, appropriated the term "phenomenology"), did not prevent him from seeing that existentialism had given needed expression to something real: a deeply felt lack of direction for man's existence as a whole, a sense of the emptiness of Europe's cultural values, a feeling of crisis and breakdown, the demand that philosophy be relevant to life. And these texts are full of Husserl's tacit admissions that his philosophy had not seemed to speak to these needs ... Husserl was convinced that his philosophy was every bit as relevant to human problems as existentialism, and he is at pains to show this in the *Crisis*. This no doubt influenced his attention to the "crisis" theme itself, his emphasis on values, on the practical aspects of existence, on "life," and hence his borrowing of many terms made popular by existentialism (*Dasein, Existenz, existentiell*, etc.). But for Husserl, existentialism, whatever its value as an expression of man's problems and Europe's problems, was precisely a symptom and the farthest thing from a proper remedy. It attacked the old rationalism, with its sterile legacy of nineteenth-century positivism, simply by turning it upside down; it was "irrationalism" pure and simple, upon which Husserl makes his most vigorous attacks.' But see also p. xxxvi: 'The concern with historicity might seem to reflect another aspect of Husserl's interaction with Heidegger,

this work frees us from the necessity of considering the development in Husserl's own ideas. He began as a mathematician, became engaged in the problem of the foundations of mathematics, was much attracted by Brentano, who spoke about intentionality, and attacked psychologism, that is, the reduction of all science to accidental events that occur in us. Gradually Husserl developed his notions of phenomenology, of *epoché*, of transcendental reduction. Those notions went through a considerable transformation as the years went on. His first philosophic work of importance, *Logische Untersuchungen*, was written in 1901.[4] His *Ideas for a Pure Phenomenology* appeared about 1912 or 1913,[5] and his formal and transcendental logic, I think, about 1929.[6] *Erfahrung und Urteil* (*Experience and Judgment*) is a posthumously published work edited by Landgrebe.[7] The works are being published by Nijhoff, but the editing is done at Louvain and at Cologne. Gradually they are getting out what can be edited, what are not simply jottings and things that are endlessly corrected.

---

considering the importance of this concept in the later pages of *Sein und Zeit*. Actually it represents Husserl's attempt to come to terms with an adversary of much longer standing, namely, the historicism he had attacked in "Philosophy as Rigorous Science." Dilthey is the key figure here rather than Heidegger.'

4 Edmund Husserl, *Logische Untersuchungen*, 2 vols. (Halle: Niemeyer, 1900–01, 1913, 1928); in English, *Logical Investigations*, 2 vols., trans. J.N. Findlay (London: Routledge & Kegan Paul, and New York: Humanities Press, 1970).

5 Edmund Husserl, *Ideen zu einer reinen Phänomenologie und phänomenologischen Phänomenologie. Erstes Buch: Allgemeine Einführung in die reine Phänomenologie*, ed. Walter Biemel (The Hague: Martinus Nijhoff, 1950); in English, *Ideas Pertaining to a Pure Phenomenology and to a Phenomenological Philosophy*, vol. 1, *General Introduction to Pure Phenomenology*, trans. F. Kersten (The Hague: Martinus Nijhoff, 1980). Volumes 2 and 3 of the *Ideen* were edited by Marly Biemel and published by Martinus Nijhoff in 1952. Vol. 2 is entitled *Phänomenologische Untersuchungen zur Konstitution*, and vol. 3 *Die Phänomenologie und die Fundamente der Wissenschaften*; in English, *Studies in the Phenomenology of Constitution*, trans. Ted E. Klein and William E. Pohl (The Hague: Martinus Nijhoff, 1980) and *Phenomenology and the Foundations of the Sciences*, trans. Ted E. Klein and William E. Pohl (The Hague: Martinus Nijhoff, 1980).

6 Edmund Husserl, *Formale und transzendentale Logik: Versuch einer Kritik der logischen Vernunft* (Halle: Niemeyer, 1929); in English, *Formal and Transcendental Logic*, trans. Dorion Cairns (The Hague: Martinus Nijhoff, 1969).

7 Edmund Husserl, *Erfahrung und Urteil: Untersuchungen zur Genealogie der Logik*, ed. Ludwig Landgrebe (Hamburg: Claasen, 1948, 1964); in English, *Experience and Judgment: Investigations in a Genealogy of Logic*, trans. James S. Churchill and Karl Ameriks (Evanston: Northwestern University Press, 1973).

Husserl spent his life working at these things, perpetually. He was constantly trying to work towards the foundations of science, and his notions were strictly scientific. His fundamental inspiration was philosophy as strict science. I will give you a brief outline of this book – it is about three hundred and fifty or four hundred pages, as I remember – and, if possible within the hour, a critique of some of the fundamental ideas in it. Then we may go on at eleven o'clock to a general discussion of phenomenology.[8] These ideas lead right into Heidegger and the existentialist movement as determined by Heidegger. And Heidegger, despite the difficulties that we will find in his thought, is the most profound of the lot.

2   Is There a Crisis in Science?

Husserl begins by admitting that it seems paradoxical to speak about a crisis in science. Its achievements are unmistakable, its labors in endless fields continue apace, what unsolved problems there may be will be solved either by existing methods or by the development of further methods, and the development of further methods will occur just as the developments of the methods of the past occurred. Still, he says, the need of new methods can be discovered only by a critical survey. And if the need exists at present, then the survey will not only serve to discover the need, but also provide a signpost to the direction in which the need is to be met.

Such a survey demands a criterion, and a criterion that can hardly be rejected is an act of recall in which we reenact within ourselves the original intentions of the scientific enterprise.

He goes on to say that these original intentions had two principal manifestations, the first in fourth-century Athens and the second at the Renaissance.

3   Fourth-century Athens

The formulation of the aim of science in fourth-century Athens consisted in a shift of meaning, an *Umdeutung*, of popular notions on *sophia*, *alêtheia*, *epistêmê* – on wisdom, truth, science. There were the popular notions that Socrates, Plato, and Aristotle found, and what they did was to shift the meaning of those words. This shift took place through the Platonic contrast of *epistêmê* and *doxa*, science and opinion, and in the contrast between

---

8 Below, chapter 12.

dialektikê and *eristikê*. It consisted in setting up an ideal of knowledge and truth that involved a sustained effort, a methodical procedure, a rigor, an attainment of evidence, and a solid, immovable basis in certainty, that simply were not contained in the original meaning of such terms as *alêtheia* and *epistêmê*. Finally, the setting up of this scientific ideal by transforming the meaning of ordinary words such as wisdom, truth, understanding, and science unfolded in the works of Aristotle, Euclid, and Archimedes, of the historians, and of Greek medicine.

4   The Renaissance

The Renaissance brought forth a far more grandiose proposal. It discovered in the ancients an ideal of knowledge and truth as opposed to merely traditional opinion, and in that ideal of knowledge and truth as opposed to merely traditional opinion it sought a principle for the transformation of society as opposed to merely traditional power. And, Husserl goes on, in the measure that that ideal and that principle are valid, Western man is the exemplar of mankind, the realization of the meaning of what it is to be a man. On the other hand, in the measure that the ideal and the principle of the Renaissance are not valid, in that measure Western man is just another anthropological classification. He is of concern to us not because of any intrinsic value or significance, but merely because he is the type or species to which we belong. You see there the radical sweep of his idea. The notions of truth, science, and so on, were the creation of the Greeks. The Greeks took the raw materials of popular conceptions and constructed an ideal out of which arose Greek science. In the Renaissance there is not merely the rediscovery of the Greek ideal of knowledge and truth; there is also the taking over of that ideal as a means of destroying merely traditional opinion and undermining merely traditional power, to put human society on a basis of truth and reason, freedom and responsibility.

5   The Criterion

This is the criterion, then, by which Husserl proposes to judge modern science. If we are to judge modern science by the criterion of its original intentions, we must ask what hope modern science offers of the attainment of knowledge and truth and of a principle that frees man from merely traditional opinion and power and enables him to set up human society on a basis of reason and truth.

6   Five Criticisms of Modern Science

In the light of this criterion, Husserl has five criticisms of modern science.

6.1 Tendency to Splinter

First of all, there is its tendency to splinter into specialties. You can look at any university catalogue or any international congress, and you will find endless divisions. Five or six years ago there was an international congress of biologists in Montreal. The list of papers read was a book about two inches thick. The same tendency to splinter has been lamented in *Deus scientiarum Dominus*,[9] which commented on all the different things that theologians are now taught as opposed to the much more unified formation that was given earlier. It is a fact: there is a tendency to splinter.

6.2 Autonomy of the Splinters

Secondly, Husserl criticizes science for the autonomy of the splinters. What counts effectively within each of the departments, sections, and subsections is what is recognized as 'good' ('good' in quotes) in that department, section, or subsection. Mathematicians will say a given book is 'good.' To find some other way of talking about it, they would have to go into the whole problem of the foundations of mathematics, which they have not solved to universal satisfaction yet. And the mathematicians, of course, are way ahead of everybody else in solidity, rigor, precision, and demonstration.

Moreover, not only does modern science splinter, and each splinter tend to become autonomous, but discussions of knowledge, science, and truth are themselves just so many other specialties, and the relevance to any other fields except the specialty that is called the study of knowledge, science, truth (that is, philosophy) is merely a matter of opinion. The department of philosophy does not rule the other departments. Each department is autonomous; it goes by its own criteria; and philosophy is just another department.

6.3 The Drift to the Criterion of Technical Competence

Thirdly, he criticizes science for the drift to the criterion of technical competence. Upon a background of merely traditional norms that are not

9 *Deus scientiarum Dominus* was an Apostolic Constitution of Pope Pius XI
  (1931) regarding Catholic universities and faculties of ecclesiastical studies.

questioned, the effective principle is technique. Husserl has a term for these merely traditional norms: the *Selbstverständlichkeiten*. It recurs in Heidegger, and it reappears in Marcel as the *tout naturel*, what is self-evident, with an ironical twist to the word: what is understood by itself, *Selbstverständlichkeiten*.

On this background, then, of things that are considered self-evident in the sense that no one bothers to question them – and they would not have much in the way of a foundation if you did question them, but they are there, they float more or less, in the background – on that background the effective principle is technique. What counts ultimately is getting results, but what counts proximately is the approved technique, doing things the way the things are done. If you write a book in history and you have the footnotes and the bibliography done just right, if that sort of thing is done well, then who is going to judge whether the contents are good or not? You have the technique; you obviously know how to do it; you know the literature in that subject; and that is all there is about it: the technique. It is similar in other fields. Each field develops its techniques, and the person who knows the technique will be accepted, while the one who does not will not be accepted.

What is the foundation for the technique, the justification for the technique? That is another question. It belongs possibly to some other department that cannot exercise any influence upon this one.

So first there is splintering, second autonomy, and third the domination of a pragmatic criterion of technique.

6.4 *The Position of the Human Sciences*

Husserl's fourth criticism has to do with the position of the human sciences. He contrasts scientific medicine with folk medicine. Scientific medicine is based on scientific anatomy, physiology, pharmacy, chemistry, physics. Folk medicine is a collection of recipes and cures that rest upon the knowledge of the people. It still existed to a certain extent when I was a boy. An aunt of mine had various cures for various things, and I was submitted to various types of medicines. But it has practically disappeared. However, as Husserl notes, while folk medicine for the individual has disappeared, what do we have for society? There arise crises and what do we have? A series of nostrums that are no better and no worse than folk medicine. There is nothing scientific about it. De facto, the unification of science and the application of science to society come about by either the totalitarian state or mass democracy, and in both cases it is a unification not by reason but by power.

6.5 *The Impossibility of Reorientation on the Present Basis*

Finally – and this is apparently his ultimate and most damning criticism – there is the impossibility of a reorganization, a reorientation, on the present basis. A reorientation demands a general view, and no general view is possible. All that can be had is a shifting set of best available opinions in more or less unrelated fields. Consequently, the best available opinions are more or less unrelated to one another. A general view is the work of a mind, and no mind can master all the techniques. Because no mind can master all the techniques, no mind can utter a pronouncement that will be acceptable to all the autonomous splinters. Thus on the present basis there is no possibility of setting forth a general view of foundations of science, a reorganization of science, that could be scientifically respectable. We are confronted with what he calls *Bodenlösigkeit*: there is no foundation, no ground, on which you can stand.

7   Diagnosis

Having set up his criterion and performed his criticism of modern science, he goes on to the question of diagnosis and remedy.

If modern science does not fulfil its original inspiration of providing man with knowledge and truth as opposed to merely traditional opinion, and a principle that will enable man to put society on a basis of science and truth, has there been some radical defect or oversight in its program? His point here is the point already foreshadowed, namely, that scientific clarity floats on popular obscurity, scientific evidence on popular *Selbstverständlichkeit*. The real basis of science has not been explored, examined, evaluated.

7.1 *Two Worlds*

What we have is the Greek ideal of knowledge and science and wisdom reached by transforming popular notions, but the validity of that transformation of popular notions has not been explored, examined, evaluated. We still have reflected the original bifurcation between science, *epistêmê*, on the one hand, and *doxa*, on the other. There exist two worlds and two truths. There is popular truth in the sense of telling the truth in the home, in business, in law courts, in newspapers and periodicals, in autobiographies. There is scientific truth in the sense of a validated set of propositions in logic, mathematics, physics, chemistry, and so on. And those two truths are just different worlds. They are just different. There is no apparent common measure between them. They simply reflect the original bifurcation of the Greek invention or shift in meaning of words like 'truth,' 'science,' and so on. Plato and Aristotle took these words of ordinary usage, gave them a new meaning that ordinary usage knew nothing whatever about, and on that basis there is floated the scientific ideal.

Again, there is the popular world of poets and men of common sense, of everyday assumption, opinion, activity; and there is the quite different world of the scientist or philosopher, a world of protons and electrons and so on, or any of the various types of universes excogitated by any of the various schools of philosophy. And there is a total difference between, on the one hand, the world of everyday assumptions of ordinary living and, on the other hand, the world of science or the world of the philosopher.

7.2 'Shovings Under'

Moreover, there has occurred a series of what he calls *Unterschiebungen*, or 'shovings under.' The scientific, philosophic world is put under the popular world, as though it were the real thing. What you see is the table, but what is really there? Electrons and protons, and so on: that is the real reality, and it is shoved underneath. There really is no connection between the two, at least apparently, but they make one world out of two by saying that the scientific or the philosophic world is what is really there and that the popular world is what the ignorant layman thinks.

7.3 *The* Lebenswelt

On the other hand, it remains for Husserl that the fundamental truth, the really basic world, is not the scientific or philosophic but the popular. He argues that one has only to take any scientific procedure or conclusion, and with a little probing one will find that the ultimate evidence lies in the popular world, in the *Lebenswelt*, with its *Selbstverständlichkeiten*. Science claims to rest on experience, but what is experienced is not the scientist's real world but the popular world. Science rests on the testimony of observers and experimenters, but the observers and experimenters are operating in the popular world and after the fashion of the popular world. For example, Michelson and Morley perform an experiment. What they see is one thing, and what they say is another. Science makes no investigation whether there is any psychophysical parallelism to justify connections between what Michelson and Morley observe, on the one hand, and what they say, on the other. There is just a leap there. There is no scientific investigation of the process between what they see and what they say. Science rests on what they say. There is nothing queer about that if you are living in the popular world. That is something that is self-evident to everyday common sense: what one sees and what one says are intimately correlated, connected in the closest fashion possible. That is all very well for men of common sense. But if the scientist does not rest his case simply on the popular world, he has to put a scientific connection between the two, and he has no scientific connection between the two. And so Husserl claims that the scientific world, so far from providing the underlying reality and the underlying truth, rests on the popular world and on popular notions that the scientist does not examine and does not criticize.

8   Remedy

That provides him with his opening to step in and show how science should be grounded. If there is a malady and a diagnosis, there is also a remedy and a cure.

8.1 The Priority of the Subject

The first step in the cure is to acknowledge the priority of the subject. The subject is the source of both truths and both worlds. There is what he calls a natural attitude, *natürliche Einstellung*, that yields popular truths from the popular world, and, on the other hand, there is a cultivated attitude, a cultivated *Einstellung*, developed in Athens in the fourth century and again in the Renaissance, and maintained in the scientific tradition. It is that cultivated *Einstellung*, with its developed notions of truth and science, that grounds the scientific truths and the scientific world. But the subject is the source of both. The subject can be both a man of common sense and a scientist. He will live in the popular world where he draws his salary, and he will live in the scientific world in which he writes his books.

8.2 Intentionality

Now what the subject is the source of is intentional. He is the source of what he means, symbolizes, represents, intends. Note that in Husserl 'intentional' has no metaphysical presuppositions. In scholastic circles, 'intentional' is a metaphysical term denoting entities of a class through which you know other entities. There are no metaphysical presuppositions or implications in Husserl's notion of the 'intentional.' There is the intending subject and the intended object. The object is nothing more than what is intended by the subject, and the subject is nothing more than what intends the object. The two are correlative, and the fact is primary, basic, undeniable, unavoidable.

Husserl has done, with enormous labor, a fine analysis of psychological process. Two of his most brilliant discoveries are the correlation of *Abschattung* and *Horizont*, and again the correlation of *Einstellung* and *Welt*.

Abschattung and *Horizont*: for example, we have here in the center of the campus a building called Gasson Hall; I have seen it from the outside, from various angles; I have seen the basement, where on one side there is a sort of lunch room and on the other side there seems to be a dining room; I have not been in any of the upper floors, but I know that they are there and that they serve several useful purposes, and I know there is an inside to the tower and probably some means of going up inside the tower. What has been my perception of Gasson Hall? Is what I mean by Gasson Hall what I have seen? No. What I have seen is Gasson Hall from a certain number of perspectives, a certain number of angles. But what I have seen has been integrated in what Husserl would call a horizon, an infinite series of all the things I could see and feel and bump into. And that is a construct. You get that construct insofar as you have an intentional act, a meaning. You mean a building, and what you mean is an integration of possible *Abschattungen*, things that you would perceive in any instantaneous or any momentary perception. What I see when I lean over one way is a different visual object from what I see when I lean the other way. The laws of perspective will establish that. But I have practically no awareness of the fact, because I automatically integrate all these visual percepts or objects into one thing, and that integration is the *Horizont*, the horizon.

Again, visual objects or any other type of sensible objects are integrated by acts of meaning into a single *Horizont* that we regard as the object, the one thing that is out there that we never see. I never see Gasson Hall from all sides at once; I can see it from above if I'm in a plane, and from the north or the south or the east or the west, but I cannot see it all at once. But Gasson Hall is for me what I would see if I were seeing it from all sides at once. That is the thing that is there, and that is what I mean by Gasson Hall: the totality of perceptibles, so to speak.

Besides *Abschattung* and *Horizont,* there are *Einstellung* and *Welt.* Just as objects are integrations of indefinite series of perceptibles, so is the world. You never see the world. There is no observation of the world. What is the world? The world is the total horizon in your knowing, and it corresponds to your *Einstellung.* The commonsense man has his orientation, his attitude, and corresponding to his attitude is his world. But his world is not any object that he ever experiences in any way. It is the field in which the totality of objects occur, are available, to the subject.

The subject, then, is prior, and the subject is the source of the intentional. These are the first two points.

8.3 *A Return to the* Cogito

The third point is that what is needed is a return to Descartes's *Cogito.* Let the subject realize that everything he thinks, believes, is certain of, whether on popular, scientific, or philosophic grounds, is just intentional: the term to an intending act that means, intends, considers, defines, posits – what you please. There are intentional acts, and everything that you know is object of intentional act. Let him first, then, grasp that everything that he knows is just a term of an intentional act. Let him ask how much he can primarily, irreducibly, immutably hold. For example, there are Descartes's 'I doubt' and 'I think thoughts.' Having reached that ground of absolute certainty, let him be careful to avoid Descartes's two wild leaps. The first leap of Descartes was to go from the term of sensible experience to Galileo's real world of things with primary qualities but not secondary qualities. That is just a leap beyond the intentional. It posits what is not given. On the other hand, let him avoid Descartes's leap on the side of the subject, from the intending subject that qua subject and intending is just the opposite pole to what is intended, from that, of which there is no practical doubt, to a metaphysical entity called a soul, which in Cartesian philosophy amounts to a ghost in a machine.

So Husserl cuts off, on the one hand, as beyond what is intended, Galileo's mathematicized world of real objects and, on the other hand, Descartes's metaphysical subject, the soul. He maintains that both of these leaps made by Descartes are disastrous. They have taken people away from the intending subject and the intended object. What has been the result? While everything comes from the subject, while the subject is prior, still science has a 'real world' (in quotes) of protons and electrons and so on, and an utter incapacity for human science, *Geisteswissenschaft.* Scientific psychology is an attempt to understand the subject, which is the source of everything, in terms of the pseudo objects of this 'real scientific world,' and it is no wonder that human science has gone on the rocks.

So the first point is that the subject is prior, the second point is that what the subject is the ground of is intentional, and the third point is a reiteration of Descartes's *Cogito*, a more strict repetition in which there is no leap to a soul and no leap to a Galilean world of mechanist determinism.

8.4 Transcendental Phenomenology, Transcendental Psychology, and Transcendental Philosophy

In the fourth place, there is Husserl's identification of transcendental phenomenology, transcendental psychology, and transcendental philosophy. They are all one. It is a matter of studying the transcendental subject. This yields what is necessary in minimal acts of intending and what follows from it: he might call it a transcendental ontology, namely, the necessary characteristics of any object that is intended. He uses 'transcendental' in the Kantian sense, namely, moving to the conditions of the possibility of the 'I think' of knowledge.

In Husserl there are the two key notions of *epoché* and *transcendental reduction*. *Epoché* is a Greek term that means 'restraint,' 'holding back.' What Husserl means by the *epoché*, at least in his later works, is a withdrawal of interest, a withdrawal of concern. A man spends his day at the office thinking about his business, but he goes home to his wife and family and forgets the business at least for a while; it just drops out of his field of concern. Again, he is very much in love with his wife and children, but he goes to the office and he is entirely occupied by his business. What Husserl means by the *epoché* is the drop of all concern with the 'really real.' The desire to have a soul, a metaphysical entity as the soul of the subject, or the desire to have Galileo's world of mechanist determinism, that is, the 'really real' on the side of the object – Husserl is saying, 'Just withdraw all concern for that, and attend simply to the intending subject and the intended object.'

The second notion, his *transcendental reduction*, is just turning upside down mechanism and behaviorism. Mechanism reduces everything you see to little pellets, hard little knobs, as Chesterton described it, or something more refined in terms of more recent science. But what you see is not the real; the real is something else to which science reduces these appearances. Behaviorism reduces everything about consciousness to observable, external correlations, to stimulus and response, and so on. But instead of reducing the subject to objects, Husserl wants the opposite reduction, the reduction of objects to the subject. The subject is what is prior. The subject is what we are certain about. The subject is what we can examine with perfect accuracy and complete certitude, without any wild leaps. If you are going to talk about anything, if you are going to have science or common sense, you are going to have intentional acts, and they are what we know and what we can be certain of.

Finally, he says, through the *epoché* and the transcendental reduction secure for science and philosophy an immovable ground. Build them not on some flimsy ideal construction that consists of taking over popular words like *alêtheia, epistêmê,* and so on, as the Greeks did, and setting up an ideal where the whole thing floats on popular notions that are more or less accurate and more or less true but simply get science into a mess. Husserl says, Put the whole of science and philosophy on the solid foundation of transcendental phenomenology, which is the essential tool in the study of basic psychology, psychology reduced to its ultimate, and which provides a transcendental philosophy, a philosophy that grounds all possible philosophies.

9   Critique of Husserl's *Krisis*

Such is Husserl's program, and we will attempt right away a few of the more obvious criticisms. In the next hour, we will go on to his key notion of phenomenology, which has of course had a far richer life than Husserl's philosophic ideal.[10]

9.1 Human Science, Philosophy, and Theology

First of all, then, there is a real problem set by science and especially by human science, and its only solution lies in going to a philosophy. Natural science can get along pretty well by relying simply on the pragmatic criterion of success. We produce the results, and everybody can see the results. They can keep going on that basis. But even so, they suffer from a neglect of basic research, of fundamental thinking, simply because it is difficult to see the necessity of fundamental research when your criteria ultimately are pragmatic, the results that everyone can appreciate. But human science cannot get along on that basis. Human science is involved in philosophic

10 See below, chapter 12.

issues from the simple fact that the human scientist is one of his own objects. He cannot be totally detached in the science without special guidance. The history of the human sciences, where the element 'scientific' has been emphasized, has been a continual flight from what is truly human to what, in man, is not properly human. That flight has grounded a great deal of the success of the phenomenological movement.

So we can admit, in the main, Husserl's strictures on the situation of modern science. In other words, science has problems that it cannot solve, that can be solved only in terms of a philosophy. Further, not only does the problem of science raise philosophic questions. The problem of human science raises theological questions. We have today a situation that is essentially different from the medieval situation. St Thomas could produce the synthesis in a vertical line: theology, philosophy, science, where philosophy was the 'handmaid of theology,' *ancilla theologiae*, because in his time science was simply a department of philosophy. It was Aristotelian science and a part of Aristotelian philosophy. In subsequent periods, as long as science did not become strictly human, that mode of organization could be maintained. But at the present time, where we have empirical human sciences, where we have sciences dealing with men as they are, men under the influence of original sin, offered God's grace, and either accepting it or refusing it, there are theological issues involved in concrete human living, and consequently science, insofar as it includes human science, cannot be simply subsumed under philosophy, where philosophy prescinds entirely from theology. The problem of synthesis today is the problem of synthesis in the form of some sort of triangle, where philosophy, in our traditional sense, is related to theology and is related to science, but where science is not related simply to philosophy. A part of it, namely the human science, is in need of a direct contact with theology.

9.2 *Science, Necessity, and Certitude*

Secondly, while we admit Husserl's problem and, moreover, point out the necessity not only of philosophy but in certain respects also of theology, we have to disagree with Husserl in his pursuit of philosophy as rigorous science in the sense in which he understood this, namely, as grounded in necessity and yielding absolute certitude. This ideal has undoubtedly its Greek and its Cartesian antecedents, but it stands in need of distinction. All human judgments rest on a virtually unconditioned, as I argued last week.[11]

---

11 See above, pp. 72–77.

They are true as a matter of fact. The pursuit of absolute necessity, absolute certitude, is the pursuit of more than man can have, and consequently it is doomed to failure because it is overshooting the mark. I think that is the fundamental criticism of Husserl, but obviously it needs considerable development and we cannot attempt that here. We do not automatically go along with a philosopher simply because he is out for absolute necessity and absolute certitude. God is absolutely necessary, and God has absolute certitude without any conditions whatever. But we arrive at our certitudes and our knowledge of necessity insofar as, as a matter of fact, each one of us reaches in particular cases the virtually unconditioned. Our knowledge is based on the knowledge of a contingent world, and our knowing is a contingent event. To demand the absolute and to be content with absolutely nothing else results in a skepticism.

9.3 The Two Worlds and the Two Truths

The correlations of *Abschattung* and *Horizont*, of *Einstellung* and *Welt*, are, I believe, valuable contributions to cognitional analysis. Still, the alleged two worlds are but one set of beings considered from two standpoints. The one set of beings, as considered relatively to us in its relevance to human living, to the practical problems of man keeping alive and keeping the peace, is the world of common sense. The same set of beings, insofar as one seeks the universal laws of their interrelations, is the world of science. The two worlds are unified in the notion of being and distinguished by considering different relations among beings. The relations of all beings, insofar as they concern the practical problem of man's living, is the world of common sense, and the relations of beings to one another according to their natural laws is the world of science.

Again, the two truths are simply the result of applying the appropriate criteria to two cases of knowledge that is sought from different standpoints. If I seek to know this one set of beings insofar as it concerns me practically, there is a set of criteria I follow, and I arrive at certain truths that way. If I seek to know what is universally and necessarily so with regard to the relations between beings objectively, then I have to proceed in a different manner, and use different criteria and more elaborate methods. And so we arrive at a distinction between the truths of common sense and the truths of science. You can see how this problem of the two worlds in Husserl moves Heidegger on to the question of Being.

9.4 *Science and the* Lebenswelt

In the fourth place, de facto science does not rest on the evidence and procedures of commonsense living, of the *Lebenswelt*. There has been a failure to attempt the phenomenology of the scientist or of the phenomenologist. If there were a phenomenology of the scientist, certainly it would note that the scientist does not fit into the 'life scene,' the ideas of common sense on human living. Thales was an oddity, looking at the stars and tumbling into the well. He was not living in the *Lebenswelt*. Archimedes running naked through the streets shouting, 'I've got it!' was not an ordinary specimen of commonsense humanity. Newton, living absorbed in his problems for days in his room, having his meals brought to him and hardly pecking at them, was not an ordinary specimen of common sense. There is a specific pattern of consciousness, a specific pattern of experience, that characterizes the scientist at work. And so I do not think that Husserl is accurate when he says that the sciences rest simply on popular notions. What is true is that, when you start questioning scientists, they usually are not very good at philosophy, or the philosophy they have is not very good, and they will not be able to give a very good account of themselves. But as Einstein pointed out to epistemologists and theorists of science, 'Don't pay any attention to what the scientists say to you; watch what they do.'[12] If you ask them what they are doing and start asking them for their opinions, all they will do is trot out some third-rate theory of knowledge. But if you watch what they do, you will be getting at the facts of scientific inquiry, and that is something quite independent of the third-rate theory of knowledge that any particular scientist may happen to hold.

Now as far as I know – I have not read all of Husserl, but no one has, as a matter of fact, because a lot of it is still in this queer shorthand of his – Husserl has not done justice to that point. And as well, the subsequent concern with engaged consciousness, consciousness as orientated upon choosing, the flight from any type of intellectualist attitude as though it involved one necessarily in positivism or idealism (which are anathema to the contemporary European), has resulted in a neglect of that field due to a subsequent bias.

12 See Albert Einstein, *Essays in Science*, trans. Alan Harris (New York: Philosophical Library, 1934) 12.

9.5 *Normativeness of Human Intelligence and Reason*

Fifthly, Greek, Renaissance, and subsequent normative accounts of truth, science, and method are not just artificial ideals floating on popular obscurity. It is true that they are nonphilosophic, or inadequate philosophic expositions. But they really are expressions, clarifications, objectifications of the immanent normativeness of the human intellect, of our *participatio creata lucis increatae.* In other words, human intelligence and human reasonableness intrinsically involve norms. There are things that are intelligent and stupid, there are things that are reasonable and silly, and there is that ultimate normativeness in our intellect that comes to us from God, the *lumen intellectus nostri* that is a *participatio creata lucis increatae.* This aspect of human intelligence comes to light, not at all perfectly but to a notable extent, in Heidegger's notion of *Erschlossenheit,* openness, what the French call *ouverture:* the openness of mind, which is the orientation of mind to Being *in tota sua latitudine.*

9.6 *Priority of the Subject*

Finally, there *is* a real priority of the subject in knowledge. The human sensitive psyche is not the animal psyche. Köhler, in his experiment with the apes, discovered that even the most intelligent type of ape, the chimpanzee, does not have free images.[13] If you put an element of the solution of a problem to the ape within its field of vision, the ape will solve the problem. But if you put the element such that the ape can only see one element now and another element later, he cannot form a free image, even though the problem is essentially the same. In man, the free image is fundamental. Children are continually imagining and pretending, and that is something essentially human. To educate people you have to give them a formation in language and literature; otherwise they lack imaginations, and without imaginations they do not have a sensitive tool of sufficient suppleness and range to provide a basis for intellectual activity. This is the theoretical ground of classical or humanistic education.

Again our *participatio creata lucis increatae* is in fact the ground of questions and of all intellectual activity.

13 Wolfgang Köhler, *The Mentality of Apes,* trans. (from 2nd rev. ed.) Ella
   Winter (Harmondsworth, Middlesex: Penguin Books, 1925; Pelican Books,
   1957).

Still, though the subject has this priority in knowledge, we cannot follow Husserl in his demand for absolute necessity and absolute certitude. What we know is true as a matter of fact, and to demand more is to move towards an impossible ideal that backfires into skepticism. In Husserl's *epoché*, as we will see in the next hour, there is involved the confusion between what Santayana calls animal faith and, on the other hand, rational judgment.

And finally, Husserl's transcendental reduction to the subject is not ultimate: the ultimate reduction is of subject and object, scientific world and world of common sense, to being.[14] The subject *is*, and if he is, then he is among the beings.

So much, then, for a very rough outline of a powerful book by Husserl. I hope I have conveyed to you some idea of its sweep and radicalness, and some intimation also of the way Husserl worked all his life long in search of foundations for philosophy and science. The more obvious limitations I have indicated, but Husserl's great discovery was phenomenology, and in the next period we shall say what comes to mind regarding its nature, its significance, and its limitations.[15]

14 Note that Lonergan's distributed notes add, 'and not to "intending," which also is.' See above, p. 183.
15 The prominence given *Krisis* in Lonergan's study of Husserl raises the question whether he was acquainted more widely with Husserl's work. In the years when he was at Regis College, Toronto, Lonergan borrowed at least four books by Husserl from the College Library. Library records provide evidence for three of these; and for the fourth, *Ideas: General Introduction to Pure Phenomenology*, trans. W.R. Boyce Gibson (London: Allen & Unwin, and New York: Macmillan, 1931), an earlier translation of a work mentioned in note 5 above, there is the evidence of a scrap of paper used as a bookmark and still there with Lonergan's handwritten references to pages in Husserl.

12

Phenomenology: Nature, Significance, Limitations[1]

1 The Nature of Phenomenology

It will be more helpful to start from the question of the nature of phenomenology. What is phenomenology? It is an account, description, presentation of data structured by insight. That is my own definition of it: an account, description, presentation of data structured by insight.

1.1 'Of Data'

It is 'of data,' of what is given, what is manifest, what appears, phenomena. It is not just of external data, external phenomena, but also of inner data. That is the basis of its opposition to experimental psychology as commonly understood, to all behaviorism, to all types of mechanism. It includes the internal among the data, the phenomena, what is manifest, what is given. But it is also not exclusively of internal data. The inner intentional act terminates at the outer datum, and the outer datum is just the term of the inner intentional act. The subject is nothing apart from the intended term, and the intended term is nothing apart from the intending subject. The synthesis of both is found in the intentional act.

Nothing is excluded from consideration. Phenomenology is not a matter of considering primitive data as opposed to derived, natural as opposed to cultural, sensitive as opposed to intellectual, cognitional as opposed to

---

1 The second half of the seventh lecture, Tuesday, 16 July 1957.

emotional or conative. It is concerned with everything that appears, everything that is given, everything that is manifest.

1.2 Data Structured by Insight

However, it is of data structured by insight. In other words, it is selective. It does not offer an exhaustive description of all and any data whatever. There is a structure to it, a selection, and the selection is of the significant. It seeks basic, universal structures. Husserl spoke of the *eidetic* (from *eidos*, form), the structure in the data. He also spoke about *Wesensschau*, an intuition of essence. I bring in the word 'insight' because what he is talking about seems to me to be quite parallel to the distinction in Aristotle's *Metaphysics* in book VII, about chapter 10, between parts of the matter and parts of the form.[2] With regard to a circle, for instance, the form is the necessary roundness of the circumference, resulting from the equality of all radii, where radii are multiplied to infinity. If all the radii are absolutely equal, you see that the curve has to be perfectly round. And if any one is not equal to any of the others, you see that this curve, because of the inequality of any one radius, involves a bump or a dent. That is something manifest, something presented, something that appears, and it is structured by the insight. Parts of the matter, on the other hand, are, for instance, the fact that the circle is white on black, that the drawing is in chalk, or that it is in almost a vertical plane, or that it is just this size and no bigger or smaller. But the parts of the form are the perimeter, the equal radii, and the center. You understand why the circle has to be round because of the inner ground of that roundness in the circle. Among the multiplicity of data, some are merely casual and are called parts of the matter, but there also are parts of the form: the center, the equal radii, and the circumference. That is a case of data structured by an insight. Husserl and phenomenology are concerned with considering all data without any exclusion, as structured by insight, as given a form, an *eidos*, from the insight.

Now to present data structured by insight, Husserl does not proceed on the basis of the first bright idea that comes along. To find what really is the proper structure of the data takes effort and time; it calls for scrutiny, penetration, contrasts, and tests. It may be necessary to overcome spontane-

---

2 Aristotle, *Metaphysics*, VII, 10. See Bernard Lonergan, *Verbum: Word and Idea in Aquinas* (see above, p. 13, note 14), chapter 4, section 1. See also *Topics in Education* (see above, p. 61, note 38) 113–14.

ous tendentiousness, systematic oversights, common over-simplifications, preconceptions arising from a scientific outlook or a philosophic position or any other source. In other words, we are not accounting for data structured by the first insight that comes along but for data structured by an ultimate insight that hits things off and meets the issues.

1.3 Not Insight as Such

On the other hand, phenomenology is not concerned with insight as such. Insight as such is something extremely elusive. If the phenomenologist had hold of the insight itself, the act of intelligence by which you grasp the necessity of roundness when the radii are equal and the impossibility of roundness when they are unequal, he would immediately be led to unity, to a unification of insights into a science, to the movement of the sciences from lower to higher viewpoints, to the integration of the sciences, and to the integration of science and common sense in philosophy. A study of insight leads immediately to a synthetic position, as is illustrated in the book called *Insight*. And there is not that kind of tendency to unity in phenomenology.[3] Husserl spent his life perpetually discovering new fields of possible investigations. He would investigate something and define the issue more and more closely. He would set aside other fields, and the fields of possible investigation in which data might be structured by insight kept multiplying. He kept filling in more and more pages of notes in shorthand.[4]

Again, there is no tendency to unity among his successors either; they do brilliant work in particular limited fields, but phenomenology does not head towards a synthesis, towards a unification.

1.4 Data, Not Concepts

Finally, it is the data as structured by insight that are the objects of phenomenology, not the subsequent conceptualization or definition or theoretic statement of the data in their essential features. What you have to attend to in the circle, the data as structured by an insight, consists in *these* radii or other particular radii that you imagine, and similarly *this* circumference or another particular circumference that you imagine: not the concept of

---

3 See below, pp. 356–57, where Lonergan illuminates this point during the question sessions.
4 Lonergan added in an aside, 'They are up to forty-thousand or some such figure in his archives at Louvain.'

radius, of which there is only one, or the concept of center, of which there is only one. The basis on which you grasp this necessity is only in the imagined multiplicity. You need an infinity of radii to be able to get the insight: only if every possible radius is absolutely equal to every other one do you get this necessary roundness.

There is a sharp distinction, then, in phenomenology between what appears as structured by proper insights and, on the other hand, the thematic treatment, the phenomenological exposition, of the data as structured by the insight. What is manifest is one thing, and on the other hand what the phenomenologist says is quite something else. Just as when you grasp this necessary roundness in the data you define the circle as the locus of coplanar points equidistant from the center, so the phenomenologist, considering what is manifest, what is given, what appears, as intelligently structured, distinguishes that sharply from his thematic treatment of the data. He is not concerned with his own statement about it; that is just his report. His report is one thing, and what he reports on is another.

Consequently, there is in phenomenology a terrific emphasis on what is called the pre-predicative, that is, what is known before you conceptualize, before you formulate any theory, before you make any judgment or any statement. It is the pre-predicative manifest, what is manifest pre-predicatively, pre-theoretically, pre-judicially, what is there, what is given, that the phenomenologist is concerned with. He distinguishes that sharply from what he calls the thematic treatment, exposition, presentation, which is his writing, his report, his observations. The etymology of the word 'phenomenology' speaks of phenomena and *legein*: to read off the phenomena. The phenomena are what are manifest, and reading them off, *legein*, is what the phenomenologist does. I make my own statement in terms of insight because I think it makes the matter very clear, and I have provided a broader context for it. But they do not speak of insight, although they are sufficiently aware of the fact.

2   The Significance of Phenomenology

2.1 Psychological Explorations

Secondly, we treat the significance of phenomenology. It has more or less swept the field in a variety of ways, and its first significance is that it provides a technique for the exploration and presentation of whole realms of matters of fact that are important but that have been neglected or treated superficially. In a psychology, for example, that calls itself scientific, there is a bias in favor of outer data, in favor of what can be measured, in favor of events that can be counted. Phenomenology, as contrasted with scientific psychology in that sense, opens up new vistas and possibilities in a manner that is comparable to Freud's discovery of significance in dreams, and far broader in its scope and implications. So phenomenology appears as a break with scientific tendencies in psychology.

It also appears as a break with some older, more traditional psychologies. When compared to the results of phenomenological research, we find in traditional psychology as represented, for example, by William James or by the Scholastics either rough and ready statements, on the one hand, or on the other hand, when precision is attempted, a tendency to bog down in a set of indefinable 'somethings.' For example, when you start talking about consciousness, just precisely what are you talking about? How do you pin it all down? When you want to draw precise distinctions and get things accurately, there is needed a technique, and phenomenology provides such a technique.

This may be illustrated by Husserl's distinction between *Abschattung* and *Horizont*, but it is found also in entirely different fields and in writers that have no philosophic connections either with Husserl or pretty well anyone else. For example, there is a very short little book written by Buytendijk, on the phenomenology of a meeting. The French translation, which is the only thing I've seen, published by Desclée in 1952, is entitled *Phénoménologie de la rencontre.*[5] The book presents a description of just what is involved psychologically when one person meets another, what appears, what is manifest in a meeting when you really understand what it is for two people to meet. The description is brilliant, and it presents us with the human event, the meeting, in a way that otherwise one could not arrive at. Buytendijk wrote an earlier work, also published in French by Desclée in the same year, I think, or perhaps earlier: *La femme*, on woman's attitudes.[6] Whether his earlier work on the essence and meaning of play, *Wesen und Sinn des Spiels* (Berlin, 1933), is of this type or not I do not know; I imagine it is but I have not seen it.[7] At any rate, he is a representative of the use of phenomenology as a technique in psychological study.

---

5 F.J.J. Buytendijk, *Phénoménologie de la rencontre,* trans. into French by Jean Knapp (Bruges: Desclée de Brouwer, 1952).

6 F.J.J. Buytendijk, *La femme* (Paris: Desclée de Brouwer, 1952); in English, *Woman: A Contemporary View* (Glen Rock, NJ: Newman Press, 1968).

7 F.J.J. Buytendijk, *Das Wesen und Sinn des Spiels: Das Spielen des Menschen und der Tiere als Erscheinungsform der Lebenstreibe* (Berlin: K. Wolff, 1933).

Again, Stephan Strasser has a recent book called *Das Gemüt*, which has been very highly praised in a review in *The Philosophical Review*, a publication from Cornell.[8] It is a study of the emotions. He also has an earlier work, the French translation of which is called *Le problème de l'âme*.[9] It is a study of the respective objects of metaphysical psychology and empirical psychology, a study of how you go about stating just what you mean by the soul and how you investigate the soul. The approach again is phenomenological; Strasser is not a member of some particular philosophic school.

Next, there is the work of Maurice Merleau-Ponty. He is a leader in the existentialist movement and a professor at the Sorbonne. In 1942 he wrote his *Structure of Behavior, La structure du comportement*,[10] and in 1945 *La phénoménologie de la perception*, published in Paris by Gallimard.[11] He is brilliant, in his account of perception, on the significance of one's own body in one's perceiving. He is, of course, engaged in an attack upon Sartre's sharp distinction between the *pour-soi*, the conscious, and the *en-soi*, the mere dead thing. The body is both *pour-soi* and *en-soi* at the same time, and it has to be both at the same time; you cannot account for perception without one's consciousness of one's body. In other words, the perceiving subject is spatiotemporal; we have a feeling of space and time, so to speak, in our bodies. This is what Marcel would call *incarnation*, the incarnate subject: not just the idealist subject or any merely observing subject, but the

8 Stephan Strasser, *Das Gemüt: Grundgedanken zu einer phänomenologischen Philosophie und Theorie des menschlichen Gefühlslebens* (Utrecht, Uitgeverij Het Spectrum, and Freiburg im Breisgau: Herder, 1956); in English, *Phenomenology of Feeling: An Essay on the Phenomena of the Heart*, trans. Robert E. Wood (Pittsburgh: Duquesne University Press, 1977). The review to which Lonergan refers was written by John Wild, *The Philosophical Review* 66:3 (1957) 428–30.

9 Stephan Strasser, *Le problème de l'âme: Études sur l'objet respectif de la psychologie métaphysique et de la psychologie empirique*, trans. into French by Jean-Paul Wurtz (Paris: Desclée de Brouwer, 1953). The work appeared originally in Dutch: *Het zielsbegrip in de metaphysische en in de empirische psychologie*. Strasser revised it when he translated it himself into German: *Selle und Beseeltes: Untersuchungen zum Verhältnis der empirischen zur metaphysischen psychologie*. The French translation is of the German edition. There is also an English translation: *The Soul in Metaphysical and Empirical Psychology*, trans. Henry J. Koren (Pittsburgh: Duquesne University, 1957).

10 Maurice Merleau-Ponty, *La structure du comportement* (Paris: Presses Universitaires de France, 1942, 1960); in English, *The Structure of Behavior*, trans. Alden Fisher (Boston: Beacon Press, 1963).

11 Maurice Merleau-Ponty, *Phénoménologie de la perception* (Paris: Gallimard, 1945); in English, *Phenomenology of Perception*, trans. Colin Smith (London: Routledge & Kegan Paul, 1966, 1989).

incarnate subject. The subject has to have a body to perceive. The body enters right into consciousness. Merleau-Ponty's treatment varies time and again, and very convincingly. He has no philosophic commitments, in the sense that he is not tied down to some particular philosophic school. His work offers very useful material, I believe. I am not at all an expert in his views, but I think that anyone interested in psychology would find Merleau-Ponty extremely stimulating and probably very helpful. Again, he presents the incarnate subject, the subject as the subject of feelings in a body, the body of the subject. You cannot have either the body or the perceiving subject as intelligible without bringing the other in; you have to have the body to understand the subject, and the subject to understand the body. What is the human body? It is the incarnation of meaning, of a principle of meaning. And of course this ties in with the old-time axiom that a person by the age of thirty is responsible for his own face.

There has been a study of Merleau-Ponty by de Waelhens in 1951 at Louvain, *Une philosophie de l'ambiguïté* (*A Philosophy of Ambiguity*).[12] It is a study of Merleau-Ponty's existentialism, but in it you get a good deal of his psychology.

So one part of the significance of phenomenology is that it has provided a tool that seems extremely fruitful and that appears in various types of investigations of a psychological character.

2.2 A Powerful Instrument for Philosophical Psychology and Philosophy

The second type of significance of phenomenology is its ranging out into different fields. First of all, there is of course its philosophic significance in Husserl's own work. But to illustrate this aspect of its significance let us present something of a preview of Heidegger.[13]

Just as the phenomenologist finds structured meanings in what is manifest, so for Heidegger a man is a source of meanings. He uses the term *Dasein* to eliminate the subject-object opposition. We might say in English that the existing subject is the origin of the meanings that correspond to

12 Alphonse de Waelhens, *Une philosophie de l'ambiguïté: L'existentialisme de Maurice Merleau-Ponty* (Louvain: Publications Universitaires de Louvain, 1951, 1968).

13 Lonergan was to go into greater detail on Heidegger in the following lecture. Unfortunately, however, most of the tape of that lecture has not survived, so that his detailed presentation is lost; we must rely on his distributed notes (see above, pp. 187–93).

the understanding of the phenomenologist in the manifest data. And just as the phenomenologist has to refrain from taking on face value the first bright ideas that come along, just as he has to scrutinize and dig and feel around and penetrate and eliminate oversights and oversimplifications, so man in his living first of all lives inauthentically, and if he begins to penetrate things a little better he may move on to authentic living. In other words, Heidegger's existentialism (if one may call it that) is parallel to phenomenological analysis. Not only is what Heidegger is doing a phenomenology of the existing man but the man is structured after the fashion of phenomenology itself. Just as the phenomenology is the meaning in data, so man is the source of meanings in a living. Just as man may be a source of meanings in a superficial manner that results in inauthentic living, so the phenomenologist may be misled by superficial ideas. And just as the phenomenologist may penetrate to the real meaning of the phenomena and read them off rightly, so the man may move from inauthentic to authentic living. In other words, there is a profound influence of phenomenology on Heidegger, not only from the viewpoint of method but also from the viewpoint of content.

Now Heidegger's existentialism was followed more or less immediately by Ludwig Binswanger's new interpretation of the dream. Freudian interpretation of the dream had been a matter of taking the dream symbols and moving up immediately to a conceptual level of interpretation; Freudian interpretation of what the dream really means is up on the conceptual level. But Binswanger considers the dreamer as the existential subject, and he interprets the dream on the level of the dream. At least that is his ideal. He has writtten a very short essay entitled 'Traum und Existenz' ('Dream and Existence'). It has been translated into French,[14] and the value of that French translation is that there is also an introduction by Michel Foucault of about 125 pages pulling out all the implications and background and significance of this thirty-page essay that had been written about twenty years previously. Binswanger distinguishes between dreams of the night and dreams of the morning. It is a fairly old distinction. Dreams of the night are

14 Ludwig Binswanger, 'Traum und Existenz,' in Binswanger, *Ausgewählte Vorträge und Aufsätze* (Bern: A. Francke, 1947), vol. 1, pp. 74–97; in French, *Le rêve et l'existence*, with introduction and notes by Michel Foucault (Paris: Desclée, 1954); in English, 'Dream and Existence,' in *Being-in-the-World: Selected Papers of Ludwig Binswanger*, trans. Jacob Needleman (New York: Harper Torchbooks, 1963) 222–48.

mainly under somatic influences. Dreams of the morning are the existential subject beginning to create his world.[15]

Then there is Heidegger's influence on Bultmann. 'The objective' for Bultmann is either science or myth. Christianity is not science, and therefore what is objective in Christianity has to be just myth. What the interpreter of the New Testament has to do is find the existential elements in Christianity. These existential elements are faith, and faith is Christian understanding of Being; the rest is myth.

As you know, of course, this has led to a terrific amount of discussion. There has been published by H.W. Bartsch a series of five volumes under the title of *Kerygma und Mythos, Kerygma and Myth*, from 1948 to 1955. There is also a supplemental volume to numbers 1 and 2, in which are collected together discussions of Bultman's interpretation of the New Testament.[16] From a Catholic viewpoint, René Marlé of Louvain, a Jesuit, has written a book on Bultman and the interpretation of the New Testament. It was published in Paris by Aubier in 1956, in the collection of our Jesuit Fathers of Lyons, *Théologie*, volume 33.[17] Marlé provides us with a very thorough exposition of Bultmann.

So much, then, for the significance of phenomenology: it is a new approach in psychology that has yielded very rich results; there is the profound influence of Husserl's philosophy on Heidegger, and through Heidegger on others, including Sartre; there is the work of Merleau-Ponty; the influence goes into depth psychology and into scriptural interpretation. It ranges pretty well over the map.

3   The Limitations of Phenomenology

We have treated first, then, the nature of phenomenology, and second its significance. Now in the third place we shall discuss its limitations.

---

15   There is a gap on the tape here. Lonergan's distributed notes (see above, p. 185) indicate that the gap is quite short, but when the tape resumes Lonergan is talking about Heidegger's influence not on Binswanger but on Bultmann. The material in the next paragraph down to 'and therefore' is an editorial construction, relying on the notes that Lonergan had distributed. The recordings and previous typescripts have to be reordered at this point, by comparing fragments on the tapes to the outline as given above in chapter 7.

16   H.W. Bartsch, *Kerygma und Mythos*, 5 vols. along with *Beiheft* to vols. 1 and 2 (Hamburg: Herbert Reich, 1948–55).

17   René Marlé, *Bultmann et l'interprétation du Nouveau Testament* (Paris: Aubier, 1956).

The first limitation is that just as phenomenology is essentially prepredicative, so it is essentially preconceptual, and prerational. Phenomenologists talk about what is there; and if that is all there is to knowledge, then they tell you all there is. But phenomenology, by its existence, is a challenge to Scholastics. Scholastics can offer a foundation of their position that is very easy to say but is not adequate. Once phenomenology arises, they cannot remain in that position; they have to seek adequate foundations and a more adequate statement of their real foundations. Otherwise they will be swept right along with the phenomenologists. That, of course, to some extent is a difficulty in Europe at the present time.

In other words, the phenomenologist provides himself and the reader with the evidence in which they can grasp the virtually unconditioned, as I would put it, and so reach the absolute on which the judgment is based. The evidence is set out. But the phenomenologist has not penetrated to the judgment itself, to the rational process within which grasp of the virtually unconditioned and the judgment occur. Thus phenomenology fails to give due weight in its psychology and in its philosophy to rationality, to affirmation, to being as known through affirmation.

That is the fundamental point. Phenomenology is concerned with what is evident. The thematic treatment of what is evident is considered secondary, merely the phenomenologist's report. But there has not been done a phenomenology of rational consciousness, of the process of asking, Is it so? and grasping the virtually unconditioned and on the basis of that absolute making your affirmation and taking your affirmation as the *medium in quo cognoscitur ens.* And that to my mind is the real basis of Scholastic thought.

Hence, phenomenology's criterion of the true is the manifest, the evident, what becomes manifest, evident, when one lets the phenomena appear, when one is not hiding the phenomena, running away from them, brushing them aside, when one is not living the life of the escapist. That is for phenomenology the criterion of the true: what is manifest. But just as affirmation rests upon the manifest, what is given, so negation also has to have its basis, and its basis is what is not, its basis is nothing.[18] As there is the manifestness of what is given so also there has to be the manifestness of nothing, and so we find in Heidegger and Sartre the insistence upon nothing. Just as there are affirmative judgments, there are negative judgments. Just as affirmative judgments reduce to positive manifestness, so negative judgment has to reduce to a manifest nothing.

18 Lonergan's distributed notes are clearer than his spoken word here. See above, p. 186.

They find the manifestness of nothing in the anxiety crisis. What is an anxiety crisis? Ordinary conscious living is a patterned flow of consciousness in which your world, yourself, and what you do have meaning. That patterned flow of consciousness can break down, and when it breaks down you have an anxiety crisis. In the anxiety crisis what you are presented with is brute reality. It is there, and there is no meaning to it. It is just as solid and hard as before but it is meaningless. The source of meaning is the flow of consciousness, the structured flow of consciousness, and that has broken down. There are descriptions of this in Sartre and to some extent also in Heidegger. But that anxiety crisis is the manifestation of nothing. It is the negation of intelligibility. It is Heidegger's *das Seiendes*, being in the sense of *what is*, as contrasted with *Sein*, which is the intelligibility given to what is presented, given to brute reality by the intelligence of man giving a meaning to life.[19]

So the first result of making your criterion of truth the manifest is that you have to have a manifestness of nothing to account for negations; and they find that manifestness of nothing in the anxiety crisis.

The second result is the possibility of Husserl's *epoché*. He says, 'Withdraw from interest in, concern with, the really real. Just attend to the intending subject and the intended terms of meaning.' Just withdraw from questions about the really real, what is really there, in the same way that one forgets about his business when he goes home or forgets about his home when he is occupied with his affairs. What is the possibility of that *epoché*? It is possible if the really real is just a term of attention, something that could be manifest if you attended to it. In that sense the *epoché* is quite possible. I can gaze upon you or upon a landscape and attend simply to what is presented, without bothering my head about whether the objects of my attention really exist or not. It makes no difference to my apprehending you here before me whether I have a veridical perception or a very lively hallucination; the presentation could be the same in both cases. In that sense I withdraw from consideration of the really real. But there is a radical difference between a spontaneous orientation upon the really real – what Santayana calls 'animal faith' in his book *Scepticism and Animal Faith*[20] – and, on the other hand, what is posited absolutely in judgment. True judgment is the *medium in quo*

---

19 Lonergan added, 'What kind of a meaning we will see tomorrow, when we discuss Heidegger.'
20 George Santayana, *Scepticism and Animal Faith: Introduction to a System of Philosophy* (New York: Charles Scribner's Sons, 1923; reprinted, New York: Dover Publications, 1955).

being is known. You have to distinguish between those two. If you do not, then you will not be uncovering the ambiguity in Husserl's *epoché*. It is merely a psychological feat not to yield to spontaneous extroversion upon what is called the 'really real' (within quotes), the term of 'animal faith.' It is quite a different matter to suspend one's judgment. One cannot play fast and loose with what one knows to be true. One cannot turn rationality off and on as one pleases. They are two entirely different things, and there is an ambiguity in Husserl's thinking, and also in the others who do not use the *epoché* to the same extent, but in whose work the effects appear.

From that *epoché* and its ambiguity there follows another point. H.L. van Breda held a meeting of outstanding phenomenologists from Germany, France, Belgium, and perhaps Holland. The meeting was held at Brussels in 1951, and the papers were published by Desclée in 1952, with the title, *Problèmes actuels de la phénoménologie.*[21] One of the speakers, H.J. Pos, points out that there is no return from the *epoché*. If by an intentional act I suspend the really real, what kind of an intentional act is going to restore the really real? In other words, in my perceptions I have dropped the really real from them, and the perceptions keep on coming. If I make another act to say that they are real, what difference does that make to my perceiving? How could it make any difference to my perceiving? So behind this ambiguity of the *epoché* there arises the problem set out in some detail by Pos in his paper in that meeting. Just as there is an ambiguity to the *epoché* itself, so there is an ambiguity to the return from the *epoché*. Through the double ambiguity you can arrive at Pos's view that while you can practice the *epoché* you cannot get out of it afterwards. You have suspended the really real, and no amount of meaning is going to give it to you again.

There is a more fundamental point in the paper by Eugen Fink, who was one of the faithful pupils of Husserl. He discusses the possibility of a phenomenological treatment of speculative issues, and answers in the negative. His point is that within phenomenology being is what appears, and nothing else. But there is no possible proof within phenomenology that that is what being is. For the phenomenologist, what is is what he can point to, *das Ausweisbare*. But that being is all that appears and nothing but what

---

21 *Problèmes actuels de la phénoménologie: Actes de Colloque internationale de phéno-*
   ménologie, Bruxelles, 1951, ed. H.L. van Breda (Paris: Desclée de Brouwer,
   1952). The papers to which Lonergan refers are: H.J. Pos, 'Valeur et limites
   de la phénoménologie,' pp. 31–52, and Eugen Fink, 'L'analyse intentionelle
   et le problème de la pensée speculative,' pp. 53–85. Fink's paper is presented in both French and German.

appears is not a question that any amount of pointing out, any amount of indicating, any amount of manifesting, is going to settle. In other words, phenomenology does not contain within itself the necessary implements to handle speculative questions. Insofar as you are in phenomenology you do not have the necessary tools for dealing with speculative questions. This is just another aspect of the fundamental point that phenomenology, as it deals with what is manifest pre-predicatively, also is by that very fact pre-rational. It is on the rational level of truth, absolute affirmation, that we know the real really.

Finally, in Heidegger, to a less extent in Sartre, but really in the whole movement, truth arises as the fundamental problem. Heidegger, in his account of concrete human living, distinguishes between being in the truth and being in untruth, as he distinguishes between authentic living and inauthentic living, genuine and nongenuine being a man. The genuine man, you might say, is being in the truth, but that is a further point.

In other words, he knows about the remote criteria of truth. Besides the proximate criterion of truth that you have insofar as you reach the virtually unconditioned, there are remote criteria that regard the way you go about investigations and inquiries, the attitude of mind you should have in an investigation. Heidegger knows about those remote criteria, and he speaks of them as being in the truth and being in untruth. But the existentialists are constantly writing about truth and cannot handle the issue, and the reason is that phenomenology is an inadequate method.

Again, Fr Lotz – he was a pupil of Heidegger and now teaches half a year at the Gregorian – in an article that was translated into French and published in *Archives de philosophie* a couple of years ago, states that on Heidegger's position it is not possible to prove the existence of God because of the method on which the position rests.[22] In other words, Heidegger's method of doing philosophy is phenomenology, and as long as he retains that method there is no possibility of arriving at a proof for the existence of God.

Again, Merleau-Ponty, is busy writing a book on the origin of truth.[23] As far as I know it hasn't appeared yet. But the people in France and Louvain know that the big question raised by phenomenology is the question of

22 J.B. Lotz, 'Heidegger et l'être,' *Archives de philosophie* 19:2 (1956) 3–23.
23 The work to which Lonergan refers was published as *Le visible et l'invisible* (Paris: Gallimard, 1964); in English, *The Visible and the Invisible*, trans. Alphonso Lingis (Evanston: Northwestern, 1968); see the editorial note there, p. xxxiv.

truth. Alphonse de Waelhens gave a series of lectures at the Sorbonne as visiting lecturer, published in 1953 under the title *Phénoménologie et vérité* by Presses Universitaires de France.[24] In his preface he says that he limits himself to discussing the question in Husserl and Heidegger and does not put the question further because it is the main preoccupation of the professors of the students to whom he is lecturing. He does not feel it is the position of a visiting lecturer to step in and tell where these professors should be going or heading. But that is what they are working on.

So this is the fundamental limitation to phenomenology. While phenomenology has got hold of something that is of great importance and has proved very fruitful in a variety of ways, since it is dealing with the manifest as structured by insight, and since it provides the evidence for judgment, still it has no reflective account of judgment, and until it gets it, it is not on the level of *verum, ens, bonum* in any Scholastic sense. Consequently we have to be aware of that fundamental difference and not feel that just because these people are talking about existence in a way that seems extremely acceptable to Scholastics because of its extreme difference from positivism and idealism and other non-Scholastic philosophies, therefore they are all with us, so that we can just join hands and all be happy. They are on an essentially different plane. What they mean by being is going to be quite different from what we mean by being. The basis they provide us with is not a basis on which you can prove the existence of God, nor is it a basis upon which you can do dogmatic theology.

24 Alphonse de Waelhens, *Phénoménologie et vérité: Essai sur l'évolution de l'idée de la vérité chez Husserl et Heidegger* (Paris: Presses Universitaires de France, 1953).

13

Subject and Horizon[1]

1   The Horizon in Human Science, Philosophy, and Theology

In mathematics and natural science the subject is not part of the object. When a physicist pronounces upon the electron, he is not saying anything that directly involves himself: he is not an electron. But when the human scientist says something about men, he says something that implicitly regards himself. Similarly, insofar as a philosophy or a theology has human implications, when the philosopher or theologian speaks about his object he is also speaking about himself.

It follows that a broadening of the horizon in a field that includes man is not merely a matter of new concepts, new principles, new methods, new

---

1 Unfortunately, there does not seem to exist a recording of most of what Lonergan had to say on Heidegger, which certainly occupied him for a good part of the eighth lecture (Wednesday, 17 July 1957). Nor is there a recording of the first part of the material on the subject, some of which was given in the eighth lecture and some in the ninth (Thursday, 18 July 1957). Only Lonergan's opening words about Heidegger are still on a recording, and they appear in this book as appendix D. On Heidegger, then, see the lecture notes, above, chapter 7, pp. 187–93. On the 'dilemma of the subject' as well as a large portion of the material on 'subject and horizon,' see above, chapter 8, pp. 195–200. The current chapter presents the end of the first half, and all of the second half, of the ninth lecture. The text presented here corresponds to the material that appears on the outline starting on p. 201, at III ('The Horizon in Human Science, Philosophy, Theology'), §3 ('The difference ...').

topics, new approaches. Just as there is a revolution on the side of the conceptualization, so there has to be a revolution within the subject himself. Just as the advance in the object of a science involves an entirely new approach, a new slant on everything, a new perspective, so if that perspective is to be realized in the subject that is thinking the philosophy, there has to be a conversion within him. He has to reorganize his living as well as the concepts that he uses in talking about his object. Otherwise he is not keeping pace with the broadening of the horizon.

Briefly, then, if there is a development, a new higher viewpoint, on the side of the objects, and if the subject is one of those objects, then the subject has to have a new concept of himself, new principles to guide his thinking, judging, evaluating, new principles guiding everything that concerns him. Such a change is a conversion in the subject. Without that conversion in the subject running concomitantly with the broadening of the horizon, the new ideas not only are inoperative in one's own living – it is clear that the conversion is needed for them to be operative in one's living – but also they are insignificant to oneself. They have no effective meaning to one; they have no vital expansiveness even in the domain of objects.

In other words, the thinker has to be on the level of what he thinks about, and when his thought moves to a higher level, he himself has to move to that higher level. Otherwise his thought will not really be on the level; he may repeat the words of the master, but he cannot give the words the meaning they had in the utterances of the master. That is why it is that original thinkers can found only a school. There are going to be only a limited number that will not merely seek a new set of concepts and a new approach but also have a revolution, a development, in themselves. The school splinters because those that follow him effect in themselves a conversion in unequal measures. The school can also decay: you can have a decadent Scholasticism as you can have a decadent Platonism or a decadent Aristotelianism or a decadent Kantianism, because the successors do not produce in themselves a conversion that is on the level of the conversion of the original thinker.

2   The Existential Gap

This point can be summarized in a phrase: the existential gap. The existential gap consists in the fact that the reality of the subject lies beyond his own horizon. We have knowledge; we have *docta ignorantia* limited by the horizon;[2] and if the field of *entia* goes out beyond that, it is possible that the reality of the subject or part of the reality of the subject lies beyond the horizon; he does not really know himself. It is insofar as the subject does not really know himself that we have the fundamental problems in philosophy, and the fundamental problem of incommunicability. Insofar as the subjects are beyond their own horizon, you cannot get at them; they have not got at themselves, and it is through their getting hold of themselves that you can get at them.

To say, then, that the reality of the subject lies in whole or in part beyond his own horizon is to say that he suffers from an *indocta ignorantia* with regard to himself. This *indocta ignorantia* with regard to himself is not any matter of biochemistry or synapses or any of the subtleties that he could well be excused from knowing. It regards his knowledge of his own intelligence, his own freedom, his own responsibility. For those are the fundamental things in human science, philosophy, and theology insofar as theology is affected by philosophy.

I gave you already the example of Hume.[3] The knowledge that Hume describes is one thing, but the knowledge that Hume uses in setting up that description is something entirely different. Hume as intelligent and reasonable is quite different from Hume as known by himself. Hume partly lies beyond his own horizon. He is intelligent and reasonable, but in his description of knowledge as a series of impressions linked by custom there is no room for either his intelligence or his reasonableness. And so on, through the whole series of philosophic differences. The existential gap, then, is the gap between what is overt in what one is and what is covert in what one is.

Again, the existential gap is not eliminated by affirming the propositions that are true and denying the propositions that are false. A man can be brought up in a given philosophy. For example, he can be taught Thomistic philosophy. He can repeat all the theses that are considered essential to Thomistic philosophy. And yet it can all be an expression of something that is just beyond him. He is not on the level of that thinking, and it will be in

2 It is clear from the notes that in his opening remarks explaining the notion of horizon Lonergan appealed to the notions of the known unknown and the unknown unknown and to the related notions of *docta ignorantia* and *indocta ignorantia*. See above, p. 198. That explanation of horizon is repeated in the summary that Lonergan provides in the paragraph beginning 'Briefly' below, p. 283.

3 See above, p. 109.

that measure meaningless to him, insignificant to him, not worth while to him. He can repeat the words and not have the meaning, because his horizon is not broad enough to give a meaning to the theses that he can repeat.

Again, if he does not know the philosophy and you persuade him to repeat the words and give him some grasp of the meaning, you are not making a Thomist out of him. He has to do that for himself. He has to broaden his own horizon, and that is something that occurs slowly.

I think you will see that this idea of horizon is an idea of great philosophic significance, that it represents a concern with the transition from the per se to the concrete subject that exists, that it is concerned with a transition from the nonhistorical to the historical, and that it involves a study of notions that are very conspicuous in existentialism. It is connected with notions such as immediacy: what is consciousness? what are you as prior to what you say about yourself? It is connected also with notions such as obnubilation and discovery. Insofar as you are beyond your own horizon, insofar as your reality is beyond your own horizon, your own reality is hidden from you, and if it is hidden from you, that is not entirely without any fault on your part (using 'fault' in the broadest sense possible). And insofar as there is this obnubilation, there is a need for dis-covery (with a hyphen between the 'dis' and the 'covery' – taking the cover off; this type of emphasis is very much in the line of Heidegger). There is a normative element to it: one ought not to be beyond one's own horizon. If one is a philosopher, one's knowledge ought to measure up to one's reality. If there are norms, there are also freedom and responsibility. And so we move right into the circle of ideas that are very prominent in existentialism.[4]

Briefly then, what we have attempted to say is that there is a distinction between the questions we raise and answer, the questions we can raise but cannot answer, and the questions that we neither raise nor answer. The first defines our knowledge, the second our *docta ignorantia*, and the third our *indocta ignorantia*. The horizon is the limit between *docta* and *indocta ignorantia*. That horizon has a significance in the natural sciences, but also and much more pronouncedly in anything that involves man, such as the human sciences, philosophy, and theology. Because in the case where man is involved the subject is one of the objects, a higher viewpoint on the object

---

4 In the notes distributed for the lectures, Lonergan included 'transcendent' and 'Existenz' among the 'notions that are very conspicuous in existentialism.' See above, p. 203.

demands a conversion in the subject. Without that conversion running parallel to the higher viewpoint on the object, you can repeat the words but you will not have their meaning. For instance, you can agree with Augustine that the real is the true, but you will not be saying that and meaning what Augustine meant when he said it. Jaspers says that the philosophers are our educators, and he means all the philosophers. How are they our educators? By our reading them and attending to the places where they are obscure. Insofar as a philosopher is obscure, insofar as he is saying things that you do not get the hang of or that you do not see any importance in, insofar as you do not see where his ideas are leading, the philosopher is providing you with evidence of the existence of your horizon. You learn to broaden your horizon by reading the philosophers and attending to the parts where they are obscure, where you do not understand them. Finally, to systematize the matter a little we spoke of the existential gap, namely, the difference between what is one's reality and the measure of one's reality that is within one's own horizon.[5]

3   Horizon and Dread

The notion of horizon is, of course, a variable: the horizon is at different spots in different people. While the obvious application of the notion has to do with the big differences in philosophy between positivists, idealists, existentialists, Scholastics, and so on, still the fact that there are disputed questions among Scholastics themselves shows that the notion has some relevance also to us.

3.1   The Stream of Consciousness

If we ask about the *ground* of the horizon, we are led to a further topic that I entitle 'Horizon and Dread' or 'Horizon and Anxiety.' ('Dread' is commonly the word used in English translations of the existentialists.)

Because the horizon is defined subjectively and occurs in a series of positions relative to different subjects, its foundation is not a matter of objective fact, of objective truth, but of some subjective phenomenon, some subjective principle. The foundation of the horizon has to be in the subject.

---

5 A break was taken at this point. The recording resumes in mid-sentence with the words '... on the notion of horizon, which is of course a variable.' Judging from the distributed notes, it seems that nothing major was lost in the recording.

Consequently we are led to ask, What is the conservative principle in the subject that keeps him at the horizon he has, that makes it so difficult for him to move on to a broader horizon? Why is it that the original philosopher does not bring about a universal and permanent difference in the history of philosophy? Why does he merely found a school? Why does even the school that he founds splinter and become decadent and undergo renewals and revivals?

If we raise that question, we are led immediately to the stream of consciousness.[6] To speak of an act of seeing involves a violent abstraction. Seeing occurs within a context of other conscious acts. There is an emotional context, a conative context, an intellectual context, an imaginative context, as well as the total context of the movements of the body. If we want to see something, we bend over, turn it over, look at it this way, and so on. Correlated with the seeing there occurs a whole number of other movements in the body, and in the eyes especially – focusing the eyes. And there are other more recondite things such as emotions, desires, conations, and purely intellectual and volitional elements. Consequently, instead of speaking of single acts, one has to consider the flow of consciousness, the orientation of consciousness.

We can draw an analogy with mechanics. In studying movements in mechanics, one begins from ideal constructs and gradually moves to things that are more concrete. For example, in mechanics you can begin from the first law of motion, namely, that bodies continue in a uniform state of motion in a straight line as long as no force intervenes; and then you can add on the law of movement in a central field of force. In this way you arrive at a first approximation to the movements of the planets. Then you can consider each planet as the center in a field of force, and this will give you a second approximation to the movements of the planets. Or in the study of heat, there is the Carnot cycle and the way in which that is used in an analysis of the flow of heat. Such things are ideal constructs that, through a series of approximations, bring one closer and closer to the concrete.

Similarly, if one asks about the flow of consciousness, one is led to obtain one's bearings by setting up a series of ideal constructs.[7] One can consider a

6 Lonergan adds, 'the sort of thing we were talking about yesterday.' See above, chapter 8, p. 197, in the section on 'The Dilemma of the Subject.' See also note 1 in this chapter, above.

7 Greater detail on the patterns of experience (here also called 'flows of consciousness') is found in *Insight*, chapter 6. It is not clearly stated there, however, that the patterns are ideal constructs.

merely biological flow of consciousness by thinking about an animal chasing its prey, or a dog chasing a cat. The dog's every movement is determined by where he sees the cat now and where he imagines the cat is going to go next. Furthermore, the movements of the dog are constantly changing what the dog sees. This whole flow of sensations, images, conations, emotions, and movements of the body continually shifts in a single flow of consciousness in the dog.

There is also an aesthetic flow of consciousness. It is a liberation from purely biological determinants. Kittens will play and snakes can be charmed, but the aesthetic liberation of consciousness is the freedom of the flow of consciousness from merely biological goals such as food. The liberation of consciousness occurs first in man in an aesthetic manifestation. Children's play has an aesthetic motive, in a broad sense of the word 'aesthetic.' Children are so active, running around and imagining things and pretending this and that and playing at this and that; and this is an aesthetic liberation of their consciousness. They are always ready to eat, but they can be completely absorbed by other things that have no direct biological reference whatever.

Again, there is the dramatic flow of consciousness. The primary aesthetic achievement of man is being himself. The child is the man at the basic level. We have here a drama, and not a drama in which we study a part and add to the words from the study of the play the appropriate gestures and the appropriate tone of voice. The drama arises in us spontaneously. We are in the presence of others. We are doing our little act in their presence, and we are thinking about what they think of our acting: not in any explicit fashion, but person to person, or person in the presence of other persons. If you are alone in your room and somebody comes in, there is a complete shift in the flow of consciousness. You are reacting to the presence of someone else. Again, insofar as people succeed in dealing with others, they tend to be extroverts, and insofar as they do not succeed in their dealings with others they are misunderstood. But because they are misunderstood they act awkwardly, and they become more misunderstood. They withdraw into themselves, and they go into fantasies and dreams, in which of course they are the hero, the successful person; it is a merely private theater.

So there is the dramatic flow of consciousness, and it is very fundamental in human life. The hardheaded businessmen and the captains of finance use advertising and other psychological techniques to stimulate the dramatic imagination of people. They want them to dress in this new way and live in this new type of home and have this new type of car, and so on right along the line. The fundamental motives are 'keeping up with the Joneses' and that type of thing, which is just an aspect of the human dramatic action.

Next, there is the intellectual pattern of experience, which we have already illustrated by Thales not seeing the well, by Newton absorbed in his study, by Archimedes shouting that he's got it. And there is the practical pattern of experience of people absorbed in getting things done.

These patterns represent different ways in which the flow of consciousness can occur. People live their lives in a number of different flows of consciousness. Now they are in one, and now in another. They shift from one to the other. A certain part of the day they will devote to study, which is more or less in the intellectual pattern of consciousness; when dealing with others, they will be in the dramatic pattern; when getting things done, they will be in the practical pattern. The different patterns will be related, but each one has its own series of characteristics.

Now I said that man's primary artistic creation is his own living. But there is a difference between the artistry involved in one's living and the artistry that can be embodied in colors and shapes, or in words, in poetry. The difference is that the amount of artistry possible in one's living is strictly limited by the fact that one's flow of consciousness also has to be an integration of one's neural patterns. The neural patterns govern one as a biological existence. In other words, if the stream of consciousness runs too freely, if it runs off in 'abnormal' fashion – you may put the word 'abnormal' in quotes, because it is very hard to define – there will be a conflict between the orientation of the flow of consciousness and the needs of the body that that consciousness informs and governs. Such conflict heads towards neurotic phenomena, the nemesis of compulsion, the invasions of consciousness, and, in the limit, anxiety crises.

The appearance of anxiety in consciousness is, as it were, the danger signal, the sign that consciousness is running on a line that is too free. Human consciousness is not in a flow that is determined by external data. It is, as it were, a free creation. It is man the artist realizing his art primarily in himself, in a spontaneous fashion. We are, of course, using metaphors when we speak about man the artist realizing his art primarily in himself, but this expresses the truth that the flow of consciousness is not determined by environment and sensible data. On the contrary man finds clues; the data of an environment are merely clues for one to think out the way in which he can dominate and re-create his environment. A modern city is not any natural environment; it is a human creation. Similarly, a lecture room is a human creation; it is not something that results from determinations in any human environment. The free creativity of human consciousness seeks certain ends and produces all the means and paraphernalia necessary to pursue those ends in a certain manner that is thought helpful and that consists in giving lectures.[8]

3.2 'World'

A world is what lies within a horizon. We speak of people having different worlds. The stockbroker has his world, the captain of industry has his, the tramcar conductor has his, and so on. People have different worlds; their consciousness is occupied with a certain totality of objects. A world is not any particular object or any determinate number of objects; it is a sphere of objects to which we attend, with which we are concerned. According to our measure of interest and concern, that is, in the measure in which things are relevant to us, we pay more or less attention to them.

A difference between a perception and a sensation is, I think, well illustrated by going down stairs in the dark and going down a step that is not there. Perception is anticipatory of a further step, and sensation is the jar you get when the step is not there. The mere sensation that is not embodied in the flow of consciousness is a break, an interruption, and it pulls you together.

Now each person has his or her own concrete synthesis in successful living. In other words, there is a freedom to consciousness, the creativity of imagination, and it is organized in a certain fashion that gives one a horizon and a world within that horizon. That concrete synthesis in conscious living is, on the one hand, an integration of underlying manifolds of neural complexes and, on the other hand, a set of modes of dealing with the *Mitwelt* of persons and the *Umwelt* of things. To change that concrete synthesis in human living is to be converted to a new world. It is to move to a new horizon. When we speak of the horizon as relative to the subject, we mean that it is relative to his concrete synthesis in human living. To change his horizon is to change his successful integration of the problem of conscious living. And any change in that concrete successful synthesis of human living gives rise to anxiety. Anxiety is the anchor that keeps you where you are, the conservative principle. If your synthesis is not successful, you will have the experience of anxiety, to a greater or less extent, and that

---

8 The notes are perhaps better, even if the account there is brief. See above, p. 204, §4.

anxiety will force you to bring about a change in your synthesis so that your synthesis will be something that works. In other words consciousness has to be the integration of the underlying neural manifolds, and to keep it tied to that job, insofar as we are incarnate souls, souls in a body, when consciousness runs off too freely one starts feeling anxiety.

3.3   Corollaries

3.3.1 Philosophy and Conversion

You cannot change yourself by taking thought. You cannot become a new man simply by taking thought. Consider the work of developing the spiritual life that is attempted in the noviceship. If people go at it hammer and tongs or too naively they have breakdowns. You cannot fool around with the concrete synthesis of conscious living that is your *Dasein*. You feel anxiety as soon as you start playing with it. The anchor of the horizon lies in the anxiety, the dread one feels whenever there is any attempt to fool around with the concrete synthesis that is successful in one's living.

I do not know if my own experience is general but I found that this concrete synthesis in living is much more tentative, much less successful, in childhood. I know that whenever I was ill when I was a child, I would very quickly go into delirium. I do not know whether this is a general thing, but it seems that children will go into delirium with a minor illness, whereas this will occur with an adult person only under extraordinary stress of illness. I would say that the reason is that in the child the organization of consciousness has not yet achieved the stability that we have in adult living.

However, conversion involves a change in that concrete synthesis. It means that living is organized in a new way, in the light of new concepts, new principles, new norms of action, new modes of response to situations, new types of interests, new orientations. It is very easy to introduce a new viewpoint, a higher viewpoint, on the side of the object, as Plato does when he distinguishes between the sensible and the intelligible, the *aisthêta* and the *noêta*, and says that the *noêta* are what really exists. But when Plato wants to *convert* people to the *noêta* as what really exists, he has quite a job on his hands. He has to write a series of highly literary dialogues and to introduce the myth of the cave in order to explain what he means. And even when he gets his meaning across, it is quite another thing for people to live according to those principles.

Again, the Stoics, although on the intellectual side they were materialists, represented a high morality in the ancient world. They described the *sophos,* the really wise man. He had a terrific series of predicates, all the things that the wise man would do and would not do. But the question the Epicureans asked the Stoics was, 'Well, we'd like to meet *Mr Sophos* some-time. Where does he live? Are there any such people?' In other words, to set up an ideal on the side of the object is one thing, and you can get an ideal from a set of philosophic principles; but to embody it within a concrete synthesis of successful living is something entirely different. And if there is a conflict between the implications for me of the philosophy I am invited to accept and, on the other hand, my conscious living grounding my horizon, what is going to suffer is the philosophy. The philosophy is going to be watered down. Its meaning is going to be contracted to something that can be fitted within a narrower horizon. That is where the phenomenon of the decadence of philosophic schools is seen. But if someone comes along who has the broader horizon, he can revitalize the original message and bring about a revival of the school.

Now the fact that conversion is connected with philosophic development means that putting a philosophy across, selling a philosophy to others, is not exclusively a matter of proofs and arguments and evidence. There is also the very complex psychological process of conversion, and in that process there is a connection between the prior situation and the later situation. There is always a pivot on which one turns. Newman said that in becoming a Catholic, he retained everything that was good in Anglicanism. There is always the pearl of great price to which one clings, but there are a lot of other things that one has to throw aside. Conversion is not any simple matter of setting down principles and drawing conclusions. It includes changing one's mode of life, changing one's concrete successful synthesis in living. And the occurrence of that change will be accompanied by anxiety phenomena, by dread. If you suddenly were to become convinced that a philosophy that you hitherto considered completely wrong were true, you would be terribly upset, and you would find all sorts of reasons, good and bad, for not making the change. A successful synthesis maintains itself, just as in the psychoanalytic situation or the therapeutic situation in general, the patient spontaneously puts up a resistance – spontaneously, not deliberately – to the moves of the therapist that would bring about the cure he needs; he finds all sorts of reasons to maintain his present position. This is just an instance of the inertia that fundamentally is healthy and necessary in maintaining the existing situation.

When a person is confronted with a new philosophy that also involves or necessitates in him a conversion for its total acceptance, he will not and cannot attend simply to the philosophic arguments. His consciousness will be concerned to maintain itself, and he will trot out what Husserl and Heidegger will call his *Selbstverständlichkeiten*, his obvious, self-evident truths, to the effect that no man of common sense would ever dream that such a philosophy could possibly be true; it is too ridiculous to merit attention; the only thing to do is to brush it aside. You get all sorts of pique and indignation and emotion and resentment in philosophic debates because philosophic issues are concerned with the horizon of the subject, and the horizon of the subject is connected with his own vital solution to the problem of conscious living.

Briefly, then, the first corollary is that, because a philosophy has implications that regard the subject, a new philosophic viewpoint is correlated with a conversion in the subject, and a conversion in the subject is also connected with his problems in conscious living, with his personal solution to the problem of living. As a result, his emotions and in particular his anxiety will be involved in any attempt to move him from one philosophy to another. To change people's philosophies is to change them, and to change them is not any simple matter of true propositions and axioms and deductions. It is changing a concrete synthesis in living, and that change necessarily involves a whole retinue of emotions that give rise to the *odium philosophicum* and the *odium theologicum*. Conversion involves something of a leap. There is a pivot on which the movement turns, but for the person to find just where the pivot is and to turn upon it is not a simple matter. It will be hard to find as honest and sincere and direct a soul as that of Cardinal Newman, but it took something like from eleven to fourteen years between the time when he first saw that Catholicism was perhaps the truth and the time at which he was converted. It was sometime between 1831 and 1833, as I vaguely remember, that he first saw the light, and it was in 1844 or 1845 that he entered the Catholic Church. We cannot expect religious conversion to occur overnight. Psychologically the same type of problem is involved in philosophic conversion.

3.3.2 The Self-constituting Subject

The second corollary has to do with the self-constituting subject. There is the freedom of the will; there are rational alternatives and free choice; the rational alternatives are expressed in propositions, and I select one and reject the other. But besides that freedom, which is the freedom with which ethics and moral theology are concerned, there is freedom in quite another and prior sense. That prior and less accurate sense is nonetheless a true sense, a sense of a different kind, of course, one that is hardly even analogous, in fact one that might even be called equivocal; but it is very significant. It is the freedom with which consciousness emerges upon the flow of neural determinants.

We may say that a sensation is determined by the object, but our perceptions are simply of clues. We do not have to see everything there is to be seen about an object; we get a clue. 'There is someone lurking behind over there': I don't have to attend to just who it is in order to know that there is another man there. We are content with minimal sensible clues, to have perceptions of men and things. There is a terrific amount of economy in our perception, and connected with that economy there is also a freedom.

It is the direction of the flow of consciousness that determines what we attend to. Unless we are interested, we do not see what is there to be seen. You have to have an interest if you are to develop the science. Again, unless we develop knowledge along a certain line, we do not see what is there to be seen. You can put an insect on this piece of paper. I'll say it is a bug, but a person trained in entomology will tell you a hundred things about the bug that I just do not see because I do not have the categories into which I could locate all the things he can tell me about it. And because I do not have the categories, I will not know whether he is repeating himself or not by the time he comes to the sixth or seventh item. The fact that a person has a developed knowledge in a field makes him see ever so much more because he has the categories into which he can synthesize the sensible data.

There is a loose sense, then, an equivocal sense, in which the flow of consciousness is free, inasmuch as it is not uniquely determined by the sensible data or by one's own biological, neural reality. There is also a determinant from the direction of the flow of consciousness itself. It is that prior freedom of which Heidegger is always speaking. He does not mean freedom in the sense of freedom of the will; that is something beyond his horizon, as far as I know. When he is talking of freedom, he is talking of this freedom of the self-constituting subject, of the subject as the ground of the flow of interests and concerns, of the subject as grounded in his *Sorge*, in his care, in his concern. It is at that point that we find the root of the horizon. In other words, Heidegger's *Sorge*, concern, care, which lies behind the whole flow of consciousness, is also what determines a horizon. It is, as it were, the nucleus, the fundamental element, in the concrete synthesis of conscious living. It determines the horizon within which lies our world, our potential totality of objects.

Now, we all develop from infancy to wherever we happen to be. To a greater or less extent we have been exercising our freedom and making ourselves what we are, making our mode of living, our own concrete synthesis of conscious living. My own is absorbed principally in speculative questions and does not fit in very easily with practical decisions; it finds them an awful nuisance; they do not fit in with the flow of my interests, with my orientation in consciousness. Everyone, insofar as he has some knowledge of himself, would have some similar awareness of what kind of a man he is. We have been making ourselves in the past, and we have been doing so spontaneously. And as we have made ourselves in the past spontaneously, so in general men can remake themselves. That is the point to a noviceship, a long retreat, a tertianship, and so on: the possibility of remaking ourselves.[9] We recognize it openly with regard to the spiritual life, but it is not only on the moral side that we make ourselves. That is where freedom lies in its full immediate sense, but to a certain extent we exercise our freedom in our choice of interests, and these have repercussions upon what our concrete synthesis in living is.

When we advert to the connection between our concrete synthesis in living, what we have made ourselves to be spontaneously and more or less unconsciously in the past, on the one hand, and, on the other hand, our horizon and its connection with our philosophy or our measure of understanding of philosophy, there arises in us the question for fully deliberate freedom: Are we going to make any change in it or not? Once the question arises one has crossed a divide. Once the question arises one can refuse any change, or one can decide upon a change. If one refuses any change, one is at that point taking upon oneself full responsibility for what one has done spontaneously and inadvertently and indeliberately in making oneself in the past. And if one decides upon a change then again deliberately and advertently one undertakes the business of making oneself what one *wills* to be.

During a first period, then, one's self-constitution is something that the self-constituting subject is not aware of. It is something that he does not do

9   Lonergan is referring here to the standard Jesuit formation of the time. An initial two-year novitiate included a 'long retreat' – thirty days doing, under direction, the Spiritual Exercises of St Ignatius Loyola. This was followed by years of study, each year including a 'short retreat' – eight days instead of thirty. The number of years varied from Province to Province, but normally included a secular degree, three years of philosophy, three years of teaching, four years of theology with ordination to the priesthood at the end of the third year, and finally a third year of probation – 'tertianship' – which included a further directed long retreat.

deliberately, with full advertence and consent. But there can come a time when he will advert to the significance of his self-constitution, and when that advertence arises then he makes the choice, and no matter what way he makes it he becomes the responsible, deliberate, self-constituting subject. He gives blanket approval to whatever he has made himself inadvertently and spontaneously, or he does not. And if he does not he changes it.

That is putting the problem of conversion from another angle.

3.3.3 The Basic Function of Philosophy

These questions regarding horizon, conversion, the self-constituting subject are concerned with a basic function of philosophy. From one viewpoint one could say – and this is the third corollary to these considerations – that philosophy is the attempt to illuminate the effort of intelligent, reasonable, free, fully responsible self-constitution. The second volume of Jaspers's *Philosophie* (1932) has the general title *Existenzerhellung*, a clarification of *Existenz*, of one's soul in the religious sense, the illumination of one's *Existenz*. Because of his Kantian presuppositions, Jaspers does not dictate to people what they should be; he feels that since he does not acknowledge the validity of any objective knowledge, he cannot tell people what they should be and what they should go about doing; but what he does attempt is a clarification of *Existenz*. He aims at making clear to people just what is the self-constituting subject, what has been the subject's non-responsibility in the past, the lack of complete liberty, the spontaneity and inadvertence in one's self-constitution, and once the question is put the inevitability of taking over responsibility for what one has done inadvertently or, on the other hand, for remaking what one has done inadvertently.

Again, insofar as philosophy is concerned with the subject in his free and intelligent self-constitution, philosophy is concerned with the good, because *that* is the good of the philosophic subject. Consequently, a point of comparison between Scholasticism and existentialism is found much more easily in the realm of the good than in the realm of being. One can be misled by the existentialists' talk about being and existence. They do not use the words in what I believe is a satisfactory Scholastic sense. But what they are really talking about, and the medium through which they reach being, is the good, and particularly the good of the subject. Marcel's thought is essentially a movement from mere existence to being, and it is the being of the self-constituting subject that he is concerned with. And so it is within the field of fundamental ethics, in the field of notions of the good, of what precisely one means by the good, that one finds more readily the point of comparison between Scholasticism and the existentialists. This concern with the good is a concern with improving my operative solution, my functioning synthesis in concrete living; with the transition from freedom of images to freedom of enlightened, responsible choice; with the question, the possibility, the possible need of some conversion that will bring with it a broadening of the horizon. It is a concern with improvement not in general, not for the other fellow, but with *my own* improvement. It is a concern not with truths in general but with the truths that I live by, the truths involved in my self-constitution. It is a concern not with what Newman would call notional apprehension and assent, but with real apprehension and real assent. The difference between the two is perhaps just that: the notional is not what you live by, and the real is what you live by. It is a concern with the solution in living, and living is never abstract but is living in this world, in this place, at this time, with these people. It is a concern with living within a technical civilization and in the particular part of technical civilization in which I live. It is a concern with personal relations within a technical society. It is a concern with the concrete possibility of that living at its highest point, the exercise of freedom in such matters as involve norms and absolutes. So, inevitably, it leads on to what we know to be God. (Jaspers in his earlier works speaks simply of *Transzendenz*, because he cannot talk about God. He does now talk about God; in his later works he considers God a necessary postulate of a philosophy.) To give to ethical norms a concrete interpretation that will be effective, you have to go beyond any mere categorical imperative. You have to step into the realm of personal relations with God. It may involve a contradiction if I live in one way and think in another, but what of it? In other words, you can set up your categorical imperatives in whatever way you please, but when a man is faced with suffering, with death, with struggle, he has to have something more. There will be the movement to *Transzendenz*, the moral argument for the existence of God.

Again, the concern with the good is a concern with history. The philosopher responds to the problems of his age, just as everyone responds to the problems of his age and the problems of his environment and location. That is what living is: responding to problems. But he responds as a philosopher insofar as he is meeting those problems at their deepest level, at the point of maximum consequence for human welfare or human disaster. In the nineteenth century, Karl Marx was regarded as an old man with a queer face and a long beard who wasted his life in the British Museum. He was looked upon simply as an oddity. At the present time a greater portion of the surface of the earth is dominated by Marx's philosophy. One cannot help feeling that what we need most of all is that in the nineteenth century we too had some queer face with a long beard that spent his life in the British Museum and was operating in the opposite direction. There is a need for fundamental thinking, a need for meeting the problems of the day on the deepest level. If we are not doing it, our position a century later is apt to be that we are arriving too late with too little.

Again, the philosopher is concerned with history. The problems with which he is concerned are the problems of his time. This is in general, more or less, the position of Jaspers, but I have been noting that a good deal of the responsibility among Catholics falls upon the theologians.

Again, the philosopher is open. He goes beyond the horizon based upon his personal anxiety, upon the accident of his personal history. He has to move to a horizon that is coincident with the field, and as I remarked in the previous hour, there is a good deal in what Jaspers says when he remarks that the philosophers are our educators. This principle can be applied excessively, and I would not advise anyone to follow it out literally, but there is a measure of truth in it. It is where the philosophers are obscure that they can help us most of all. Where they are obscure for us is where they can help us because their obscurity is relative to us, and their obscurity relative to us may be a revelation, an indication, a signpost to something like a blind spot in us, a question of horizon, a difference between horizon and being.

The philosopher also has to be genuine. In other words, if he has a horizon that is not as broad as the philosophy he is representing, then if he is genuine he cannot really speak that philosophy. If he is not genuine he can speak it, but he will be devaluating the currency. He will be making things mean less than they really do. He will be cutting things down to his own size.

From a practical viewpoint, as far as results are concerned, there is a good deal of truth in that position of *Existenzerhellung* of Jaspers, not for his reasons but for other reasons. Insofar as the objective change in philosophy is connected with a conversion in the subject, the free will of the subject is going to be involved, and insofar as free will is involved, you can give principles and you can provide motives but you cannot guarantee results. You can clarify the issues but you cannot impose the solution. It is up to the free will of the people you are endeavoring to convert to bring about the conversion within themselves.

Finally, as the word 'conversion' suggests, simply de facto will is not within a state of pure nature but in a world with original sin and with the offer and the acceptance of divine grace. Philosophic issues on this level are very closely related to theological issues. In other words there is a shift in meaning, an *Umdeutung*, in the meaning of the word 'philosophy' as it is employed among Catholics. We have the meaning of 'philosophy' as concerned with truths that can be known by the natural light of reason, but when the Catholic philosopher moves to the consideration of things existentially, he finds that things as they exist in all their concreteness cannot be completely illuminated without a consideration of the supernatural. One finds in existentialism, I think, a suggestion that philosophers and theologians need to cooperate perhaps more than they have done in the past. As long as everyone regarded philosophy as what is known by the natural light of human reason, there may still have been many reasons for collaboration between philosophers and theologians, but at the present time, when philosophy is moving to existential considerations that have immediate theological repercussions, there is need for even closer collaboration between philosophers and theologians.

14

Horizon, History, Philosophy[1]

1 Horizon and History

1.1 Enlarging the Significance of the Existential Gap

We have been asking, What seems to be the chief significance for us of existentialist studies of man in the concrete flow of consciousness? I am speaking of that concrete flow as it is analyzed by Heidegger and treated in different manners by other existentialists such as Jaspers, Marcel, and Sartre. What seems of fundamental importance is contained in the word 'horizon.' This is not to be taken here in the more restricted sense used by Husserl, in which the meaning of a concept defines the horizon of an indefinite series of visual objects. (Our example was the indefinite series of visual objects all of which we referred to the single building, Gasson Hall.) We mean 'horizon' rather in a broader sense that includes the total field of what one is concerned with. Heidegger took *Sorge* as the fundamental existential, the fundamental category of the existential subject. It was in connection with that idea that we worked out yesterday a more general tool that we named 'horizon.' There are the questions that you can answer, and that totality constitutes what you know. Then there are the questions that one can raise, find important and worth knowing about, understand what they mean, and know how to solve at least in some remote fashion, even if one does not know the answers to them. That forms the second, surround-

---

1 The tenth lecture, Friday, 19 July 1957.

ing circle of *docta ignorantia*. Finally, there are the questions that cannot be raised at all, or that if raised are not understood, or that if understood in some fashion are not considered worth while, or that if considered worth while have no apparent method of solution. This constitutes the still further region of what may be named *indocta ignorantia*. The limit between the last two, between *docta* and *indocta ignorantia*, might be called the frontier or the horizon.

A horizon manifests itself differently in the natural sciences from the way it does in the human sciences, philosophy, and theology. You can have a totally new mode of conception of the object of the natural science without it having any effect on your concrete living, except in some very external fashion: you use electric light instead of gas light, for example. On the other hand, there can arise a new mode of conceiving man, whether that new conception of man arises implicitly in philosophic development or explicitly insofar as a philosophy is directed towards man. There are many illustrations in the history of philosophy. Consider Plato's emphasis upon the distinction between the *aisthêta* and the *noêta*, or Augustine's transition to insistence upon truth and judgment. Really to assimilate such a philosophic advance calls for a conversion in the subject, and such a conversion in the subject is difficult. It is difficult, first of all, because philosophers are apt to think that no conversion in the subject is needed, that everything is merely a matter of objective argument: if you put forth the right argument, everything will go fine. Though they have controversies that last over centuries, still at least some of them do not ever advert to what the real trouble is. Such a conversion is difficult for another reason, that is, because one's horizon is a consequence; one's horizon is grounded in one's concrete synthesis in conscious living, in one's flow of consciousness, in the modes in which one's consciousness can flow. Any tampering with such a successful solution causes anxiety phenomena. It makes one feel very uneasy, and it spontaneously gives rise to various defense mechanisms, whether emotional or otherwise, that tend to block off the possibility of this transformation.

Such is the general idea of horizon. You can see how it is rooted in the existential subject. It is not based upon a notion of man per se. It is based upon a study of men as they concretely are.

Now that notion of horizon is significant not merely in the field of philosophy in the strict sense. It is important not merely in studying the differences between the different philosophies and the various series of schools with their periods of flowering, their periods of decadence, and their later periods of revival. It also has reference to the historical process, and this dimension too is important in existentialism. It is true that the nature of Heidegger's work makes it impossible for him to have any general views on history, and that Marcel by his nontechnical approach can hardly attend to historical considerations. But the historical aspect comes out very prominently in Jaspers's thought. The difference between the horizon that a man may have and, on the other hand, the field that is defined objectively in terms of the totality of beings that exist, the difference that we have called the existential gap, is not merely a call to the authenticity of the subject in his private existence.[2] It is also a call to authenticity in all subjects, an invitation to understand something about the process of human history, and a summons to decisiveness at a rather critical moment in the historical process.

1.2   Dialectic

To help us grasp that I had best say a few words about the notion of dialectic.

First, it is a notion that has several very different meanings. It appears first in Plato as a mode of reasonable dialogue opposed to eristic. It appears in Aristotle as a review of the opinions proposed by others in the hope of selecting the elements of truth in all of them. In the Middle Ages, it became the technique of public disputation. In Hegel it has to do with the triadic process through thesis, antithesis, and higher synthesis. It is turned over in Marx into a dialectical materialism as opposed to a mechanical materialism.

What *I* mean by 'dialectic' is the fact that if you start from an inadequate position and follow its results out logically or let its natural consequences arise, then the deficiencies of the initial position become more and more obvious. Our study of mathematical logic last week provides an illustration of the point (perhaps not too clear an illustration). They started from an ideal of what a deductive system should be or had to be and what it could do. They felt that there is a set of little nuggets named truths. From that set you are to pick out certain fundamental ones, and the problem is to pick out the right ones. On that basis, it is thought, you can deduce everything that lies within a given field. That is the spontaneous notion of a deductive system. But the mathematical logicians ran into the difficulties known as

---

2   See Lonergan's more precise expression of the meaning of 'field' above,
   p. 199. On dialectic, see *Insight*, the index under Dialectic. The words, 'the
   difference that we have called the existential gap,' are an editorial addition,
   for the sake of harmony with the notes that Lonergan distributed.

Gödelian limitations, difficulties of various kinds that seemed to show that that ideal could not be achieved.

Let us put the matter more concretely. If you begin by finding fault with Euclid's *Elements*, if you point out that he is constantly using casual insights that are not contained in his axioms and are not deducible from them, you proceed to a rigorous axiomatic system as the expression of a geometry. But the mere fact that you are excluding the little insights that come all along the way when you proceed by Euclid's methods changes entirely the nature of your project. If you are proceeding with Euclid, you start off from his set of axioms, which is the first big step, and you go along having casual insights all the time. Consequently you are really moving to a fuller basis of thinking; your mind is developing all the time, and you are stepping up with it. But if you lay down a set of axioms and proceed by a rigorous symbolic technique, you may start off at a point somewhat higher up, but you will not move straight along that level. When you get to the end of Euclidean geometry and want to go further, you will have to take a jump.[3]

What is the dialectical element in this? We say Euclid is wrong: he lays down axioms, and then he keeps constantly adding insights onto them, and in that sense he is wrong. However, when you apply a rigorous symbolic technique, you discover that you are depriving yourself, as you go along, of the little insights, and then you have to make a big leap when you get to the end of the row. Consequently, while you have eliminated one defect in Euclid, you have given rise to problems of a new kind.

Hegel, in his *Phänomenologie des Geistes*, is constantly using such a technique. He starts off from a very simple notion and raises the obvious difficulties that involve a deepening of the notion. Then he goes further, and finally by the time we are around the circle he has given us a fully nuanced notion. That is an excellent device in teaching as well as in writing. In teaching, you do not presume at the start any more than the class's outlook will allow, and you derive gradually all the reasons that lead to a transformation of that original outlook. Finally, you get across what you consider to be the right idea of the subject.

Now in those cases we are considering the dialectic of an idea, of a concept, of an insufficient concept: give it enough rope and it will hang itself. That is a very common idea. But there is also a dialectic of man, and that dialectic regards man not insofar as man is nature, not insofar as man is reproduced without a transmission of acquired characteristics. It is on the

3 See above, p. 95.

technical, social, and cultural levels that the dialectic occurs with regard to man. It occurs precisely insofar as he is not merely nature, insofar as we do not get the whole of civilized man merely by having a newborn baby. In the technical, social, and cultural fields man is the maker of man. It is important to grasp that fundamental statement. In the field of the technical, the social, and the cultural, it is man's ideas upon man that will determine the fabric of human living.

Man is technical. As using tools he is not merely satisfying immediate animal necessities. He creates a human environment. The city and the state, with their endless array of artifacts of all possible kinds, are entirely a human creation, a human product. Again, as social, man creates the state precisely as an *ens iuridicum*. The state is not just a function of geographical features. A given state may coincide with well-determined geographical features, but it is a great deal more than those features. The family, too, is a social institution, not a matter merely of casual sexual relations. Again, there are the economic system, the political system, the systems of alliances between states;[4] and there is the cultural level, where human living depends upon man's ideas upon man. Culture in the anthropological sense is the current effective totality of immanently produced and symbolically communicated contents of imagination, emotion, and sentiment; of inquiry, insight, and conception; of reflection, judgment, and valuation; of decision and implementation. In these fields man presupposes nature but also makes himself by taking thought.

1.3   The Objective Functioning of the Dialectic

We are considering the field of human history, and within that field there functions a dialectic. Some brief analysis of that dialectic is our immediate purpose.

At any given moment, then, there is some technical, social, cultural situation. It gives rise to a variety of problems that become the concern of men of various types and various walks of life. They think these problems through to a greater or less extent, they get ideas about them, and they discuss these ideas. As a result of their ideas, they finally change policy, and there follows a new course of action to deal with the problems in situation *A*

---

4   There is an indeterminate gap, probably quite short, on the tape at this point. The remainder of the paragraph is assembled from Lonergan's distributed notes. See above, pp. 209–10.

and to guide the production of the next situation, *B*. So the prior situation is always acting upon man as a stimulus to modify his current ideas about the technical, social, and cultural aspects of human living. From this reflection there arise ideas that will result in the new situation.

The situation, then, is both a product of man in his previous stage of development, his previous set of ideas about himself, and an occasion for change. Man technically, socially, culturally, in 1956 depended upon man as he was in 1955, plus the ideas that arose from man's reflections upon his situation in 1955. A similar difference occurs between 1956 and 1957. Now think of this without taking into account precise temporal limitations. Because these changes come along at any time, there is, as it were, a continuous process. The general picture is one of the situation influencing man's thoughts about man and leading to a successive transformation of the situation.

There is a further point. If the situation is defective or unsatisfactory, this can have two causes. Either the previous situation had those defects and nothing was done about it, or there was some defect in the thinking that added to the defects of the previous situation. So defects in *A* result from defects in the prior situation or from defects in the thinking. Similarly, defects in *B* derive from defects in the thinking that immediately produces it or from the prior situation with which this thinking had not coped.

Now one can say that each successive situation provides a concrete and almost visible *objectification* of what man has been feeling, thinking, and deciding about man. The objectification is not visible, of course, because the technical, social, and cultural as such is not visible. Still, it is the thing that men most willingly talk about. People will discuss economic systems and political questions without end. They are the concrete goods that people know, and the situation as it concretely exists provides the objective evidence, the objective revelation, of what men are thinking about mankind.

Again, not only does the concrete situation objectify and reveal what man's thought about man has been, but it also suggests and motivates changes in man's thinking about man. It suggests changes insofar as there are evils in the concrete situation: the existence of the evils moves people to do something about them. And it motivates changes because, again, nothing is more effective to move people to good than the concrete vision of evil.

Such is the general process, but there is a further point that arises when we bring in the notion of horizon.

Insofar as all our thinking about man is under the limitation of a horizon,[5] then, first of all, the suggestions and the motivations that arise from the situation are given a twist by the limited mentality of that horizon. People will see what they want to see, what can fit within their horizon, and they will omit the rest, or at least they will omit the significance of the rest. Insofar as thinking, reflecting, deciding, and policy-making are under the limitation of a horizon, there is the recurrence of overemphasis and oversight in the consequent situation. The succession of situations, then, will reveal the cumulative effects of the limitations of this horizon in the mentality of men.

Insofar as people's horizon is limited, the situation can be as bad as you please and they still will not see in the situation its real significance. They will be looking for all sorts of remedies and cures and ways of fixing things, but the one thing necessary is what they will miss, and they will miss it because their thinking is within the limitation of a given horizon. The result is that the situation progressively deteriorates. We have had from the liberals a doctrine of automatic progress, and there is a sense in which that is true: namely, the better man understands techniques and social relationships and the nature of human culture, in that measure there is an improvement. But at the same time there has been a continual degradation, and it is very easy for Catholics to acknowledge it. In the Middle Ages there was a synthesis of the natural and the supernatural, of society and church. That synthesis shattered on the conflict between church and state, and the conflict ended with the breakdown of the unity of Christendom at the time of the Reformation. That breakdown of unity was followed by the religious wars, and the conclusion that people drew from the religious wars was that we have to keep religion out of life, out of practical affairs, because nothing has been more atrocious than these religious wars. Man has to be guided by reason. So, first of all, in the Middle Ages there is acknowledged supernatural guidance of human life. In the second place, there is a split in the supernatural guidance because of the conflict between church and state ('cuius regio eius religio'). Next, because of the wars of religion men thought that the obvious thing to do was to base practical affairs on reason alone and keep out the supernatural. There is the deification of reason in the French Revolution. However, they very shortly discovered that just as the different religions differed from one another and were irreconcilable,

---

5 In the notes that Lonergan distributed there is mention of the 'existential gap.' See above, p. 210.

so too the one goddess Reason spoke with several voices: there were as many opinions of Reason – with a capital 'R' – as there were men. And so they moved to a doctrine of tolerance, where each one has to follow reason as he sees it and each one has to respect the other's opinion. And so we move from reason as a universal unifying principle to tolerance, in which every man follows the light of reason that God gave him. But tolerance proved unequal to dealing with concrete social problems. There existed economic problems, problems of social classes, and all the rest of that retinue, and in states that did not have the elbow room and raw materials and possibilities for increasing population that existed on this side of the Atlantic, those problems called for a method of effective action that was not possible within the limitations of the parliamentary game. Thus tolerance was followed by the totalitarian state. People had to deal with social problems in the manner that they thought would be effective, and so tolerance was succeeded by the state taking over all cultural functions and subordinating them to the state's own ends, economic and political.

This continuous deterioration of the objective situation illustrates at once the idea of dialectic and, at the same time, the limiting effect that horizon can have within the social process: from the medieval unity through the breakup of Christendom to a rule of reason, and through tolerance to totalitarianism. This is just a very sketchy expression of the matter, but it is perhaps rich in its suggestiveness. There is an intelligibility to the historical process. That intelligibility has its basis, first of all, in the fact that man as technical, social, and cultural is the product of human thinking, a result of man's ideas about man, and, in the second place, in the fact that man's ideas about man are shaped, on the one hand, by the situation in which he finds himself, and, on the other hand, by the horizon within which he thinks about himself.

1.4 Resolute and Effective Intervention in the Dialectic

So much for the general idea that there is a dialectical process in history, and that there is a responsibility of man for man. Now the existentialists believe in intervening in this dialectic. And they do not write simply for professional philosophers; they write novels and plays, and they are ready to use those techniques of human communication that can have a maximum diffusion. This is particularly true of the French existentialists: Sartre, Camus, and a number of others use the literary field as a vehicle for the communication of their ideas.

Just as each individual can choose to be himself or, on the other hand, merely drift, choose to be like everybody else, so there is a historic authenticity. Men can disregard the fact that they by their living are inevitably making their own personal contribution to the historical process. Or they can advert to the fact that they and everyone they know and with whom they live and endless numbers of others similarly are constantly making their contributions to the historical process. And they can raise the questions, What is to be done about it? How should one go about it? In other words, if there is this objective dialectic of history the question arises, Can we get enough knowledge of it to be practical, to exercise some control over this historical process? That is the question of human action applied to the point where it can exert a maximum of influence.

History has been influenced enormously by individuals, but by and large the influence seems to have been a chance influence. A given individual puts forth the ideas and gains the adherents that led to these results, but he did not know and could not know that he was doing it; he could hardly appreciate that it was going to have all these effects. It was left to the historians who came centuries later to investigate the matter and say, 'Well, so-and-so is responsible for this.'

Is it possible to do something with regard to the dialectic of history, the historical process? That question, of course, is more a question for the age than for an individual to answer fully. It calls for an enormous amount of investigation. But perhaps a few points can be made.

First of all, insofar as there is to be a resolute and effective intervention in this historical process, one has to postulate that the existential gap must be closed. In other words, one has to postulate that the people who are seeking to influence history, to put their lever at the vital point in the historical process, are not operating, not doing their thinking, planning, and policy-making, from within the pair of blinkers of a personal or communal horizon. They have to be people in whom the horizon is coincident with the field. If they are not, then all they possibly can do is increase the confusion and accelerate the doom.

That is one side of the picture. On the other hand, insofar as there are those who (we have reason to believe) have a philosophy in which the horizon and field are coincident – in other words, their philosophy is large enough to take into account the universe of being, they have a true philosophy of being – they are the ones who should be making whatever effort in this direction is to be made. If those that live within a narrow horizon are all 'ahoy' for changing the historical process while those whose horizon is coincident with the field retire into an ivory tower and exert no influence upon society at large, then we are in the situation where the people who can do the most harm are doing it and the people who could do the most good are not.

In the second place, heightened, effective intervention consists in a heightening of the effects of the dialectic itself. Men at any given moment have a set of ideas about men. Those ideas are expressed in the social, technical, cultural situation that proceeds from them, and that situation manifests them in their limitations. It manifests them in their strong points, but it also brings out the limitations of their thinking. All is there in the objective situation. What is meant by heightening the process of the dialectic is drawing attention to the objective features in the situation that manifest the defects in current thinking about man. Current thinking about man produces the situation; defects in the thinking result in defects in the situation.[6] The defects are there, but they will not be noticed by people within this horizon. Insofar as you want to make use of the historical dialectic as a tool for manifesting these defects in people's mentality, you pick out the points in the situation, in the objective field, that reveal these limitations. You make use of the objective manifestation of men's minds in the situation to make manifest through the situation the defects in the minds. The effective writer is the one who draws attention to what everyone knows, to what everyone feels is true with regard to the current situation, to what everyone has an uneasy feeling about, but which no one can articulate. For example, in the situation prior to *Sein und Zeit* and to Marcel, people were not satisfied with idealism, and they were not satisfied with positivism and scientism. As soon as these new ideas came out, there was a terrific reaction in their favor. Why? Because these people were picking out concrete features of the current situation and giving vivid expression to defects in the situation, defects in current modes of thought, that everyone felt in an inarticulate fashion to be true; but precisely because it had been an unanalyzed feeling, it could not produce results. Heidegger's description of the *Umwelt*, of man using tools, is almost a caricature of technical civilization, and because technical civilization was the thing that was really getting the Europeans down, Heidegger had a terrific following. You can utilize the situation produced by current mentality to manifest the defects in current mentality. Insofar as you are doing that, you are heightening the action of

---

6 The topic is dealt with in *Insight*, chapter 7, §8, and in *Method in Theology*, chapter 2, §7, and chapter 3, §10.4.

the dialectic that produces unsatisfactory situations as manifestations of unsatisfactory minds. It produces those manifestations so that we may react to them and correct them.

Again, intervention not only clarifies the situation but also drives home the point. You can show that the evils of a situation are due to these errors, and that the evils of another situation are due to the same errors again, and that to retain these errors is to perpetuate the evil. The argument will be entirely in the concrete. You will be making manifest to people what everyone knows is bad, what they do not want, and you will be showing the connection between what is bad in the situation and its ground in the limitations of their thinking.

1.5   The Essence of the Dialectic

What is the essence of this historical dialectic? Note that it is not a conflict between philosophies. The conflict lies, rather, between, on the one hand, what man is and is to be and, on the other hand, what man thinks he is and is to be. The conflict is not between two schools of thought but between human reality and human thought about human reality. It is human thought about human reality that produces the technical, social, and cultural situation. It is the limitations in human thought about human reality that produce the evils in the technical, social, and cultural situation. And it is the revolt of human reality (as distinct from human thought about human reality) against those evils in the situation that provides the lever for correcting the defects in the human thinking about man.

The dialectic, then, does not operate within the field of concepts and judgments. It operates between the field of concepts and judgments, on the one hand, and human reality, on the other. It is the conflict not between philosophies but between the defects in any philosophy and, on the other hand, the reality of man. The verdict of the dialectic is not a label of approval or disapproval on any given philosophy. It lies in the facts of the concrete situation, in its tensions, and in its basic hopefulness. And the facts of the situation can be properly grasped only by one whose horizon is totally open, whose horizon is coincident with *ens*, with being.

Starting, then, from existential concern with the concrete flow of individual consciousness, we proceeded yesterday to a consideration of horizon in its philosophic import, its relevance to revivals and decadence within philosophic schools, its connections with the emotions and anxieties of the particular subject and so of the philosopher. Today, in a rather rapid sketch, we proceeded from the same center of the subject as a concrete flow of conscious living and considered the totality of human subjects in their concrete living, that is to say, we considered history. The totality of human subjects has a natural aspect, the aspect of what is reproduced by birth and by transmitted characteristics. On the other hand, it has the aspect of the historical process that lies on the technical, social, and cultural level. Insofar as there is that component of human living – and it is a component that is of maximum importance in modern life – it depends upon what man thinks about man. It depends, for example, on whether he thinks of man as a creature subject to God or as almost a god working out his own destiny. Man's concept of man determines fundamentally what kind of technical, social, and cultural situation people will produce. The results of that thinking will be cumulative. If there is a horizon limiting the understanding, then unless there is some intervention each successive situation will constantly increase the evils in the situation until finally the civilization vanishes.

We considered the general nature of such intervention first of all in its preconditions. You only cause harm if you are operating from within a limited horizon. On the other hand, if you have a true philosophy of being then it is you rather than others that should be trying to influence the historical process. Secondly, that influence can take place with the greatest possible concreteness, by way of a critique of the existing situation in its evils, in evils that are recognized by all to be evil; at least they know it is so, although they may not have the means of saying it is so and may not have the concepts through which they can express that it is so. All men are reacting to the historical process, as all are contributing to it, but one is a philosopher precisely in the measure that one's reactions occur on the most fundamental level. (Incidentally, the same point is made by Jaspers.) And so it is by manifesting the realities of the situation that one manifests the defects in current thinking, and it is by manifesting the defects in current thinking that one changes that thinking.

There is one further point, and it has to do with communication. The thinking of people generally, at the present time, has been the result of communication in the past, communication through education and through the media of communication. Their thought comes from that source. They may change their ideas to a certain extent, but by and large their ideas correspond to the communication. As people's minds as they are at present are the result of communication in the past, so what they ought to be in the future can be a result of further communication. The notion of communication is a category of maximum importance among existentialists, and particularly among those who follow Jaspers. While there is an exaggerated element to Jaspers's talk about communication – insofar as he does not recognize any objective truth, he obviously has to struggle in a special manner to have communication – still communication from man to man and the art of effective communication provide an important topic worthy of the philosopher's attention, though just how he should go about it I am not prepared to say.[7]

2   Horizon as the Problem of Philosophy

I have been providing an account of man, of concrete human living in its significance for the solution of philosophic problems and particularly of the problem of being. Perhaps I have not sufficiently emphasized during these two weeks that this course of lectures is not aimed at equipping you to write on either mathematical logic or existentialism. I am simply throwing out some ideas that I have found helpful myself in understanding what these people are talking about, ideas that may help you when you get down to the job of reading them. That, of course, is the hard work, and it cannot be omitted if you wish to make public pronouncements on these issues.

We have considered the fact of horizon and the multiplicity of horizons. This multiplicity may be considered in three ways. It may be considered first of all as a mere matter of fact, and then it provides material for a history of culture, a history of thought, a history of opinion. In the second place, the multiplicity may be considered as a problem to be explained, and then one gets as a solution to the problem a book such as Jaspers's *Psychologie der Weltanschauungen*, presenting a psychology of worldviews.[8] In the third place, this multiplicity may be considered as an issue calling for judgment and decision.

It is on the third level that the multiplicity of horizons is a philosophic question. The multiplicity is a philosophic question when we ask whether some horizon is the field, whether some horizon is coincident with the limits of all that there is of the universe of being. If we answer that question affirmatively, if we say that some horizon is the field, then how can that horizon be determined? The positivists, the pragmatists, the skeptics, and the relativists would deny that any horizon is the field.[9] The positivists

---

7 A coffee break was taken at this point.
8 See the bibliography, below, p. 371.
9 'Field' is a term that Lonergan borrowed from existentialism. The field, of course, may be simply identified with the object of the open heuristic of mind. It is thematized in *Insight*, but a more sophisticated thematic of the field emerges from Lonergan's later work.

would say to the philosopher's question, 'Let's do science.' The pragmatist would say, 'Let's perform some experiments and see what happens.' The skeptic would want to inquire some more. The relativist would say that all answers are very interesting but none can be definitive. But if we affirm that some horizon is the field, we are involved in the second question, How do you determine which horizon coincides with the field?

That question is simultaneously ontological and epistemological. It is ontological consequently. Insofar as a given horizon also defines the field, what is outside that horizon is nothing and meaningless and insignificant in the absolute sense: not merely nothing for me, insignificant for me, meaningless for me, but nothing absolutely, meaningless absolutely, insignificant absolutely. Thus to select the true horizon is to lay down the basis of metaphysics, to lay down the criteria of what is and what is not. It is to answer the question, What is being? in the concrete fashion that says that being goes so far and there cannot be anything beyond it or there is nothing beyond it.

This consequent ontological aspect is simply consequent. It is not the real issue. The real issue is the antecedent issue that has to do with how you pick out the right horizon. The antecedent issue is epistemological and also ontological, though ontological in a different sense. It is epistemological insofar as you ask for the definition of the true horizon that coincides with the field. You ask why that definition is true, and how you know that definition is true. But it is ontological in an antecedent sense, insofar as that truth depends upon evidence of some sort or other and that evidence is evidence of some reality, in some meaning or other of the term 'reality.' Thus the epistemological aspect of the issue and the antecedent ontological aspect of the issue are as it were simultaneous. To determine which is the right horizon is a determination in virtue of an evidence, and the evidence is evidence of some reality. It is epistemological in terms of evidence, and it is ontological in terms of reality.

That antecedent ontological aspect Heidegger names *ontic*. Ontology, a *logos* of being, occurs when you start talking about being, but the ontic is the being that is prior to your talking about it.

Now let us consider very briefly what are the general characteristics of a possible answer to the question that we have raised.

First of all, any determination, justification, evidence for a horizon as the true horizon arises within some stream of consciousness. Consequently, it arises within the horizon that is set up by that stream of consciousness.

Secondly, the justification of a horizon cannot rest on the *consequent* ontology, on the realities known within the horizon. For if that were so, then every different stream of consciousness could claim that the realities known within its horizon were all the realities to be known, and we would be no further. We would be justifying all the horizons. If we let each one justify itself by the realities of its consequent ontology, we would be in the relativist position. You cannot justify a horizon by saying that from within this horizon these are the realities that we know. Someone else may have another horizon that knows these and other, or fewer, realities, consequently differing, and he can claim that his realities justify his horizon just as you can claim that your realities justify yours. One is slipping back into a relativist position if one justifies a given horizon by the realities known from within that horizon.

Another difficulty of justifying a horizon by the realities known within that horizon is that you can justify any horizon in that fashion, because every horizon will have some determinate philosophy and that determinate philosophy will be its horizon and coincide with its realities.

Again, the justification of the horizon cannot rest on the norms, the invariants, the principles that de facto characterize, determine, constitute any given horizon. If *I* appeal to the way *I* think to justify the horizon that *I* happen to have, there is nothing to prevent the next man from appealing to the way *he* happens to think to justify the horizon *he* happens to have.

And so we have blocked out two types of answers to the question, How do you pick the true horizon? How do you pick the horizon that coincides with the field? It is not going to be based on any consequent ontology, because consequent ontologies will vary with horizons and there are not several true ones. It is not going to be based on de facto modes of thinking, because everybody has de facto some mode of thinking, and if yours justifies your horizon, his justifies his equally well.

Consequently, the justification of the horizon has to involve the discovery of the evidence or norms or invariants or principles that naturally, ontically, possess a cogency, inevitability, necessity, normativeness. You have to examine the self-constituting subject and find in the self-constituting subject norms or evidence or invariants or principles that have a natural basis, an ontic basis, that are a reality there to be discovered apart from particular attitudes of particular men, and so that recur in everybody – norms or invariants or principles that have some cogency, inevitability, necessity, normativeness independently of the horizon of any given particular thinker. Because they possess their cogency independently of the horizon of any particular thinker, they thereby constitute a self-justifying horizon. If they constitute a self-justifying horizon, they determine a horizon that is coincident with the field.

Moreover, if one's account is to be fully satisfactory, it must not only discover the norms or invariants or evidence that recur in every subject independently of his horizon. It must also account for the fact that de facto people do have different horizons, despite the existence of these norms that recur in everyone. It must account, at least in principle, for the existence of the multiplicity of horizons, and it must have some capacity for explaining the varieties of philosophic opinion that have existed for thousands of years.

Finally, not only must it pick out norms that recur in everyone independently of horizon. Not only must it find such norms that, though they recur in everyone and have the full force of norms in everyone, still are not necessarily effective in everyone, so that there can be the many horizons that do de facto exist. While it must account for all the horizons that occur, it must do so in such a way that it also discredits them. It accounts for them as violations of the norms, humanly intelligible violations of the norms, so that it is plausible – not only possible but plausible – that many philosophers should think in these erroneous ways.

So much for the general prior question, What kind of thing are we to look for when we want to determine which is the true horizon? We are not to start from a metaphysics, an account of reality, because there will be different accounts of reality from various horizons, and if one man can justify his horizon by his account of reality, there is no reason why another should not justify his. We are not to appeal to de facto principles, evidence, norms, invariants, because everyone has his own de facto principles, evidence, norms, invariants. We have to reach to a more fundamental type of principles, evidence, norms, invariants that recur in every subject despite the variety of horizons. And we will do a good job insofar as we find principles, evidence, norms, invariants – whatever you want to call them – that recur in every subject despite the variety of horizons, that account for those people who have correct horizons, and that admit a hiding, a covering up, of the truth (in the Heideggerian sense of 'being in the truth'), so that the norms will not be universally effective. You can have people with limited horizons or mistaken horizons. You can account for all these mistaken horizons. Finally, you can account for these mistaken horizons in such a way that you will, by explaining them, also discredit them.

Now what is this prior reality, this ontic, evident, normative something that grounds horizon, the critique of horizons, and the determination of the true horizon? It is easy enough to proceed to the conclusion that that reality is the reality of the subject as subject.

What is meant by the subject as subject? When I look at the paper before me, there is present to me black on white. That could not be present to me without my being present. But my being present is 'present' in another sense. The word 'present' here is equivocal. I am not present in the manner in which this paper is present to me. I am not present in the sense of being presented to myself. I am present in the sense of necessarily being there if anything is presented to me. And that which is not present as presented but present as what other things are presented to, that is the subject as subject.

Again, I can reflect on the fact that this paper is presented to me and that I am present not as presented but as that to which presentations are made. And when I think of myself as that to which presentations are made, there is the subject that thinks this, and what he is thinking about is himself, namely, the one who is present in the manner of one to whom presentations are made. In thinking of himself in that way, the subject objectifies himself in a set of concepts. He conceives himself as the one to whom the presentations are made. But insofar as he posits that conception, he posits an object. He knows what he means, but you have to distinguish between 'him meaning it' and 'him meant.' 'Him meant' is the subject as objectified. I am still the one to whom presentations are made, the one who understands and brings forth concepts, when I conceive myself. But when I conceive myself I am conceiving and conceived, and I am the subject as subject only insofar as I am conceiving, meaning, intending. Insofar as what I conceive is myself, what I mean is myself, what I talk about is myself, I am subject merely as object. I am on the side of the object. I have made myself something that can be presented to me.

Again, I affirm my existence, I affirm that I exist. Insofar as I am affirming, I am subject, and insofar as I am affirming myself what is affirmed again is the subject, but not the subject as subject, not the subject as affirming, but the subject as affirmed.

The prior ontic reality in which one is going to find the norms and invariants that are common to all horizons, that recur in all subjects, the reality that provides the real norms on the basis of which one can select the true horizon, lies in the field of the subject as subject. It lies in the subject that does the talking, and not in the subject he talks about. It lies in the Hume that does the intelligent exposition of what knowledge is, and not in the knowledge that Hume talks about. The philosophers will differ in their accounts of knowledge, in what they say of themselves in their account of the subject as object, but that will not prevent them from being the same type of being, where the word 'being' is used not ontologically but ontically, in the sense of the reality you know in some fashion, in some, within quotes, sense of the word 'know' prior to conceiving it and affirming it.

Why do I say that we will find this ultimate criterion in the subject as subject? Because all objectification takes place within a horizon. All conception, all affirmation occurs within the horizon of some stream of consciousness. And if you are not to presuppose your own horizon in arriving at your own results, then you must appeal to something prior to your own horizon, something that can occur in all subjects independently of the diversity of their horizon. That something that is independent of the particular horizons of particular subjects is the subject as subject, the subject as prior to any objectification of himself, the reality that is present as what is presented to, as what presentations are presented to, the reality that is present as understanding and conceiving as opposed to understood and conceived, the reality that is present as affirming rationally as opposed to rationally affirmed.

Insofar as you move in upon the subject as subject in that fashion, you step within the field that Heidegger enters in order to set up his *Fundamentalontologie*, his fundamental ontology. When you say that Heidegger did not succeed in finding a fundamental ontology on the basis of which he could set up an ontology in the proper sense of the word, what are you saying? He discovered *Sorge, Verstehen, Rede, Sprache,* and so on, but he did not find anything normative, anything common to all subjects, on which he could determine which is the true horizon or pick out what is the horizon that is coincident with the field and account for the mistaken horizons.

Again, that same reality of the subject as subject is what Jaspers is talking about, and he talks about it with all the more conviction because, due to his Kantianism, he does not believe that any objectified knowledge is simply valid. However, insofar as he gets to the subject as subject he gets to something that does not fall under the strictures of the Kantian criticism. The subject as subject, because it is prior to objectification, is not under the limitations of being mere appearance in the way in which Kantian philosophy would argue every object of human knowledge to be simply an account of appearance.

What is the reality of this subject as subject? It is reality in a very prior and probably conceptually incomplete sense, but nonetheless in a very real sense. The subject as subject is reality in the sense that we live and die, love and hate, rejoice and suffer, desire and fear, wonder and dread, inquire and doubt. That is what the existentialist is talking about: the subject as subject, the subject that is the victim of all these things and the origin of all these things. That subject in his own living has a certain presence to himself, in some queer sense of the word 'presence' – it is not the same as the presence of objects – that experiences the terror of death, the agony of suffering, and the torture of guilt. That subject is reality in that ontic sense prior to any ontology, prior to any conception of himself, as *there.* 'Here we are': it is true of all. The subject as subject is Descartes's *Cogito* transposed to concrete living. It is the subject present to himself, not as presented in any theory or affirmation of consciousness, but as the prior non-absence prerequisite to any presentation, as the a priori condition of any stream of consciousness.

For the existentialists, one cannot say that within that prior reality there are to be found the norms, invariants, principles, evidence that will enable you to select the true horizon and institute a critique of all mistaken horizons. However, the argument as I have presented it – and it does not coincide of course with the exact mode in which the various existentialist philosophers have approached the question – shows that if there is an answer to the question, *there* is the field in which you have to look for it. *There* is the something that is real, the something that contains norms that in some sense at least are not conditioned or dependent upon modes of thought or upon the realities that are thought and affirmed within any given horizon. When you move to the subject in that prior and fundamental sense of the subject as subject, you are at the root of the horizon. You are dealing with something that is not merely determined by the horizon within which it happens to think.

That presents the problem, the issue, into which existentialism takes us. The existentialists have not got beyond that approach to the problem. Heidegger attempted his analysis of the subject as subject. He did not call it that. He objects to the distinction between subject and object. For him, subject and object are simultaneous, and the proper name for them in his work is *Dasein.* He wants one word for *consciousness* and for the emergence of consciousness in its *world.* Consciousness and world are simultaneous. Just as the dreamer and what he dreams about are simultaneous, so *Dasein* and the world in which *Dasein* is are simultaneous. The existentialists do not move on to that further issue and find a solution for it. Heidegger has been waiting for thirty years. Sartre took what Heidegger expressly considered merely as preliminary and attempted to fashion from it an ontology. And his ontology is completely unsatisfactory for reasons that are fairly obvious.

But he has at least the value of showing the insufficiency of the work that has been done so far by the existentialists.

If you want to go on to a development of how one could find in the subject as subject the foundations for a metaphysics, you will find that *Insight* is structured along the lines of the requirements I have just exposed. The fundamental argument occurs in chapter 11, where I treat the subject's self-affirmation of himself as a knower. The argument in every case turns upon what no subject can avoid. You cannot avoid experience. You cannot avoid trying to understand, even if you wanted to play the fool. In Italian churches one of the techniques they employ is to have a person who simulates great stupidity ask the questions that the preacher answers. It is known as *fare lo stupido* [to play the fool]. But if anyone of us were called upon to *fare lo stupido* we would want to do it in an intelligent way. You would want to do a good job of being the stupid fellow. You cannot avoid that tendency to ask *why*, to ask *what*. And even if you did set about trying to avoid it, you would want to do that intelligently. Again, you cannot abolish the exigences of your own reasonableness. If you want to play the fool and pretend to be silly, still you would want very good reasons for it. Or if you were to renounce your reasonableness you would find yourself asking reasons for it. De facto we are so committed to reasonableness that when we are unreasonable we try to rationalize our unreasonableness. Those three fundamental aspects of human experience lead immediately to a metaphysics in which matter corresponds to experience, form to understanding, and act to reasonableness and judgment. So you can see that you move automatically from an analysis of the subject as subject to the fundamental categories of a metaphysics.

Appendix A
Two Diagrams

The two diagrams to follow may be regarded as simply an editorial aid to reading the text. The first diagram is especially relevant to pp. 101–14, and is close to Lonergan's blackboard work during the lectures. The diagram itself is photocopied from the typescript of the six lectures given at University College, Dublin, Easter 1961, 'Critical Realism and the Integration of the Sciences,' with footnotes added for present purposes. That diagram was probably copied directly from Lonergan's blackboard work, and as the General Editors have commented in their Preface, the same diagram is relevant also to *Understanding and Being* and *Topics in Education.* Some years later I began using a simplified version of this and the second diagram. I would not have dared invent the detail of the diagram produced here. Fr Crowe and I searched the archives for a further occurrence of the diagram, but were unsuccessful. There are, of course, other sketchings at various places in Lonergan's scribbled writings that resemble the structure, but nothing comparable in detail. So the reader need not go beyond the cautious view expressed here, that this first diagram is probably copied directly from blackboard work.

What of the second diagram? The reader can easily see that it is almost a repeat of the first diagram, but on the level of action. It is constructed, then, on the basis of the first diagram, but as a way of suggesting the dynamics related to decision and action. The notes to the diagram relate it to the relevant sections of *Insight.* It seemed pedagogically important to include it here, since in both sets of lectures there is an emphasis on decision, and so there may arise the question, What are the dynamics of decision?[1]

It is extremely important to note the manner in which this issue lurks in the lectures on logic. It may well have occurred already to the reader because of the peculiar use of the word *Entscheidungsproblem*, the decision problem, for problems related to the tradition of Hilbert and Gödel. The issue there is the issue of action possibility, expressed with descriptive accuracy by Lonergan as 'the possibility of setting up automatic machines to solve problems.'[2] Lonergan's analysis of the dynamics of decision offers definitive light on completeness and incompleteness theorems,[3] on modal and deontic logics, on the logic of future contingents and the logic of operational definitions: in general, on problems of the formulation of action possibilities.

Such problems, of course, are problems of existential phenomenologies. Above all, there is the central problem of what may be called an axiom of the 'existential gap,' Fodor's thesis that 'the central cognitive processes are not amenable to investigation; while the input systems have problem status, the central systems have mystery status; we can handle problems, but mysteries are beyond us.'[4] In many contemporary universes of discourse, this is either a hidden axiom or even explicitly grants the old obviousness of the parallel axiom in Euclid.

Two other points should be made with regard to the diagrams and their meaning.

First, the distinction between the two types of what-question is modal. This distinction is conveniently caught in the ambiguity of the phrase 'might be' in the question, What might that be? This is of consequence in coming to grips with the relation of the analysis to Lonergan's later discovery of eight functional specialties.[5] The what-question in the first mode

---

1 For the historical context, see Frederick E. Crowe, 'Universal Norms and the Concrete *Operabile* in St. Thomas Aquinas,' *Sciences ecclésiastiques* 7 (1955) 115–49, 257–91; 'Complacency and Concern in the Thought of St. Thomas,' *Theological Studies* 20 (1959) 1–39, 198–230, 343–95.

2 Above, p. 55.

3 An essential supplement to clarification here, and indeed throughout contemporary theories of particle physics, is the distinction between primary relativity and secondary determinations, in *Insight*, chapter 16, §2.

4 Quoted in Robert Carston, 'Language and Cognition,' in vol. 3 of *Linguistics: The Cambridge Survey*, ed. Frederick J. Newmeyer (Cambridge: Cambridge University Press, 1988) 57.

5 See note 36 above, p. 114.

characterizes the second functional specialty of interpretation; the what-question in the second mode is the dominant ethos of systematics, the seventh functional specialty. Systematics, then, is focused on action possibilities: it reaches for ever fuller creatively implementable 'closed options.'[6]

Secondly, the diagram and its meaning call for the fuller context of an axiomatics of intentionality. Lonergan did not offer such an axiomatics, but it is continuous with his view of the possibility of a Scholastic or metaphysical axiomatics.[7] The elementary characterization of the 'position' in *Insight* and the historical analysis of *Verbum* need to be lifted into a full thematic of knowing and being.

6 The title of section 2 of the chapter on Systematics in *Method in Theology*. It is worth noting that the diagrams display one central option focused on in the lectures on logic: 'The key issue is whether concepts result from understanding or understanding results from concepts' (*Method in Theology* 336, note 1).
7 See above, pp. 121–33.

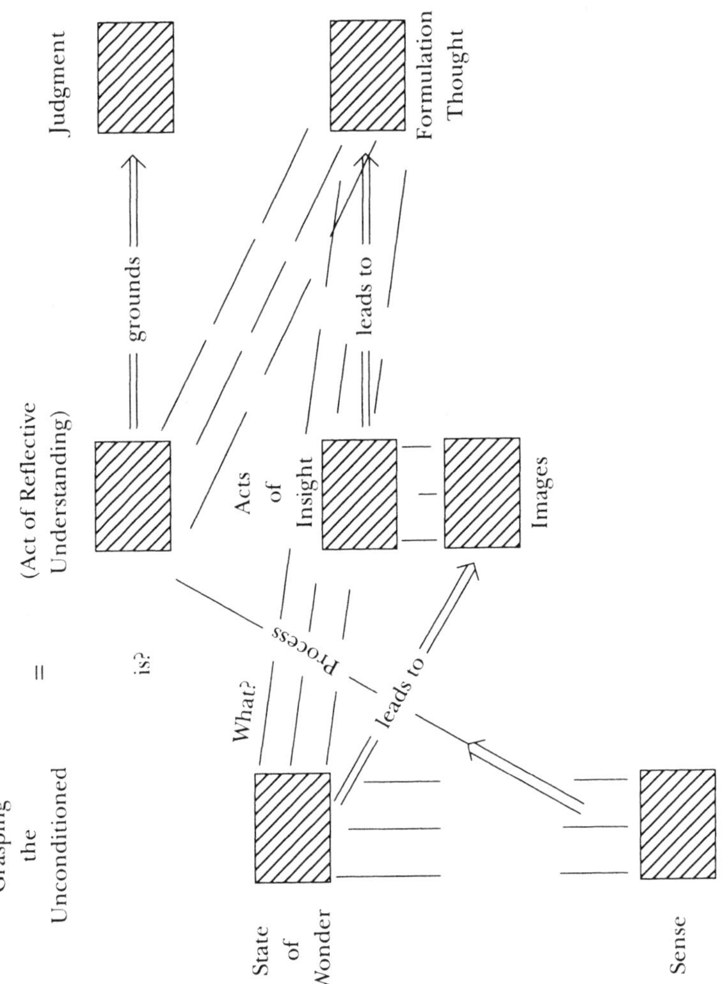

Dynamics of Knowing
Judgment
Formulation
Thought
grounds
(Act of Reflective Understanding)
Acts
of
Insight
Images
leads to
Grasping
the
Unconditioned
=
is?
Process
What?
leads to
State
of
Wonder
Sense

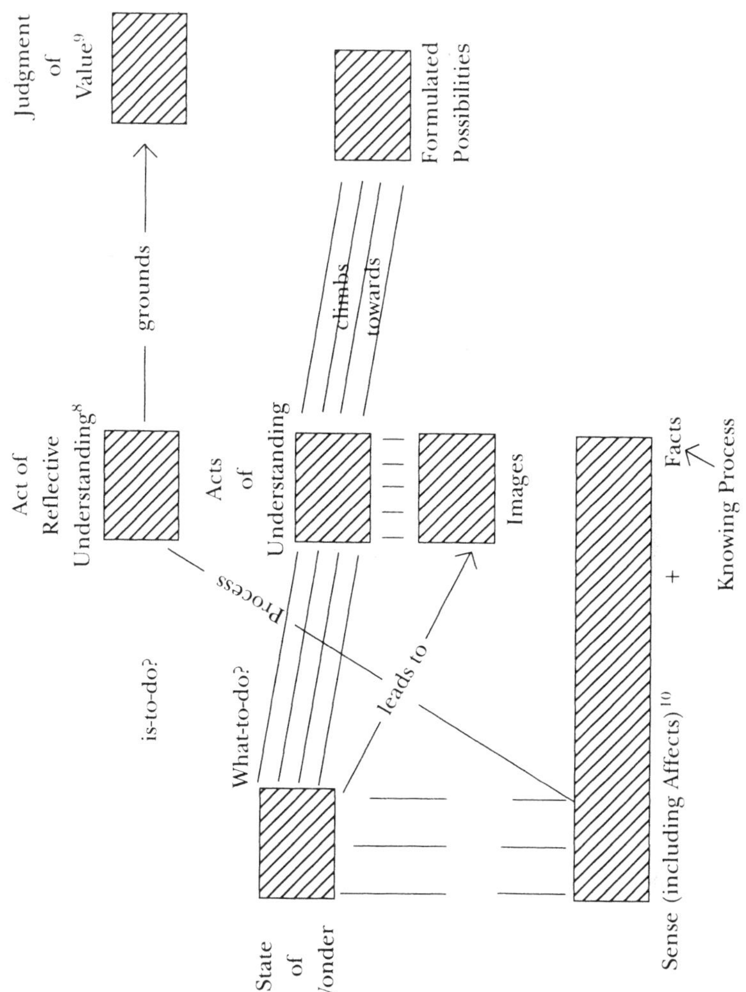

8 See Lonergan, *Insight*, chapter 18, §2.2.
9 Ibid. §2.4.
10 Ibid. §2.5.

Appendix B
The Experience of Science[1]

What is meant by 'science'?

1 Question arises from problem of integration.

    a Not any answer will do: the answer must

        a) account for the Ar[istotelian] concept of science

        b) account for its transformation into a modern concept of science

        c) provide a norm that will make possible a critique both of the ancient and the modern concepts

        d) provide a key for the problem of integration

    b no use consulting authorities

        a) anonymous authority of English usage: obviously unequal to task

        b) genuine authorities, with names and reasons for what they say: *La crise de la raison dans la philosophie contemporaine*, Desclée 1960, E Barbotin J Trouillard R Verneaux D Dubarle S Breton

        c) in any case we should need reasons for accepting or rejecting what authorities say; best find out what reasons are.

---

1 The title is editorial, indicating a concluding homing in on the topic of the volume. The remarks are related to a series of six lectures that Lonergan gave at University College, Dublin, 23–25 May 1961 on 'Critical Realism and the Integration of the Sciences.' They are included here because of their relation to the issue of science raised in chapter 5 above. It was in these years (late 1950s and early 1960s) that Lonergan worked out his articulation of the distinction between classical and modern notions of science. See above, p. xv.

2 Procedure / Annomic[2]
   a) experience: what happens when one learns science, makes discoveries
       = (1) learn something (2) attend not only to what is learnt but to the learning process, the decisive event in learning
   b) formulate the experience (1) descriptively (2) theoretically
   c) objective theoretical context: logic method metaphysics metaphysical-psychology[3]
   d) subjective theoretical context: subject, acts, structures of acts

3 Experience
   a) why is this plane curve perfectly round
   b) Euclidean scandal: not non-Euclidean geom but defective proofs first problem: equilateral triangle theorem: external angle greater than interior opposite
   c) why mathematical, symbolic logic
   d) Gödel and parallel theorems: unless trivial, either incomplete or incoherent. J. Ladrière, Les limitations internes des formalismes, Louvain about 1957.

4 Aristotelian-Thomist formulation of experience
   a) Appeal to experience exists
       Sum theol 1 84 7: quilibet in semetipso experiri potest ...[4]

2 This word, underlined, and the preceding stroke, are handwritten by Lonergan.
3 The autograph has 'met-psy.'
4 It seems useful to give here a loose continuous translation of the five Latin passages in the text: 'You and I surely notice the experience of image-dependence in ourselves?' 'It is an experience of our way of knowing, of course, that fits right in with Aristotle's talk of it.' 'For we gather creatively and collect from images, by our minding, our actual understandings: and how otherwise might we come to appreciate these activities if they were not part of our experience?' 'What reveals perfectly our creative energy and our nature is that natural activity of understanding *through the understanding of which* we get that perfect revelation.' 'And that's what the Philosopher did: he scrutinized the nature of the capacity to understand by attending to his understandings and their natural reach.' So: '*Admiramini enim subtilitatem Aristotelis*' (Lonergan, *De Deo trino: Pars systematica* [Rome: Gregorian University Press, 1964] 283: the entire appendix 2 there is relevant), 'You just have to admire Aristotle's subtlety.' The decision problem is to reach for a Bell-curve translation of that admired subtlety into industry, commerce, classrooms, governments, religions, homes.

Sum theol 1 88 1: secundum Aristotelis sententiam quam magis experimur ... secundum modum cognitionis nobis expertum

C Gent 76 §17: homo enim abstrahit a phantasmatibus et recipit in mente intelligibilia in actu; non enim aliter in notitiam harum operationum venissemus nisi eas in nobis experiremur

b) Hence not analogous but proper knowledge; not as a blind man's knowledge of color; human knowledge of angelic intellect; we know our own intellects, intellectual souls, by their acts.

Sum theol 1 88 2 3m: anima humana intelligit se ipsam per suum intelligere, quod est actus proprius eius, perfecte demonstrans virtutem eius et naturam

In III de Anima, lect 9, §724: unde et supra Philosophus per ipsum intelligere et id quod intelligitur scrutatus est naturam intellectus possibilis

c) Technique formulated for introspective study

In II de Anima, lect 6, §§305–308 (lect 1 ss, definition of soul; now definition of specifically different souls)

A. Objects B. Acts. C. Potencies. D. Essence of soul.
Regularly employed by Aquinas: ST 1 87; De Ver 10 8; C Gent II 75 III 46

– TS VIII (1947), 61–73

d) Object defined (1) logically (2) metaphysically

Logically: Post Anal II, 1 ss: 4 qq, ti & dia ti; syllogism equivalence:
what is eclipse;
why is sun darkened in this manner

Metaphys Z 17: what is a man, house: dia ti ti estin. ousia.
physis. Proton aition tou einai.
Difference between knowledge of material and immaterial

Z 17: announced 1041a 7ss; given 1041 b 9s.

Proper object: quidditas sive natura in mat corp existens 1 84 7

Only analogous knowledge of angels, God: Sum theol 1 88

Solution to ontological argument 1 2 1: we don't know quid sit

e) Aristotelian and Thomist Logic Metaphys Rational-Psychology[5] are conceived in functional relation with experience of insight, knowing cause

Ar physics (a) brilliant (b) ag[ain]st Ar's own principles.

---

5 The autograph has 'Rat-Psych.'

Appendix C
Question Sessions

Question sessions did not occur after every lecture. The following is an almost complete account of the sessions that occurred and were recorded. The questions are given in italics. Regularly, questions were inaudible or indistinct on the recording; in those cases some editorial comment is given in brackets before Lonergan's answer. Almost all the answers are given in full, with occasional editorial reduction or re-expression. A few notes are added, relating the answers to the text or to other works of Lonergan.

1   Questions after the First Lecture (Chapter 1)

[The tape begins in the middle or toward the end of a response to a question that had something to do with the isomorphism of mathematics and physical reality.]

To say that at such and such an hour there will be an eclipse of the moon involves a similarity in structure between the way his equations work out and, on the other hand, the way the moon moves.

The example you gave from Hilbert on point and straight line: does it involve any additional descriptive material?

Well, it's a type of definition that is not satisfactory from the descriptive viewpoint. On that business of definitions: you can define a straight line as what lies evenly between its extremes, and there you are just giving a descriptive definition of a straight line, and it doesn't help you to solve problems in geometry. After he gave that definition Euclid had to introduce a postulate that all right angles are equal. You can define a circle as a perfectly round plane curve, and you would have the same sort of useless definition of a circle as you have of a straight line. If you define a circle, though, as a locus of points equidistant from a center, then you don't need any postulate that all radii of the same circle are equal. You have it in your definition. So here [in the first case] you have just description of what you look at; here [in the second case] you have what's equivalent to a description, because it enables you to pick out the proper use of the name 'circle,' and at the same time it gives you a postulate about circles. Now the implicit definition pulls a fast one. It drops out all descriptive reference and saves only the postulational elements.[1]

Is it all right?

Oh yes! It works, all right! But the reason why it's all right is this, that in your thinking you need a descriptive technique to link your explanations with concrete experience. Implicit definitions are quite all right and work splendidly, super-splendidly, within the purely explanatory field. However, you need something else to link them up with concrete sensible data, and there's where your descriptive definitions are necessary and useful.

However, when you're using implicit definitions, you're solving problems with perfect generality. For example, geometers discovered that in projective geometry there are parallelisms: what will hold for a point and a plane will hold for two other things. Well, instead of working out all their theorems twice, they get some more general definition and work the whole thing out once. That's the importance of this implicit definition: it is a perfect generality. And again, mathematical logic has that advantage with regard to mathematics. When you do your mathematics in this more general logical field, you bring together a whole lot of different departments of mathematics that formerly were separated. They have the same logical structure, and they all come together. Consequently, one of the advantages of mathematical logic is called the unification of isomorphic fields. All mathematicians recognize that side of symbolic or mathematical logic. It

---

1 A context for the discussion here and in *Insight* is Lonergan, 'A Note on Geometrical Possibility,' in *Collection* (see above, p. 13, note 14) 92–107.

brings together logically similar departments that previously were thought to be simply disparate.

Is there a danger in the isomorphism of mathematical, physical, and logical relations that relations will be made into something they are not?

Well, that's the sort of problem these people run into in the long run, but if they didn't start out somewhere they'd never discover the problem. In other words, the dangers are all there. The question of the relation between concepts and realities, and so on, crops up as a question of semantics, as they call it – the relation between object language and, on the other hand, the objects you're talking about. They study that under the name 'semantics.' But I think you're jumping ahead. In other words, what we want to do is see what they do, and afterwards make a judgment on it. A piecemeal examination doesn't lead anywhere. I brushed aside questions of terms, propositions, inferences, because the interesting thing is the system, the ability to study systems. And we will see tomorrow how the topic is developed, and the things they've discovered about deductive systems and the results it has had on them, and so on. The naive ideas with which they started out have been simply transformed. The transformation isn't quite complete yet, but a lot of things are going on. And the discoveries they make are also interesting to us. What we're seeking now is to understand somebody else's mind, what they're doing. Later on we'll see, 'Well, there may be something in it for us or there may not.' That's a further question.

[The next question had something to do with the further epistemological questions and philosophic difficulties.]

They are forced by it to discover the necessity of epistemology, as we will see – insofar as you can force anyone. It's starting off again in the Middle Ages. In general the people in this field, and in logical positivism and so on, are unacquainted with the philosophic tradition. They are not formed in *any* philosophic tradition. A lot of what they are doing is making discoveries of things that are very familiar to philosophers. They are discovering them for themselves in a new way. Peter Lombard's first book of the *Sentences*, the first distinction: signs are signs either of things or of signs. We've heard that from these people. They're making the discovery again. They have put it somewhat differently, but what concerned the Middle Ages is being done in a new way, and with proper help it can be brought on. However, it has implicit tendencies towards pragmatism and empiricism, and so on. There's no doubt about that.

Is it possible that there is some overall logic that embraces these others?

We'll come to that tomorrow perhaps. But it's a very good question, quite a good question. Very briefly, my answer to that, quite independent of these people, is that you have your *intellectus possibilis* and you get *species intelligibiles* in it and have acts of understanding which are expressed in concepts and systems of concepts, and the more you understand, the more your concepts shift. So the development of mind is through a series of logical planes. Every time you make a new distinction that you hadn't thought of before, you move up a step, and you're in a different strictly deductive system. And the philosophic problem is not one of following along intelligence as it develops but getting hold of the general shape of the development. What's true no matter what plane you're on? That's the level of proper philosophic questions. In other words, let these people discover the necessity of an infinite series of strata. What are they doing? Well, they're saying that human intellect is potency, and you can keep on actuating it indefinitely in this life without arriving at the vision of God where you understand everything about everything. For those people to discover that is a big thing, and they're moving towards it.[2]

2   Questions after the Second Lecture (Chapter 2)

Could you say a few words on what is meant by a functional calculus of the first order?

A functional calculus of the first order: the basic thing is the classical propositional calculus that deals just with sentences or propositions: P, Q, R. The units are sentences. First of all, you do the classical propositional calculus. Then you are dealing with sentences, and you don't divide the sentences up into parts at all. And you have various symbols, and various ways of combining these sentences, and that gives coherence and completeness as demonstrated. The next step is, instead of simply dealing with sentences deal with the argument X that has the predicate P. And then

2 See above, p. 66, note 50; p. 111, note 33; p. 125.

you're in functional calculus. [Lonergan went on at some length with a great deal of board work, speaking of such things as the introduction of quantifiers, etc. Because of his reliance on the board work it is impossible to recover his presentation.]

What is the historical and theoretical relation of Descartes's deductive ideal more geometrico *and these people?*

Well, it's the same sort of thing, you know. The deductivist ideal is that the way to do a science is to set down certain initial propositions and deduce everything from them. There is a measure of truth there, but it isn't the whole truth.

[The next question had to do with the completeness of a system, but the actual wording of the question cannot be ascertained.]

It's complete when, if P is any proposition that can be constructed in the system, you can always prove either P or not-P. That's one way of conceiving it. You can conceive it also in weaker forms. They can show how the functional calculus of the first order is complete, or at least so they say.

[The next question had to do with Gödel.]

Gödel's limitation: when you have a formalization elaborate enough to handle a theory of arithmetic, then you get the limitation. You can't express a theory of arithmetic in the functional calculus of the first order. As a matter of fact, in order to handle something like the *Principia mathematica* – I'm not so sure about arithmetic – you need the functional calculus of *any* order. They go on from predicates to relations. And the relations are conceived as relations that define classes, and they have intersection of the classes and all the different ways they can come in, and that's what they are really interested in.

[The next question had to do with concepts.]

Concepts occur within a heuristic structure. I conceive *ens* not as some aspect of this thing before me but as everything that is to be known about it. I have an ideal or an anticipation of a totality to be known in every object. My descriptions are just exterior specifications of this totality that I'm trying to get hold of. My series of explanations of this object are just partial knowledge of the object, but the idea of *ens* as applied to this object is something that admits indefinite progress. And every concept falls within such a structure.

You'll find a very simple illustration of it in something I put out this year on the Trinity, where I dealt with the notion of person.[3] The word 'person' was used by Tertullian, and finally toward the end of the fourth century they persuaded the Greeks to use the word *hypostasis* in approximately the same sense. In the fifth century Augustine asked, 'What does "person" mean? What do we mean by "person"?' And he said, 'Well, we have to have some means to talk about the Father, Son, and Holy Spirit, without saying "Father, Son, and Holy Spirit" every time, so we call them persons. Well, these persons are related to one another, but "person" means something absolute, so I really don't know what on earth this "person" means.' It's a purely heuristic notion of person in Augustine.

In Boethius you get a definition of person: *individua substantia naturae rationalis*. In Richard of St Victor you get another definition, this time not of a person in general but of a divine person: *incommunicabilis existentia naturae divinae*. You get another definition from Thomas: *subsistens distinctum in natura intellectuali*. And then the theologians had a number of things to compare: *subsistens, distinctum, individua, substantia, natura, divina, intellectualis, rationalis*, and so on, and they had to go into metaphysics. And you get Scotus's theory of the person and Cajetan's and Capreolus's and Suarez's and Tiphanus's, all of them trying to give a metaphysical account of what the person is. To be able to compare these definitions and pick out which is the right definition of the person and what exactly it implies they have to go into a metaphysical theory of the person.

After Descartes there is a move to psychology – unless you begin anthropologically you never know where you are, you have to get back to something closer to us. They started doing philosophy on a psychological basis, the basis of inner experience. Everything else is thought out in terms of inner experience, and you can think of a person that way too, and so you start getting psychological theories of the person, and it comes to grief very quickly. As you can see, the person has consciousness. With the Trinity in

---

3 The reference is to *Divinarum personarum conceptionem analogicam evolvit Bernardus Lonergan S.I.* (Rome: Gregorian University Press, 1957) 131–39. The same material is found in the 1964 revision of this work, *De Deo trino, Pars systematica* (Rome: Gregorian University, 1964) 153–61.

God, are there three consciousnesses? If you say yes, you're in trouble. You get into difficulties with the Incarnation where you have one person in two natures, and in the Trinity where you have three persons in one nature. Consciousness alone won't serve you to define person in any sense that satisfies theology.

Then there is the more recent existential-personal approach. A person is someone you talk to and who talks to you. You have respect for one another. A person is quite different from a thing. You're on the concrete level. They feel you can get no further with the psychological approach than you can with the metaphysical, so they're right down to the distinction between person and thing: a person is something you don't call 'it.' All along the line, they're asking the same question; they're looking for a definition. It's a pure heuristic structure, pure inquiry from metaphysics to psychology to concrete living. It's always the same question. The notion of person is developing all along the line, becoming enriched, but it's also introducing elements of corruption: human development is never just pure development, there's always a mixture of regression in it.

What do you mean by 'person'? Well, at what stage of the process are you? Are you thinking with Augustine, Thomas, Capreolus, Cajetan, the nineteenth-century psychologists, or the phenomenologists of the present time? And how do you put them all together? You need a heuristic structure that will hold all these differentiations of the notion of person as you go along. Have we all got exactly the same notion of person in this classroom? Well, we might agree on *subsistens distinctum naturae intellectualis* but we probably want a little richer notion of person than that; we want to get down to things exactly.

Now 'person' may be a notion that has developed greatly and in a conspicuous fashion. But with any of the ordinary notions there's a series of scientific explanations. There are different societies and different contexts and so on, involving a lot of variation in any concept. That fact, I think, gives to ordinary everyday logic a complexity that's not of the same type as mathematical infinities, but that makes a deductive system just as difficult as the infinities involved in arithmetic.

3   Questions during and after the Third Lecture (Chapter 3)

[The first questions in this section were raised at a break period about halfway through the lecture. The first question had to do with how an analytic principle differs from an analytic proposition.]

It transforms the analytic proposition into something that applies to the real order.[4]

[The second question was about the interpretation of Aristotle; it seems to have been a complex question, treating both the difference between two different texts attributed to Aristotle and the meaning of abstraction. Only part of Lonergan's response to the question about abstraction could be recovered.]

... by *ens* I do not mean anything abstract. By potency, form, act I do not mean anything abstract. Those three terms are always concrete for me, and they are defined by their relations to one another. The element of universalization lies in these relations, or in the possibility of providing a single concrete ground for the total series of the concrete relations. Even the relations need not be abstract. In other words, when you are dealing with a relational structure, to grasp the structure is to grasp something that may be filled in by various particulars. But that structure, insofar as it determines the nature of the particular, entirely insofar as it is nature, penetrates right into the concrete. However, the question of abstraction is enormous. By abstraction I mean the omission of the irrelevant. It's selecting what is significant, essential, important, to the point, and dropping what is beside the point. And that dropping what is beside the point occurs differently in the sciences and in philosophy. But to discuss the matter further, I think, would take us a little far afield.

[The third question was about *unum per se* and *unum per accidens*.]

I spoke of the *unum per se* and the *unum per accidens*, and distinguished two meanings of the *unum*. In one case, the *unum per se* means what is substantial, and in that sense a man is *unum per se* and a machine is not. In another sense, an *unum per se* is an intelligible unity, and in that sense a machine is an *unum per se* and a heap of stones is not. And the relational structure is *unum per se*, like the machine, but not like the man.

[The fourth question had to do again with the issue of when one has an analytic principle.]

4 See Lonergan, *Insight*, chapter 10, §7.

If your terms in the sense in which they are defined occur in true concrete judgments of fact, such as 'Something happened when I went back to my room.'

[So too with the fifth question.]

You have to have something universal before you can have either analytic propositions or analytic principles. The analytic principle goes beyond an analytic proposition insofar as you add concrete judgments of fact, such as 'This piece of paper is white.' In other words, if you have your definition of the term 'white' and other definitions which go together to form an analytic proposition, and you want to determine if your analytic proposition is also an analytic principle, then one of the concrete judgments to which you might appeal is that this piece of paper is white. But the analytic principle adds on to analytic propositions as many concrete judgments of fact as are necessary for all terms that occur in the analytic proposition to be shown to occur in concrete judgments of fact, just as defined.

Then the analytic principle is really a conclusion?

Everything is a conclusion if you analyze your judgments as based on the grasp of the virtually unconditioned. But although everything is a conclusion after you do the analysis, it doesn't follow that you have to do the analysis to reach the judgment in the first place. That's the psychological fallacy. I think there are some people here who hadn't heard before this analysis of analytic principles that I gave this morning, but I think they *did* it before.

[The same issue was asked about again, with reference to experience.]

Strictly, you have an analytic proposition when you reach the virtually unconditioned by definitions and rules of syntax. You're prescinding from experience at that point, and you go on to analytic principles by bringing in a field of experience, and not just experience but true judgment. The issue is true judgment: we're not empiricists.

[A question was asked about the distinction of science in potency and science in act.]

God knows the universal along with the particulars. I'm giving another slant on the matter than the one that's more common. With regard to science as of the universal, note that, insofar as your science is of relational structures, it admits a far greater approach to the concrete than does science that consists simply of universals that are abstracted from particulars. In other words, theory of history has to be science in this sense of relational structure.

[This was followed by a question regarding the synthetic element in propositions.]

We'll say more about that on the fifth day.[5] It's a quite complex question. You get your synthetic element on your first level, asking the question, *Quid sit?* You have in every true judgment the analytic element of the virtually unconditioned.

[The final question in this break period relates to the context of Lonergan's analysis of analytic propositions and principles.]

This distinction I've drawn is in response to the claim of these people that you're talking either tautologies or empirical generalizations. The analytic proposition pins down the tautology – it isn't exactly what they mean by tautology, they define it differently – and the analytic principle takes you beyond the tautology insofar as you're dealing with something that occurs in concrete judgments of fact.

Automatically, it's been taken for granted that when you propose an analytic principle you are proposing something that has reference to the real world. So what I say about analytic principles isn't anything new. But I'm making explicit with regard to the analytic principle something that has to be explicit if we're going to be able to tell in what sense they are right and where we differ from them. Your analytic principle is an empirical generalization only in the case of a provisional analytic principle. But there is also something that is connected with our metaphysics that can be expressed in analytic principles.

5 Lonergan mentioned a conflict between Quine and Feigl on the matter. He probably meant to treat the question in the section on Logical Atomism and Logical Positivism in the fifth lecture. But in fact he did not present that material.

[The next set of questions was raised after the lecture. The first questioner asked if he understood correctly that the fragments of truth that Lonergan was talking about had to do, not with the content of the logical systems but with 'the type of understanding.']

Don't bother about content. They have no content. The nearest they get is saying, 'X is Y' or some relation of some kind between two things, but we don't know what, call them X and Y, and so on. Each of these systems has axioms and rules of derivation, and everything in the system comes out of that.

[Next there is a question about 'x or x.']

'x or x implies x.' 'x or y implies y or x,' where x is always the same proposition and y is always the same proposition, and they may be any proposition.

[The same questioner asked if this is not trivial.]

They aim at being trivial. The first rule of derivation is that you can put in any proposition for x provided you put in the same proposition wherever there is an x. And the other rule is that, if you have premises and a conclusion you can drop off the premises. But it's trivial, and consequently it's easy enough to see that everything that really follows from them has to be admitted if you admit the starting point, and that gives you the truth parallel to the type of truth you have in the analytic proposition. Can we go a step further and say we have the type of truth that belongs to the analytic principle? We can, insofar as we can say that people can't avoid using propositions and procedures of the type you have in the axioms and the rules of derivation. People can't avoid using some propositions, but we can't find fault with this. People can't avoid using negative propositions and combinations of propositions. Therefore, such a system, as far as it goes, is absolutely true. However, while they list their axioms and their rules of derivation, they don't write in at the bottom, 'And nothing else.' And it's in view of that implicit 'and nothing else' that this system is closed. There's no question of strict implication, and no question of three-valued logic, no question of dropping excluded middle in any particular case, and so on. That 'and nothing else' is the reason I say it's a fragment of truth. It isn't a total logical system that handles all questions of logic. It's a system that handles all questions of logic that fall within this restricted domain. It's the sort of thing that you can use with confidence whenever your thinking is of this type. However, if your thinking involves strict implication, then you need to go on to one of Lewis's systems in which strict implication can be expressed. There is no room for the expression of strict implication in this system. If you say, 'If the flag is at half mast the mayor is dead,' and 'If bananas grow on trees, Eisenhower is President of the United States,' both statements have exactly the same representation in this system, and there is no possibility of expressing an intelligible nexus between propositions in the system. All you have is combinations, conjunctions; you have compound sentences, not complex sentences.

[A question was asked as to what is the factual reference that gives the truth of an analytic principle, in the case just being talked about.]

The factual reference is the fact that anyone who talks is going to use propositions. Insofar as you can't single it out as making implications, you have to go beyond it and add on another one, such as a Lewis modal system. And if you talk about contingent future, you go on to a three-valued logic.

[The same question is pursued.]

The relevant judgments of fact here are judgments of fact regarding the occurrence of propositions, negations, strict implications, and so on. In other words, the facts are mental facts. The relevant facts that provide the transition from logic considered with the value of an analytic proposition to logic considered with the value of the analytic principle are universal mental facts.

[The next question was about the truth of an analytic proposition.]

It's true. It's virtually unconditioned. The criterion of truth is the virtually unconditioned. When the mind grasps the virtually unconditioned, it's rationally bound to affirm it. But it isn't rationally bound to bring it within the context of what is called knowledge. To get *that* rational obligation, you have to have concrete judgments of fact in which terms in their defined sense occur.

[The next question is not clear, but it is also about the analytic proposition.]

It's essentialist. If there were only analytic propositions and no analytic principles, you'd have to be essentialist. Because there are analytic principles as well as analytic propositions one can go beyond essentialism to an existential philosophy (not in the sense of existentialism, which we are going to treat next week).

[The next question is about the relation of this to traditional logic.]

Among Prior's thirteen axiomatic systems, formalized syllogistic is one. In other words, you can express the whole of Aristotelian syllogistic in one of these mathematical systems.

Does it involve strict implication?

Yes, it does involve strict implication.

You said the formally unconditioned is God, therefore, a being. Is the virtually unconditioned a being, a proposition, or either, or neither?

It is, first of all, the content of an act of reflective understanding. Secondly, the proposition, the judgment, that follows from that act of understanding. Thirdly, the being that is known through that judgment: *verum est medium in quo cognoscitur ens.* You can apply it to all three. You can say the virtually unconditioned is the evidence you grasp, you can say it's the utterance of what you grasp, namely, the proposition; and it's what you know through the judgment. Being starts out as a conditioned, but when you get this far you have a virtually unconditioned. And you can also apply it to the instruments, the words, that you use to express it. It can move all along the line. In every case, you have what I call a virtually unconditioned, a conditioned whose conditions are fulfilled: they all satisfy that same definition.

Is having sense experience what the fulfilment of conditions is about?

Having the sense experience is the fulfilment, but if you just have the fulfilment without any idea of a conditioned or the link between conditioned and conditions, you're an animal, you're not yet on the level of judgment. In other words, the sense experience is decisive insofar as it is within the context that's thinking of a conditioned, a link between conditioned and conditions, and the fulfilment of the conditions. Sense experience as such has nothing to do with it. Sense experience becomes fulfilment of conditions within a certain context. It acquires that significance within the context. That context is absent in the animal and present in the human being. You come to your room and say, 'Something happened.' The fulfilment there is the present experience of the room and the memory of the past situation. Again, a pencil – what do you mean by a pencil? Something with which you can write, something that's of a certain shape, and so on. You can try and see if you can write with this, see if it's the right shape, if you can hold it in your hand, and so on, and if so you will say that it's a pencil. The fulfilment of the conditions is the set of experiences. The idea, the concept, of the pencil is the list of things that you have to have to fulfil it. That's your link between conditioned and conditions. Your judgment is, 'This is a pencil.'

What is an axiom?

An axiom is a statement that isn't proved. It has to be useful for setting up a system.

[There followed a question about two-valued and three-valued logics.]

Most of these systems are two-valued. It's ever so much simpler to make up a two-valued system than a three-valued system.

[Then there was a question about the contingent future.]

'There will be a naval battle tomorrow.' 'Tomorrow morning I will take breakfast.' Contingent future. Call it P. Is P true now? If P were true now, there would have to be before me the evidence from which it would necessarily follow that P would occur. If P is contingent, there can't be that evidence now, and consequently P can't be true now. Is P false? For P to be false now, I'd have to have the evidence now from which it would necessarily follow that I won't take breakfast tomorrow. And if there were that evidence now, my not taking breakfast tomorrow wouldn't be contingent. Therefore the proposition P is not true, and it is not false.[6]

[The next question was about functors.]

6 See above, pp. 36, 84.

The stricter way of defining them is by their internal relations, but they're also defined by what they call truth functions. For example, the function *Apq*: *A* (*A* roughly means 'not neither') will be true or false according as *p* and *q* are true or false. *p* can be true, *p* can be false, *q* can be true, *q* can be false. You have the four combinations: both true, both false, one true and the other false, one false and the other true. Then you can define what you mean by *A* by picking out the combinations for which *A* will be true. *A* will be true if both *p* and *q* are true. So *A* is false only when both are false. In the other cases *A* is true. And that's what you mean by *A*: *p* or *q* with 'or' in that sense. Equivalence: *both* are true or both are false, and so on down the line. Disjunction: not both. So you can have disjunction in every case except the first one. But it's a mere matter of combining true and false. It's not a matter of nexus between propositions, and if you want to go on to a three-valued logic, instead of four combinations, you have nine to define your *A* by. It makes things much more complicated.[7]

Would you explain science in potency and science in act in relation to mathematical logic?

Ens is a concrete reality. When you know, you're knowing a concrete reality. When you're knowing the universal, you're not knowing yet a concrete reality, unless you're a Platonist. It's only when you apply your universal to particular cases or to a totality of concrete particular cases that you will be knowing reality, and that's science in act.

[Another question about contingent future]: *Is it not true to claim that it will be either-or?*

If you hold excluded middle.

[Questioner continues.]

If I have just a cup of coffee tomorrow morning, will that be breakfast?

Now you're into content.

Well, if I'm getting into content I'm getting out of either-or. Under certain

7   See above, pp. 36, 84.

conditions you have to say either-or, if you specify them closely enough. But the question in contingent future is not whether it *will* be true or false but, Is it true now? What is true now is that it will be either-or. That proposition is all right now. But the other one remains neither true nor false. Besides the proposition, Either *P* or not-*P*, there is the proposition *P*, and if you have a logic in which there isn't room for my proposition *P*, you haven't a complete logic but merely a fragment.

[The next question moves from the issue of the virtually unconditioned to ask how knowledge starts and how the fulfilment of conditions happens, and so on (in other words, what am I doing when I am knowing?)]

'Know' has a series of meanings. You have knowledge in the full sense. Knowledge in a *man* is experiencing, understanding, judging, and you have knowledge when you have all three. If you have experience alone, you're on the way to knowledge; knowledge is *in fieri*. If you have experience and understanding, you have more than you had before but still your knowledge is *in fieri*. It's when you come to judgment that you are coming to knowledge in the full and proper sense. You can consider this from the perspective of being. Matter alone isn't a being, matter and form isn't a being, it's the matter and form actuated by the *actus essendi* that gives you being. You have being in the full sense, *in actu*, only when you have all three. Similarly, you have knowledge in the full sense only when you have all three. But when you have understanding, you have the capacity to grasp the link between conditioned and conditions, and when you have experience you have the capacity to undergo, to feel, what can be the fulfilment of the conditions.

Where does reflection enter in?

Reflection enters in when you start to judge. And you can answer the question for reflection at the moment you grasp the virtually unconditioned. And when you are reflecting, you're moving toward grasping the virtually unconditioned. That's what you're trying to grasp.

Is there any judgment at all involved in understanding?

No. Understanding is just getting the point. There's no judgment in the proper sense. If you define judgment as nexus, then you don't mean by judgment what I mean by judgment. By judgment I mean *positing* the nexus. It *is.* Plato conceived judgment as linking ideas together, and that's not surprising; he was an essentialist. Aristotle spoke in the same way, and it's somewhat more surprising. Thomas often speaks in the same way, and that's still more surprising. But understanding is at the level of synthesis, while judgment is at the level of positing synthesis. Judgment is a virtually unconditioned, an absolute, the true, what's eternally so, what de facto can't be otherwise: that element is the element of judgment, and it's not nexus, it's positing.

[A question follows about implicit and explicit.]

It isn't really like explicit and implicit. You can take it in so many different ways according to the viewpoint. Usually what you'd be talking about is explicit advertence to processes that are going on, and just performing the processes: the distinction between *exercite* and *signate.*

A definition can be on the level of understanding?

Yes. A whole *book* can be on the level of understanding!

Are certitude and truth exactly the same thing?

They're pretty close. Certitude is your state insofar as you have grasped the virtually unconditioned. However, you can go on on that indefinitely. Roughly, certitude is more the state, while truth is the object. The subject is certain; what he's certain about is what is true. That's a first approximation; we could go on indefinitely.

4   Questions after the Fourth Lecture (Chapter 4)

[A first question, inaudible, asked something to which Lonergan's immediate response was, 'No, not that I know of.' Then he went on to say the following.]

What I do think is that we can surrender the Aristotelian technique, but also that what we have to hang onto are the fundamental Aristotelian ideas that are the soul of the thing. These ideas are inadequately represented by the technique, and probably, in the history of Scholasticism, they have been lost at times, in certain circles. But they are the really significant thing. Also I think it's true (but I probably have not succeeded in putting it across) that the foundations of logic that I have offered provide a basis from which one can judge the utility and real advantages offered by techniques, on the one hand, and, on the other hand, their limitations, what one is not to expect of them, and why one isn't simply to adore them.

[The second question had to do with the interpretation of Aristotle and Thomas on metaphysical questions. The questioner seemed to be calling attention to possible inconsistencies in the work of each and between their respective writings.]

The fundamental thing, I think, is to realize the complexity of the job of writing. St Thomas, when he was alive, wasn't St Thomas yet! He didn't have the idea of himself that subsequent people have had. He didn't write as though he were going to be read by people down through the centuries. He wrote for an immediate audience. I'm inclined to believe that there are differences in the mode of expression in St Thomas according to his audience. He tends to use one type of terminology in his *Quaestiones disputatae* and another type in such a work as the *Contra Gentiles,* which was aimed more or less at the cultured Aristotelian group, whereas the other writings would be talking more in the Scholastic language. Just as you, for example, if you were teaching a group of Scholastics, would present your ideas in one set of expressions rather than another, but then talk in a different manner to another group of philosophers, similarly St Thomas adjusted himself to his milieu.

In the second place, Thomas went beyond Aristotle. He went beyond his predecessors in theology. He went beyond them tremendously. But when a man is performing a tremendous synthetic work of that nature, he can't be continually fighting out the whole issue on every occasion. If it makes no difference in a given case if you distinguish between form and act, then follow the line of least resistance and write it out as Aristotle would. In historical research there is the *per accidens.* There is an ultimate residue of things we can't explain. Again, there are a further number of things that are explained by general tendencies, and these are understood by people who do some writing themselves, and they are very difficult to explain otherwise. It's quite clear that Thomas posits potency, form, and act in the first question of the *De potentia,* where he distinguishes potency and act, and then both of them in two senses, where second potency is the same as first act, that is, form. You get the same thing in the [commentary on the] *Metaphysics*, with an Aristotelian basis. Where Thomas differs from Aristotle is in his extension of that triple division from the accidental order to the substantial order. As Aristotle distinguished eye-sight-seeing, Thomas distinguished matter-form-existence. You do not have in Aristotle the triple distinction in the existential order.

[The third question had to do with foundations.]

The question of foundations to me, insofar as it's *philosophic* foundations: the problem in philosophy is not the same as the problem in science. In science there is one tradition, and you join on, accept science in its present situation, make your contribution, and die and disappear. In philosophy there is a series of different schools, and the question in philosophy is always the personal question, fundamentally: on what level of appropriation of your own intellectual and rational nature are you, and can you move up to another one? I think that is the fundamental issue in philosophy, the fundamental problem of philosophy, that moving up. And foundations in philosophy are from the subject in his grasp of his own intelligence and rationality grounding logic, epistemology, metaphysics, ethics, natural theology, right along. That is the real principle. In other words, you can use 'principle' in a logical sense, as 'analytic principle,' but Thomas very frequently says that 'principle' means *primum in aliquo ordine*, and it means very frequently in Thomas a *real* principle. He speaks, for example, about the reduction of judgments to their principles. What are their principles? Sense and the light of intellect. And it's foundations in that sense, *real* principles. The *formulation* of those foundations is secondary. In other words, the same point that I am empirically, intellectually, rationally conscious can be expressed in a thousand ways, and has to be expressed in a thousand ways, to help a thousand different people, because it's an *ascensio mentis per intelligibile et verum ad ens*. Bringing about that *ascensio* is the fundamental task of the philosopher with regard to his pupils, and bringing it about first of all in himself so that he will be able to help others. And once he has brought it about, what counts is not the expression. It is not some part in some technique, but a development in the man himself who is going to handle the technique, a development that enables him to pass judgment on the technique.

[The fourth questioner asked whether metaphysics should not precede epistemology.]

You want to have being prior?

Yes.

We'll perhaps say something about that tomorrow. In Thomas and Aristotle being *is* prior, and that was by necessity. In other words, the first exposition of this material has to start in the simplest way; and that is the simplest way. The fundamental thing in Thomas is *sapientia*, wisdom. *Sapientia* is a gift of the Holy Ghost, a mystical experience; but, on a lower plane, it is Aristotle's metaphysics. And it is by *sapientia* that you effect the transition from analytic propositions to analytic principles. *Summa theologiae*, 1-2, q. 66, a. 5, ad 4m: he doesn't put it that way, but he's showing that *sapientia* is superior to *intellectus*. And if the *sapientia* that is the ground of everything is Aristotle's metaphysics, then either you accept Aristotle's metaphysics or you don't. Once philosophy becomes separated from theology, as it does in Descartes and the subsequent tradition (not merely distinguished but separated), there arises a problem of the genesis of *sapientia*. How does one get to wisdom? It's all very well to say that everything is sapiential, that the solution to all these epistemological and ontological problems is sapiential. How does one get there? If you want to answer that question, then you do not start with being, because it's a highly ambiguous question. You start with the subject as experiencing, understanding, and judging, and you get him up to the level of being. The notion of being is not a notion that has been given just one interpretation. It has been given an interpretation by Parmenides, by Plato, by Aristotle, by Avicenna, by Averroes, by Thomas, by Scotus, by Suarez, and by Hegel – at least those, if not more; there are probably others. Which one are you going to choose? You won't choose in terms of some ontological principle, because all your principles vary with the meaning of 'being.' If you want to deal with that epistemological question, I think you have to go with the other approach.

[The fifth question is inaudible.]

The philosopher's mind has to be a unification of the sciences and of human experience and human life, insofar as possible. The practical problem of teaching leads to conventional divisions of different parts, with different professors handling different parts. And that's a more practical question, and such practical questions, as St Ignatius says, lead to a considerable amount of dispute.

[The sixth question is inaudible.]

The first thing is the subject taking self-appropriation of himself, and that by all means is hard to do. The real problem is not a logical one. In your question there is some sort of a presupposition that philosophy is one of these deductive chains, and we want to have the right starting point from which we deduce everything else. But de facto philosophy and all science is a matter of mounting through these planes and producing a real development in the subject.

5   Questions after the Fifth Lecture (Chapter 5)

[The first question had to do with the relation between what Lonergan had said about 'substance' and the physics of light.]

The physical treatment of light – there are different theories about it. If you hold the corpuscular theory of light, then you'll be saying that it's a substance. But if you hold the wave theory of light, you will say, 'Well, we've got a set of equations, and just what they apply to isn't too clear.' The treatment of light comes within the general treatment of electromagnetics, and what you're dealing with there, in the earlier physics at least, are Clerk-Maxwell's equations, which are a series of vectorial differential equations. There is no mention of substance in the interpretation of those equations. What you get from the equations is this: if you get pointer readings of this type here, you'll get pointer readings of that type there. Whether there are substances or not doesn't enter into the physical question. You get into the question of substance in physics when you start talking about electrons, protons, and that sort of thing.

[The next questioner asked for clarification on the relationship between the question of substance and the distinction between explanation and description.]

The business about the explanatory and the descriptive is the sort of argument to use against a scientist who has to acknowledge both and may be trying to deny the idea of substance. That's the reason I use that argument. What I was proving was that the notion of substance is required if you have a science that has gotten as far as explanation, if you are going to link explanation to description. I'm not talking about merely the kind of description you have in botany, but about explanatory science as in physics, chemistry, biology as now conceived, and so on (and depth psychology is also explanatory). Explanation and description have to be linked, and both change, and still they are talking about the same object that they're investigating all along the line. When you have a science that is explanatory, the key idea that provides the link is the notion of substance.

Now I'm not saying that's a proof of substance, the simplest that can be had, because obviously it isn't. It presupposes someone who's in science, and I offered that because, once people get into science they're apt to say, 'Oh well, substance is just a queer idea that the old people needed, but after all we much wiser people of today get along perfectly well without it.' An element of justification in that attitude is the fact that a lot of people that talk about substance do so in an unsatisfactory manner. They don't distinguish sufficiently between mere extroversion of animal consciousness and, on the other hand, a grasp of an intelligible unity in sensible data. You have an example of the reaction to that in Cassirer's book *Substance and Function*,[8] in which he argues that the notion of substance is passé.

[The next question was about vertical finality in relation to the natural desire to know God by God's essence. The question was the first of a series of questions about the same issue. It would seem that the issue had arisen in an unrecorded evening session the previous night. See below, at note 9.]

I said this morning that I thought Scotism tended to be a deductivist thought, from a set of self-evident principles. They got into difficulties with the intuition of the existing and present as existing and present. When you objectify, when you consider the ontological correlate of such an intellectual system, what do you have? You have a universe that consists of a series of non-communicating strata. If your objective universe consists of a series of non-communicating strata, you have nature at one end and the *finis* at the other, and you have an exigence of the *finis* in the nature. If you

---

8 Ernst Cassirer, *Substance and Function* and *Einstein's Theory of Relativity*, trans. William Curtis Swabey and Marie Collins Swabey (Chicago, London: Open Court, 1923; New York: Dover, 1953).

conceive *finis* in that fashion, as what the nature that is proportionate to the *finis* has an exigence for, then there is no possibility of a vertical finality. There is no possibility of the lower being for the sake of the higher. Objections arose around the time of Cajetan against the continuous doctrine of earlier Scholastics that man has a natural desire for the beatific vision, and they arose because of that conception of the universe: that the universe consists in a set of horizontal strata corresponding to a set of deductions from natures to ends for which the natures have an exigence.

If, however, your notion of the universe is of a series of levels, with the lower being for the sake of the higher, then you have a vertical finality as well, and while the lower is for the sake of the higher, it has no exigence for the higher. In other words, finality is not defined in terms of the exigence of the nature.

One can put that point differently. Take God's knowledge of the possibles. One can conceive God knowing all the possible natures, and all the possible natures are within the divine mind the way the animals were in Noah's ark. Or you can conceive the divine mind as grasping the series of possible universes, the total series of possible worlds. God sees in his essence every universe that might be created in every one of its details – the whole universe. The first thing, then, that is known is the universe, and within the universe are the natures. If you hold this second conception, then the ultimate end is the divine perfection, and what is for the ultimate end is the order of the universe, which is the external glory of God. According to Thomas the most perfect thing in the universe is the order of the universe, which is the manifestation of divine wisdom. *Sapientis est ordinare.* Divine wisdom is represented by this order. If this order is the highest of created ends, then everything within the universe is for the sake of this order. Consequently, the ends of the particular natures are for the sake of that order. The end of a nature is not for the sake of the nature, as is implicit in the other notion that a nature is an exigence for its end. The end is not for the sake of the perfection of the nature, but the nature is for the sake of its end. *Id quod est ad finem est propter finem,* not *Finis est propter naturam quae habet exigentiam finis,* which is a nonsensical notion.

If natures are for their ends, and the ends are for the order of the universe, then there is no possibility of any nature having an exigence for the end towards which it is directed. You can say that there are certain ends that would be insufficient to be assigned as the ends of these natures, but then you're arguing again from the proper order of the universe, in accord with divine wisdom.

Another way of putting the notion that I was trying to make last night[9] is the following. There is a notion that man cannot have a natural desire for the beatific vision because, if the beatific vision is an end that man naturally desires, he must have an exigence for it, and if he has an exigence for it, it cannot be supernatural. If he *does* have an exigence for it, it cannot be supernatural, true. But it is only by making false presuppositions about end, about nature, about the order of the universe, that you can possibly defend the notion that a nature has an exigence for its end.

[The next questioner asked Lonergan to comment on Henri de Lubac and the issues surrounding the latter's *Surnaturel.*]

De Lubac was just mixed up on the point. He isn't satisfactory on it. De Lubac says it *has* an exigence, and it does not have an exigence. The trouble with de Lubac was the following. In 1934 or 1935 he wrote a series of very erudite articles about theologians of the beginning of the sixteenth century. They remained published in theological periodicals for a dozen years without anyone paying the slightest attention to them. About 1946 he published these same articles in a book and added an epilogue. The epilogue was the mistake! While de Lubac is a man of extraordinary erudition and also respected as a very holy man by people who have lived with him in his own Province, he is not a competent speculative thinker. At least I don't find that in him. The way he tried to handle this question of exigence shows that.

Can there be vertical finality without an exigence?

Yes, yes, yes. Prime matter is *for* form, and it's for all the forms. Prime matter is an exigence for *some* form, but it hasn't got an exigence for the most perfect form. There's no exigence in all prime matter to be informed by rational souls.

What about Humani generis?

Humani generis simply states that you must not deny that God could create a universe in which man was not destined for a supernatural end. There's no

---

9 This reveals that there was at least one evening discussion that was not recorded.

difficulty whatever about that. God is free. He can create natures capable of more than de facto they attain, as is clear from this world in which many do not attain the beatific vision. To say that there is an exigence for the supernatural is a contradiction. If you say that God could not create man without giving him grace, you say that on Catholic principles it would be a contradiction for man to exist without being offered the gift of grace. And if there is such a contradiction, then grace must be due to man. And if grace is due to man it's not grace. But *Humani generis* is not a rejection of a fundamental position in St Thomas.

Can you say that the end has an exigence for the nature?

Well, that's true, yes. If God wants a universe in this order, he has to create the things that are demanded by this order. That's the argument, *Necessitas ex praesuppositione finis*. The *finis* is not necessary, but if this is the end, if you are going to have creatures destined to the beatific vision, you have to have a rational nature, because a nonrational creature is incapable of it.

Can you say that the lower has no exigence for the higher, but the higher has an exigence for the lower?

Yes. Man couldn't live without vegetables and animals to eat.

Can you say that the lower exists for the sake of the higher?

That, I think, is sound Thomist doctrine. God created the universe and gave it all to man. Material creation is for the sake of spiritual creation.

[The next questioner asked for a clarification of the meaning of the term 'exigence.']

'Exigence' is just another name for necessity. If your universe is an abstract universe, and the first thing God knows is this Noah's ark of possible natures, and you have to deduce from them, then you'll have to give these natures a necessary relation to ends, what you call their natural ends. And the natural ends are what are naturally desired. On that system there is no possibility for a natural desire for an end higher than the nature. They're contradictory statements. But, while they're contradictory statements on that set of presuppositions, that set of presuppositions, I believe, is false, and in any case not demonstrated. And it's contrary to the facts.

Isn't there a desire for the end?

Oh, it *wants* it. But wanting is one thing, and having a right to it is something else.

Doesn't the wanting have a possibility of fulfilment?

It has the possibility, remotely, not proximately.

Would that be the same as obediential potency?

It is obediential potency, yes.

[The questions were to come back to this issue, but first other matters were covered. In the first of these, the questioner asked about Lonergan's use of the expressions 'first act' and 'second act.']

Look up any book on morals: whatever happens to suit this context. *Actus primus* in St Thomas means the habit.

What about metaphysics?

Oh, you want an example *in* metaphysics. Well, defining act as determination and potency as determinable, and following out those definitions. Wherever you have something more determinate in your words, and something less determinate in your words, you posit, corresponding to that, act and potency. Potency corresponds to the capacity for an intelligible form, and intelligible form is first act. It is the potency for a further act, the further perfection that is not definable except in its relation to this intelligible form. Sight and seeing have exactly the same definition, but what sight is in potency, seeing is in act. If you use 'potency' and 'act' always in those senses, you are dealing with closely defined notions all along the line, and those notions are much more closely defined than if you simply say that potency is what is determinable and act is what determines it, and wherever I have something that I think is more determinate, I have an act, insufficiently defined – or at least one is more closely defined than the other.

Isn't there a natural motion to universalize potency and act?

I'm not saying that you can't do it. My point is that, besides living in a straitjacket, as this axiomatic method would imply, and on the other hand mere sprawling, where things tend to take on almost any meaning at all, I think there's a middle course between the two. Now on the question of potency and act, I have given the terminology that I think has historical grounds and can systematically be justified. But I'm not capable in the next two minutes of giving my reasons for that preference. However, my point in giving this course is that, between the extremes of mere sprawling terminology that can be so indeterminate as not to fix anything and, on the other hand, the straitjacket of an axiomatic system, there's a golden mean of pretty tight thinking.

[Further clarification was asked for.]

What I would say is that potency, form, and act are three terms that denote in the metaphysical order what in the cognitional order are, respectively, what is known as possibility of understanding, what is known insofar as you're understanding, and what is known insofar as you affirm what you understand. However, that presupposes a whole system.

Given what you say about the value of mathematical logic, should we be teaching it in our teaching of logic?

Depending upon the milieu. If, for example, your pupils are saying, 'This lad is teaching logic, and he doesn't know anything about these books,' you at least have to maintain your prestige and show that you know something about it. In Europe they wouldn't dream of bringing it in in most places. They're so much in an entirely opposite current. But over here you're in a different situation, and it's for the people on the spot to make these practical judgments. It does no harm to know something about it, and it's better technique than 'Barbara, Celarent, Darii, Ferioque prioris.' But the difference is that very easily it can become too complex, depending on the level of the course you're teaching. A very good book is this book of Prior's put out at Clarendon about 1956, *Formal Logic*.[10] It throws you immediately into it, and it exhibits a whole series of systems. It's a bit difficult, but it's not extravagantly difficult. It doesn't presuppose a knowledge of mathematics in the way Church's *Introduction to Mathematical Logic*[11] does; Church is thinking of mathematical logic as the most general branch of mathematics.

---

10   See above, p. 15, note 17.
11   See above, p. 19, note 19.

[The question came back to obediential potency, and asked about the difference between obediential potency and negative disposition.]

A negative disposition isn't anything. What is the natural desire to know God by his essence? When you ask, *Quid sit?* you ask to know the essence of the thing, the substantial form of the thing. You want to know the form that makes these phenomena into a thing, the *causa essendi.* That's the meaning of the question, *Quid sit?* You want to understand this thing. When you prove the existence of God, you know *Deus existit.* And naturally you say, *Quid sit?* And every book on natural theology tries to tell you something analogically about the divine essence. They're trying to answer the question, *Quid sit Deus?* as far as we can in this life. The question is there. The question is the expression of a desire to know the essence of God. And that question is natural. It isn't in virtue of some acquired habit or infused habit that you ask that question. It's in virtue of the light of intelligence that God gave you when he created your soul. So the natural desire to know God by his essence is a simple concrete fact that's exhibited by every Catholic philosopher. And Thomas puts it that way. He treats the natural desire to know God by his essence in the *Contra Gentiles,* book 3, chapters 25–63. It's the basis of his whole treatment of the end of man. In the *Summa theologiae,* it's *Pars prima,* q. 12, aa. 1 and following; in the *Prima secundae,* q. 3, a. 8; also confer q. 4, a. 4, or question 5, article 5, where similar questions arise.[12] But it's fundamental in his system. *Omnia Deum appetunt.* And intellect isn't an exception to it. *Omnia Deum appetunt: Prima pars,* q. 44, a. 4, where he's talking about the different causalities of God in regard to the universe.

[The next questioner asked about *beatitudo naturalis* and *beatitudo supernaturalis.*]

The Thomist distinction is between *beatitudo perfecta* and *imperfecta. Beatitudo perfecta est soli Deo naturalis.* If we attain the beatific vision, we're participating God's beatitude. The *beatitudo imperfecta* of which Thomas is thinking is the beatitude of the philosophers, in which they understand the whole universe, in which they get Aristotle's metaphysics perfectly. It's the sort of beatitude that you can have in this life. When Scotus came along and set up his theology of knowing what was necessary in all possible worlds, he

---

12 Thus Lonergan. Questions 4 and 5 in general are concerned with related issues.

changed the theological situation. Thomas is not talking about all possible worlds. Thomas is trying to get the amount of understanding one can have of this world. The intelligibility of this world is the order of divine wisdom as de facto chosen freely by God and realized in this world. To understand that order in this world is Thomas's concern. And as Thomas knows, he doesn't possess divine wisdom. He doesn't try to discuss all possible worlds. He has one little question on God's potency, *Prima pars*, q. 25. Consequently, when Thomas is discussing this world and understanding this world, he isn't talking about all possible worlds. He's talking about this world and what you can understand about it and what are the connections between things de facto in this world. Thomist inquiry is trying to understand this world the way the empirical sciences understand this world. Just as the law of gravitation isn't a necessary law of gravitation but the law that de facto is true, similarly the order of this world isn't what God must do but what God de facto has done. And it's intelligible just as the law of gravitation is intelligible. But it's not necessary, just as the law of gravitation is not necessary. The necessary concrete intelligibility is the divine essence, and that's the only one that necessarily is. The rest *can* be, and exist if God happens to choose it. You get a necessity of this other division – *beatitudo naturalis* and *supernaturalis* – when you enter into the Scotist context and think on Scotist premises and ask Scotist questions.

[The final questions pressed the same issue.]

The end of man is God, in any case, but the mode in which man attains God may be natural or supernatural. It would be natural, for example, in limbo, and supernatural in heaven. In both cases God is the end. Just as God is the first cause of everything, so he has to be the last end. And attaining the last end in a rational creature has to be knowing and loving God. But whether that knowing and loving God is something that is within finite possibility of attainment, or something that only God can give by communicating his own beatitude, is a different question. The beatific vision is supernatural because it's God giving to us a beatitude that is proper to God.

6   Questions after the Seventh Lecture (Chapters 11 and 12)

[There were no questions after the sixth lecture (our chapters 9 and 10). After the seventh lecture Lonergan asked whether there were any questions on the material of the past hour (our chapter 12, on phenomenology) or of the previous hour (our chapter 11, on Husserl). The first question had to do with the possibility of moving phenomenology forward to meet up with the type of work that Lonergan himself is doing in *Insight.*]

If phenomenology were pushed all along the line, you would be brought to the evidence for affirming that there is a rational process called judging, and so on. But insofar as you're merely describing what happens, you're presenting what's manifest, you're providing people with materials for doing philosophy. You'd require a transformation of phenomenology to move into Scholasticism. In other words, just as the phenomenologist is presenting data structured by insight, so a phenomenology of insight would be an insight into insight, and we'd be in an entirely different field, as you can figure out from what happens when one seeks an insight into insight.

[The questioner went on to ask whether phenomenologists do not acknowledge such acts as understanding.]

Yes, but they're not talking about them. In Heidegger one of the existentials or categories of the existential subject is understanding (*Verstehen*), but as someone remarked (I forget just where it was) the phenomenology of *Verstehen* has not yet been done, and I would say that when the phenomenology of *Verstehen*, of understanding, is attempted, then what will you be doing? You will be seeking understanding as structured by understanding, and that will be insight into insight, and it will bring you into an entirely different world from that of the phenomenologists.[13]

13 Lonergan's reflections here might well be enlarged to a phenomenology of the twentieth century's surge in geometry and topology. A further enlargement would attend to the geometric drive in contemporary physics, where 'phenomenology' has entered the vocabulary in a massively challenging way. 'The mid 1980s have brought a further succession of dramatic developments – some of the highlights being hexagon anomaly cancellation, the emergence of a new theory with the exceptional group $E_8 \times E_8$ as its gauge group, the beginnings of phenomenology, and perhaps the beginnings of an understanding of the symmetry structure of string theory, at a level deeper than hitherto possible' (Michael B. Green, John H. Schwarz, and Edward Witten, *Superstring Theory*, vol. 1: *Introduction* [Cambridge and New York: Cambridge University Press, 1987] 17–18). It is of interest to note that the subtitle of vol. 2 is *Loop Amplitudes, Anomalies, and Phenomenology.* The relevant phenomenology pivots on mega-machines subtly observational of micro and macro phenomena. A control of the deeper level of understanding requires the phenomenological luminosity of which Lonergan speaks in his short answer. See above, p. 114, note 36.

Take it this way. You have your structured data and your insight. You can attend to the data that are structured, and your attention centers there. Or you can attend to the insight, and it's a different focus of attention. When you center on the insight, you move to a consideration of the reflective insight that bases judgment when you grasp the virtually unconditioned. And then you'll be right in with what I consider Scholasticism. But insight is an elusive thing. You get hold of insights properly only by considering the history of science, the history of philosophy, and so on. Just as if you just center on what is experience, in any given mode, it's so elusive that it tends to vanish. You put insights together insofar as you say, 'Well, a geometer understands the whole of Euclid, he can tell you where the key propositions are, and prove all the propositions that follow from a given set of axioms. He's got the whole thing right in his intellectual paws, so to speak.' But that comprehensive grasp of the whole subject is not some phenomenon that you can pin right down and describe the structure. When you're seeking insight into insight, not only have you a different term of attention, but your methods of procedure have to differ if you're going to get anywhere.

[The next set began with a question whether there has been any phenomenological study of causality and moved on first to the issue of intentionality and metaphysics. Lonergan answered to the causality question that he did not know of any such study. Then, as the question was pursued, his comments were as follows.]

Well, they've been working on intentionality since 1900. Husserl talks about everything that is intentional that is a datum [*sic*]. A metaphysical theory is what you know through a set of true judgments. And what you know through a set of true judgments is not any datum. There is a sense in which metaphysics enters into phenomenology, but it's the sense in which *all* metaphysics indifferently enter into phenomenology. In that sense they're all data. The true metaphysics is not a datum until you get onto this rational level. In other words I don't think that phenomenology is a completely rounded whole, a completely satisfying system. It has yielded a lot of good, but if you start pushing it to its logical conclusions you probably could transform it into something else.

[The final question had to do with just how phenomenology moves onto the rational level.]

Well, the grasp of the virtually unconditioned. We did that last week. We can hardly do it before 12 o'clock today. *Insight,* chapter 10.

7   Questions after the Ninth Lecture (Chapter 13)

[Either the eighth lecture and the first part of the ninth were not recorded or the recording did not survive. There is no record of questions after the eighth lecture. The following questions were asked after the ninth (corresponding in part to our chapter 13). A first question had to do with the need of grace.]

Concretely in this life grace is necessary. There is a need of grace for two reasons: insofar as grace brings us a new type of perfection that we can't have naturally, but also insofar as grace perfects nature. The perfection of nature that comes from grace has the added point that, when man is in a fallen state, he needs some extrinsic help to attain even his natural ends. And that extrinsic help is de facto, within this order, supplied by grace. In theology the need of grace with regard to doing good is much more developed than the need of grace with regard to knowing the true. One cannot say that man needs grace to know natural truths except in some *per accidens* sense. But the extent to which grace de facto will be needed, I think will vary with historical contexts. The practical solution to the problem, as I conceive it, is to build up your philosophy in a series of steps in which you are constantly moving to higher levels that take more and more of the truth into account. Ultimately you are on the theological level. The philosopher goes so far, and the theologian carries on. That is a practical solution to the problem as far as exposition goes. With regard to what happens in concrete individuals, it is very difficult to say.

[The next question had to do with the emphasis on the subject in relation to objective knowledge.]

I will try to say something on that tomorrow.[14] It isn't a clear-cut issue. If you are merely on the side of the objective, you will be omitting the critique of the subject. And if you omit the critique of the subject, you will be running into disputed questions, and you won't know what the real cause of the trouble is. On the other hand, if you attend simply to the subject, you will be

14 See above, pp. 313–17.

with the existentialists; that's all they can think about. You can't take just the one or just the other; you need both: one from a more critical viewpoint, and the other from the viewpoint of objective knowledge.

[The next question had something to do with the critique of existentialism in *Humani generis.*]

With regard to the name 'existentialism' it is one of the things condemned in *Humani generis.* Just what that 'existentialism' means is difficult to say. It is safe to say that Sartre is condemned. But the name is an extremely ambiguous thing. It isn't recognized by all these philosophers. What is true to say, I think, is that there are elements in the writers named 'existentialist' that contain invitations to us that it would be foolish for us to ignore totally. I think that is safe.

[The next question had to do with the relation between philosophical conversion and the Christian context.]

It may occur only within religious conversion. But that is not true in a simple way. We have an element of philosophic conversion in a Socrates, a Plato, an Aristotle. It does not go the whole way, and de facto the ulterior steps in a complete philosophic conversion occur with Augustine and the Arab philosophers and finally Aquinas; at least that is my reading of the history of philosophic development. De facto the ultimate element occurs within the Christian context. But I don't know that those questions of possibility can be solved or that it is extremely important to solve them. If God gives his grace, he'll give it, and one won't be able probably to detect his presence until after the results come.

[The next question had to do with the relation of Kierkegaard with the material of which Lonergan has been speaking.]

I couldn't very well bring Kierkegaard into the thing. He is not exactly an inspirer [of the movement]. Kierkegaard was attracting attention from about 1900 on. He influenced theology in Karl Barth, who begins about 1912. But Kierkegaard hit the front page in philosophy after Heidegger. His own thinking is within a strictly religious context, and it is closely bound up with Lutheran ideas upon faith, upon natural knowledge of God, and so on. The elements in Kierkegaard that are incorporated in existentialism are a laicization of the religious ideas of Kierkegaard. Kierkegaard speaks about being a Christian. We started off with 'being a man,' 'being oneself,' but all this concern with freedom and *my* self and *my* responsibility for myself is connected with ideas that came to light originally with Kierkegaard. Kierkegaard is just as big a subject as the existentialists. For example, in Bochenski's abbreviated bibliographies, he has one fascicle for Kierkegaard.[15] He planned at one time at least to have one for Nietzsche. And he has one for the French existentialists.[16] Existentialism is a contemporary phenomenon. Kierkegaard is a very important influence but not the sole one. We are not saying anything about Nietszche either, but he also has a tremendous influence upon the existentialists.

8   Questions after the Last Lecture (Chapter 14)

[A first question asked about the relation of subject as subject to consciousness.]

Well, it *is* what is conscious. It is not, as subject, what one is conscious of. I see you asking me this question. You are present to me. But if I were not here, if just my eyes were here, you could not be present to me; that would be meaningless. Still, I am not present to myself the way you are to me when I gaze upon your features. I am present in some way, but not in that way. And that some way that is not that way is the subject as subject: the one to whom presentations are presented, the one who is present in the sense that presentations are made to him, but the one who is not present in the sense that he is an object presented to himself. That is the fundamental point. And that recurs all along the line. No matter how far you go at introspection, there is always the 'introspector' and what he introspects. You can always draw the distinction between the subject as subject and the subject as object. Now the point to the distinction is that the subject as object will be conceived in a certain way, in the light of a certain horizon, while the subject as subject is not open to any manipulations of that type. He is the ground of the horizon but he is not within it. Again, the subject as subject is an inexhaustible topic. My reality as that to which other things are present is a thing that I can describe in this aspect and that aspect, and so on. But this series of descriptions will not add up to the totality of my reality. Every

15 See above, pp. 50–51.
16 See above, ibid.

being is something that is known exhaustively only by God. Similarly, my own reality for myself, insofar as I am subject as subject, is something that is only partially known objectively insofar as I introspect and make analyses of consciousness and form theories and make judgments about myself.

Is this subject as subject known?

It is known, and it is not known. It depends upon your sense of the word 'known.' First of all, to take the simplest case, I will start from sense knowledge. You can have just sense knowledge, hypothetically at least. Or you can have sense knowledge plus an understanding of sense knowledge: sense and understanding, or imagination and understanding. And then in the first case you can say you have knowledge in potency, and in the second case you have knowledge in potency and first act. And you can have knowledge in a third and more complete sense, in which you not only sense the object and understand it but also affirm that it is. Then your knowledge is in second act, and you have three senses of the word 'know': as what is mere experience or as experience and understanding or as experience, understanding, and affirmation.

Now when I am just sensing, *I* do the sensing. I am present in the sense of one who has to be present, has to be non-absent, for things to be presented to me. And so when I sense, I have presence to myself. When I sense and understand, when I sense myself and understand that I am clapping my hands I have presence to myself but also the self presented. The first is subject as subject, and the second is subject as object. Again, I understand actively, I am doing the understanding, I am not *fare lo stupido*. I am doing the understanding, and what is understood is myself: the active voice (the understanding) and the understood. Again, on one side I have presence intelligently, and on the other I am present intelligently and I am also understood. You have subject as subject on one side, subject as object on the other side. And you can move up to the third level. I judge, and I am judged – the passive voice: subject as subject and subject as object. Now I can repeat that and make the divisions, and then I know myself as a subject of empirical, intelligent, rational consciousness. Insofar as I conceive a subject of empirical, intelligent, and rational consciousness, insofar as I affirm a subject of empirical, intelligent, and rational consciousness, on the side of the object I have the subject as objectified. Still, I do that conceiving, I do that affirming, and the *I* that is there doing the conceiving and affirming is the subject as subject. And it is the *I* that is doing the conceiving and affirming that cannot get away from its own intelligence, that cannot get away from its own reasonableness. And it is insofar as you invite subjects to try and put off their intelligence, to try and put off their reasonableness, that they will find in the subject as subject that they cannot do so. Discovering that impossibility in themselves, they will discover a normativeness that they cannot dodge no matter what horizons they move to. They will find a basis from which they can select which horizon is true, by a further development of the argument.

[The next questioner asked whether there are any horizons on the sense level.]

Well, de facto, sensation does not occur except within perceptions, and perceptions occur within a horizon. So the perceived is involving a horizon. But I can avoid colors and shapes only if I close my eyes. I can't avoid sound unless I build soundproof walls or move to the desert where I live all alone. I can't avoid feeling: my feet will have to be resting on something. The category of experience cannot be evaded totally. You can discover the impossibility of avoiding all experience. And that impossibility is not within a horizon. And it is the impossibility in the subject as subject that reveals the invariants, the fundamental criteria, from which you decide what the true horizon is.

[The next questioner asked how other horizons can be developed out of the structure thus disclosed; for example, how does this structure illuminate the horizon of Hume?]

Well, we will move to Hume more slowly. First of all, consider the problem of judgment, of saying, 'It is so.' Unless you already know it is so, you do not mean anything when you say, 'It is so.' And so judgment has to be reduced to the prior levels, so that, unless you know the real on that prior level, your judgment is no good. But if you already know it, your judgment does not add anything at all, and if you wash judgments out, then you can be a relativist. In other words, what is the function of understanding? Well, any given understanding is only partial. There is always more to be understood. On the relativist position, you arrive at the unconditioned only insofar as you understand everything about everything. And consequently any given understanding is true as far as it goes, and it is false insofar as it is not complete. Again, take sense and understanding. You can have Heidegger on that basis: *Seiendes und Sein.* You can have Kant on that basis, because Kant does not acknowledge the unconditioned except as an ideal towards which we tend. You can have Aristotle on that basis, where sense and understanding correspond to form and matter. And the empiricist will say that understanding is just a subjective phenomenon. It is not part of what is out there. It is a subjective activity that is useful to perform with regard to the realities that are really out there. All it does is give you useful links between these data that are out there. Insofar as that sort of thing works, science is a good thing, but the scientist insofar as he is understanding is not knowing anything about reality. When the scientist knows the law of falling bodies, he does not know something about this brush that falls. He knows how to handle the brush and get the best results out of falling brushes. That is all. But he does not know any intrinsic constituent of the falling brush. There is not intelligibility out there, it is only in my head, and once you have wiped that out you can have positivism, materialism, pragmatism, and all the rest of it. Then there are various ways of mixing them up to get more complicated philosophies.

[The next question asked (in a somewhat convoluted fashion) about the relation of subject as subject to subject as object. After clarifying the question, Lonergan responded as follows.]

The analysis of the subject as subject I conceive as breakthrough. The subject is forced to acknowledge himself as a subject of an empirical, intelligent, rational consciousness. Whether he knows anything at all is a further question. But at least he performs these three activities, and he cannot get away from these three activities. And he cannot be himself without performing them. Myself as unreasonable is just not myself. It is a denial of myself, and I cannot attempt the denial without wanting to have reasons for it. To go on to the unlimitedness of the human mind [which featured in the question] is relevant to determining just how far the horizon extends, but it is a further step. It is upon that basis, but it is a further step. You get the unlimitedness of the human intellect from the fact that no matter how many questions you answer there are still more you have to answer.

[The next questioner asked how we know that there are only these three steps. Are we not limited by a horizon in saying that there are only these three steps?]

Well, there are further steps, and you can show that they are not cognitional. Beyond judgment there is decision, and that is free will, and that is a further department. But it follows on judgment. These three [experience, understanding, judgment] you can show are perpetually recurrent because any possible revision involves them. Supposing someone were to say, 'That is all right, that is as good as we can say now, but how do you know that someone is not going to come along in five centuries and give you a much better analysis?' All right, but will he appeal to the facts of consciousness? And if he does he will have to be empirically conscious. Will he understand the facts of consciousness better than I do? If he does he will have to be intelligently conscious. And will the theory he proposes be verified in the facts of consciousness? Will there be reached a virtually unconditioned? There will have to be if it is going to be preferable to the one set forward at the present time, if you are going to say it is true or it is better. And if he does he will be rationally conscious. And if you say, 'Well, the revision might be entirely different by then,' you are running away, you are wanting to use the word 'revision' in an arbitrary sense. It is another way of putting the argument from retortion. Aristotle says, 'Let the skeptic talk; if you can get him to talk you can already have him admitting various principles.' This is another way of going about the same thing in a different mode.

[The next questioner asked about different meanings of the term 'reasonableness.']

By reasonableness, fundamentally I mean the demand for the virtually unconditioned. Without the virtually unconditioned you refuse to affirm or deny, and if you do get the virtually unconditioned then you are rationally compelled to affirm or deny. That is reasonableness in a fundamental sense. And as a syllogism is just a particular case of the virtually unconditioned, it includes that case. What has produced the judgment is the act of reasoning. Well, the act of reasoning is a discursive process. You put out successive propositions, but there is not any computer in the mind that ticks along and arrives at a certain point by the time you reach the last phrase. What the discursive presentation through words and concepts and propositions does is provide the rational subject with a compact expression of the virtually unconditioned. And insofar as he grasps that, he is compelled to assent. And his reasonableness is manifested in that fashion.

When you clap your hands is it an insight or a judgment?

Well, when I say I clap my hands obviously a judgment is involved. But what I feel is not a judgment. What I feel is localized in my hand. It is de facto a use of my past experience, the localization of sensations and so on, and it is more or less a routine use of habitual intelligence. It is not the emergence of any new insight. Maybe when I was very small I made the discovery that I had two hands, that one could bang against the other, but it is not a discovery for me now.

[The next questioner asked about the meaning of 'freedom' in Heidegger.]

Insofar as Heidegger is dealing with understanding and sense in the flow of consciousness, the flow of consciousness organizing itself in the light of its understanding of its own potentialities, the flow of consciousness is free in the sense that it is not physically, chemically, biologically determined. It is not free in the sense that you have a rational judgment about what is good and an act of will consequent upon that judgment within a consciousness of responsibility, moral obligation, and so on. This level is just missed out in Heidegger, as far as I know.

[Lonergan's concluding remarks were as follows.]

Well I wish to thank you very much for your very kind attention and the kind words you have made about this course during my stay. I may be permitted, perhaps, to add on a brief story that will express my very sincere admiration of you all. When I finished my own teaching as a scholastic and went to theology, my first week sitting on the benches listening to professors of theology found me in a state that normally is described as being 'fit to be tied.' To find so many of you, who have taught for so many years, coming and listening to anyone talk to you for two weeks steady, the way I have done, demonstrates in a very clear manner what I have referred to in my book as 'the pure desire to know.'

Appendix D
Fragment on Heidegger[1]

I propose to talk about Martin Heidegger today. The fundamental work on him is by Alphonse de Waelhens, *La philosophie de Martin Heidegger*, published in Louvain, first edition about 1941 or 1942.[2] The bibliography in that book only goes up to about that date, but there have been four reprintings or subsequent editions, the last about 1953 or 1954. Also, no one else has done the kind of systematic work on Heidegger that de Waelhens has done. He is bending over backwards to be objective all the way through. I would be inclined to say that he is apt to agree a little too much with Heidegger and his tendencies. For example, he seems to think that there is more in Heidegger's doctrine of being and his openness to Scholasticism than really there is.

However, it is true that Heidegger does offer us a challenge to clarify our own notions of being, our notions of truth, and similar fundamental questions. He starts off from Husserl. He was one of Husserl's pupils. As you can see from the bibliography, though, his first work was on psychologism in 1914, and his *Habilitationsschrift* was on Scotus.[3]

Some people lay into Heidegger, saying that he started off as a Scholastic and then became a Nazi. They have very little use for him on first principles. You will see that in French writers occasionally. And apparently he did not

1 The beginning of the eighth lecture, Wednesday, 17 July 1957.
2 See below, p. 374.
3 Lonergan clarified that the *Habilitationsschrift* was a 'writing that a German professor has to do to take over the chair.' For Lonergan's bibliography of Heidegger's work, see below, pp. 369–71.

show up well under Nazism, while Jaspers did. And, of course, Sartre showed up very well in the French Resistance movement.

However, Heidegger does seem to me to be the most profound and the most challenging of the lot. Phenomenology is his method, and simply phenomenology. He considers that the method, and one that is perfectly adequate to handle the problems. Phenomena are whatever is manifested, whatever appears. There is no contrast between appearance and underlying reality, between sense as opposed to art, culture, sentiment. There is no favoritism for what is outer over what is inner. And he does not mean, of course, what is immediately grasped by anyone at all, but what is manifest only after effort, time, attention, scrutiny, penetration.

Besides 'phenomena,' there is *legein*, to read off, let appear, dis-cover, unveil: to let phenomena appear, to let be manifest what is manifest. Not only are the phenomena to be read off, but there is also the hiding of them, the dodging of them, the escapism. And truth is based on evidence, on letting the phenomena appear. What is true is what is manifest, has been discovered, unveiled, revealed.

He goes in a great deal for etymological stunts. This makes him difficult reading in German unless you know your German quite well. As a matter of fact, de Waelhens has two translations, one of *Kant und das Problem der Metaphysik* and the other of *Vom Wesen der Wahrheit*, and in both cases he has had a collaborator, Walter Biemel, who is a German. Unless your German is first-class, you are much safer controlling your reading of the German with a French translation that has been done in this style. Also note that those French translations have a long introduction and explanation. It is more explanatory in the case of *Kant und das Problem der Metaphysik* and more expository in *Vom Wesen der Wahrheit*.[4] Those French translations are important for that reason.

His principal work is number 4 on the bibliography, *Sein und Zeit*, which appeared originally in Husserl's *Jahrbuch* and then went through several editions. It was up to about the fifth or sixth edition in 1934. For a philosophic book that is quite exceptional. His *Kant und das Problem der Metaphysik* is a study of Kant in which he contrasts the first and second editions of *The Critique of Pure Reason* and argues that in the second edition Kant deflected from his original and sounder intentions. The point is that in Kant there is, on the one hand, the empirical ego, the subject of experience, the ordinary me, and on the other hand, the transcendental ego, which is the condition

4 Lonergan is referring again to his bibliography of Heidegger's work.

of the possibility of 'I think,' of the 'Ich denke,' and that that duality becomes very pronounced in the second edition. In the first edition there is much more attention paid to a third factor, the transcendental imagination. According to Heidegger, Kant saw or felt that his first edition was heading to the position where the transcendental imagination would be the fundamental thing on which was based both the Aesthetic and the Logic. The Logic depends on and proceeds from the transcendental ego, while the Aesthetic depends on and proceeds from the empirical ego. The synthesis of both is in transcendental imagination, that is, imagination as the condition of the possibility of both. Kant, with his rationalist antecedents and probably with his tendency towards a metaphysic, was rather upset by finding out that they depend upon the transcendental imagination, and so he steered away from that. Consequently the division is much more emphatic in the second edition.

That is Heidegger's opinion. But Heidegger's opinion on those things is pretty good. Very few people attempt to argue with Heidegger, because he is extremely learned in the history of philosophy. His writings are full of allusions, and unless you know your Hegel inside out, and so on right along the line, in general the history of philosophy, you are in trouble. People are afraid of him, and they know that, if they start attacking him, he could probably make mincemeat out of them. At least I am inclined to suspect that there is something to Heidegger's knowledge of the history of philosophy that more or less protects him. People do not sail into him because he is right up there and quite able to take care of himself.

Kant und das Problem der Metaphysik and *Sein und Zeit* are rather lengthy works. On the other hand, *Was ist Metaphysik?* is short. *Vom Wesen der Wahrheit* is longer, perhaps about 120 pages. Other things are all quite brief. *Holzwege*, number 14 on the bibliography, is a collection of conferences and speeches. I have not seen the last item, *Zur Seinsfrage.*[5]

5   The tape ends with Lonergan commenting on umlauts that should be
    added to several items in the bibliography.

Appendix E
Lonergan's Bibliographies

[Evidence indicates that Lonergan distributed two distinct bibliographies, along with his lecture notes. In the Lonergan Archives in Toronto, four pages are stapled together to form item A1840. The first two are mimeographed pages of bibliographies of works by Heidegger and Jaspers. The other two are dittoed pages containing references pertinent to both of the two weeks of the Institute. The presentation here follows the same order: first Heidegger and Jaspers, then the general references. The materials are listed as Lonergan typed them, with italics or inverted commas added as appropriate. For Heidegger and Jaspers, editorial comments, a few additional bibliographical notes, and references to English translations are placed within brackets. No effort has been made here to supplement the more general bibliography, which is left just as Lonergan typed it. Further information on many of the works listed there is provided in the footnotes to the text. In other words, these are the bibliographies that Lonergan distributed for his lectures in 1957, with some editorial additions to the bibliographies of the works of Heidegger and Jaspers.]

Martin Heidegger   B. 1889 Doct. Freiburg 1914.

1 *Die Lehre vom Urteil im Psychologismus.* Leipzig, 1914. [Dissertation presented at Freiburg, 1914.]
2 *Die Kategorien- und Bedeutungslehre des Duns Scotus.* Tübingen, 1916. [Habilitation Dissertation, Freiburg, 1915.]

3 'Der Zeitbegriff in der Geschichtswissenschaft.' Zeit. f. Phil. und phil. Kritik, 161 (1916). [pp.173–88. Trial lecture at Freiburg, 1915.]

4 *Sein und Zeit.* Halle: Niemeyer, 1927, many editions. [*Being and Time*, trans. John Macquarrie and Edward Robinson (New York: Harper and Row, 1962); trans. Joan Stambaugh (Albany: State University of New York Press, 1996).]

5 *Kant und das Problem der Metaphysik.* Bonn: Cohen, 1929. [*Kant and the Problem of Metaphysics*, trans. James Churchill (Bloomington, IN: Indiana University Press, 1962); trans. Richard Taft (Bloomington, IN: Indiana University Press, 1990, 1997.]

6 *Was ist Metaphysik?* Bonn: Cohen, 1929. Nachwort added in 4th edition, 1943. Einleitung added in 5th ed., 1949. ['What Is Metaphysics?,' trans. R.F.C. Hull and Alan Crick, in *Existence and Being*, ed. Werner Brock (Chicago: Regnery, 1949) 353–92.]

7 *Vom Wesen des Grundes.* Jhrb. f. Phil. u. phän. Forschung, Festschrift E. Husserl, Halle, 1929. Sonderdruck, 1931.

8 *Die Selbstbehauptung der deutschen Universität,* Breslau: Korn, 1933. ['The Self-Assertion of the German University,' trans. Karsten Harries, *The Review of Metaphysics* 38 (1985) 467–502.]

9 *Hölderlin und das Wesen der Dichtung.* München, 1937. ['Hölderin and the Essence of Poetry,' trans. Douglas Scott, in *Existence and Being* (see 6 above) 291–315.]

10 *Platons Lehre von der Wahrheit.* 1940, 1942, Bern 1947. ['Plato's Doctrine of Truth,' trans. J. Barbour, *Philosophy in the Twentieth Century*, ed. William Barrett and Henry D. Aiken (New York: Random House, 1962) 251–70.]

11 *Vom Wesen der Wahrheit.* Frankfurt: Klostermann, 1943. *Anmerkung* added in 2nd edit, 1949. ['On the Essence of Truth,' trans. R.F.C. Hull and Alan Crick, *Existence and Being* (see 6 above) 317–51.]

12 'Brief über den Humanismus.' 1947. Appendix to 10. ['Letter on Humanism,' trans. E. Lohner in *Philosophy in the Twentieth Century* (see 10 above) 270–302.]

13 *Erläuterungen zu Hölderlins Dichtung.* Frankfurt [Klostermann], 1950.

14 *Holzwege.* Frankfurt [Klostermann], 1950.

15 *Einführung in die Metaphysik.* Tübingen: Niemeyer, 1953. 153pp. Completely revised edition of lecture with same title delivered at Freiburg i. B., 1935. [*An Introduction to Metaphysics*, trans. Ralph Manheim (New Haven: Yale University Press, 1959).]

16 *Vorträge und Aufsätze.* Pfullingen [G. Neske], 1954.

17 *Zur Seinsfrage.* Frankfurt, 1956.

There are Italian translations of 4, 5, 6, 7, 9, 11, 14.

M. Heidegger, *Qu'est-ce que la métaphysique* (Paris: Gallimard, 1938) is a French trans. by H. Corbin of 6, 7, 9, of about 100 pages of 4, and 20 pages of 5.

A. De Waelhens and W. Biemel, French trans. with long introduction of 5 & 11.

De l'essence de la vérité, Louvain, Paris, 1948.

Kant et le problème de la métaphysique, Paris, 1953.

Anteile. Martin Heidegger zum 60 Geburtstag. Frankfurt a. M.: Klostermann, 1950.

Karl Jaspers   B. 1883

1 *Allgemeine Psychopathologie.* Berlin [J. Springer], 1913, 1923 (2nd ed.); French, Paris, 1928. [*General Psychopathology*, trans. of the 4th edition, 1946, by J. Horning and M.W. Hamilton (Chicago: University of Chicago Press, 1961).]

2 *Psychologie der Weltanschauungen.* Berlin [J. Springer], 1919, 1926 (2nd ed.); Italian trans., *Le Visioni del Mondo.*

3 *Die Idee der Universität.* Berlin [Springer], 1923. [*The Idea of the University*, trans. H.A.T. Reicher and H.F. Vanderschmidt (Boston: Beacon Press, 1959).]

4 *Max Weber, Gedächtnisrede.* Tübingen, 1921.

5 *Die geistige Situation der Zeit.* Berlin [W. de Gruyter], 1931, 1933 (2nd ed.); London, 1933. Louvain-Paris (Ladrière & Biemel) 1951. [*Man in the Modern Age*, trans. Eden Paul and Cedar Paul (London: Routledge and Kegan Paul, 1951, 1953; Garden City, NY: Doubleday, 1957).]

6 *Philosophie: I. Philosophische Weltorientierung; II. Existenzerhellung; III. Metaphysik.* 3 vols. Berlin, J. Springer, 1932. Pp. 340, 441, 237. [*Philosophy.* 3 vols., trans. E.B. Ashton (Chicago: University of Chicago Press, 1969–71).]

7 *Max Weber: Politiker, Forscher, Philosoph.* Oldenburg, 1932 [Bremen: J. Storm, 1948]. ['Max Weber as Politician, Scientist, Philosopher,' trans. Ralph Manheim in Jaspers, *Three Essays: Leonardo, Descartes, Max Weber* (New York: Harcourt, Brace and World, 1964) 187–274.]

8 *Vernunft und Existenz [Fünf Vorlesungen].* Groningen [J.B. Wolters], 1935. Milano, 1942. [*Reason and Existenz: Five Lectures*, trans. of 3rd edition by

William Earle (New York: The Noonday Press, 1955, and Milwaukee: Marquette University Press, 1997).]

9 *Nietzsche: Einführung in das Verständis seines Philosophierens.* Berlin [W. de Gruyter], 1936. [*Nietzsche: An Introduction to the Understanding of His Philosophical Activity*, trans. Charles F. Wallraff and Frederick J. Schmitz (Tucson: University of Arizona Press, 1965).]

10 *Descartes und die Philosophie.* Berlin [W. de Gruyter], 1937. Paris, 1937. ['Decartes and Philosophy,' trans. Ralph Manheim in *Jaspers: Three Essays* (see 7 above) 59–185.]

11 *Existenzphilosophie* [*Drei Vorlesungen*] Berlin [W. de Gruyter], 1938. Milano, 1940. [*Philosophy of Existence*, trans. Richard F. Grabau (Philadelphia: University of Pennsylvania Press, 1971).]

12 *Vom lebendigen Geist der Universität.* Heidelberg [L. Schneider], 1946. [Full title: *Vom lebendigen Geist der Universität und vom Studieren: Zwei Vorträge*, von Karl Jaspers und Fritz Ernst.]

13 *Philosophische Logik; Von der Wahreit.* 1947. [*Truth and Symbol*, trans. (of excerpt) Jean T. Wilde, William Kluback, and William Kimmel (New York, Twayne Publishers, 1959)].

14 *Der philosophische Glaube*, München, [Zürich: Artemis, 1948]. *The Perennial Scope of Philosophy*, trans. Ralph Manheim, New York [Philosophical Library], 1949 (trans. said to be unsatisfactory, Heinemann). *La foi philosophique*, Paris: Blon 1953.

15 *Vom Ursprung und Ziel der Geschichte.* München: Piper, 1949 (8–11 thousand [sold]), 1950. [*The Origin and Goal of History*, trans. M. Bullock (New Haven: Yale University Press, 1953).]

16 *Rechenschaft und Ausblick.* München: Piper, 1951. *Bilan et Perspectives*, Desclée 1956. [*Existentialism and Humanism: Three Essays*, trans. E.B. Ashton (New York, R.F. Moore Co., 1952).]

17 *Vernunft und Widervernunft in unserer Zeit;* 3 conferences at Heidelberg [München: Piper, 1950]. *Raison et Deraison de notre temps.* Desclée, 1953. [*Reason and Anti-Reason in Our Times*, trans. Stanley Godman (London: SCM Press, and New Haven: Yale University Press, 1952).]

18 *Schelling: Grösse und Verhängnis.* München: Piper, 1955.

– *Way to Wisdom: An Introduction to Philosophy* [New Haven: Yale University Press, 1954. [Trans. by Ralph Manheim of *Einführung in die Philosophie: Zwölf Radiovorträge* (Munich: R. Piper and Co., 1953).]

– *Introduction à la Philosophie.* Paris: Plon 1952.

– La mia filosofia (anthology) Milan, Einaudi, 1946.

– There is a Festschrift for his 70th birthday.

For French Existentialism, see bibliography by Jolivet in Bochenski, *Bibliographische Einführung in das Studium der Philosophie.*

Lonergan's General Bibliography for the Two-week Course
[in many instances, full bibliographical information is provided in footnotes to the text]

I.M. Bochenski, *Bibliographische Einführung in das Studium der Philosophie,* Bern, 1948, A. Francke. Fasc. 1, 2, 3, 4, 9, 11.

Actualités scientifiques et industrielles, Paris Hermann nos 837, 1134, 533, 535.

Proceedings XI Intern. Cong. Phil. Brussels 1953, vols. 3, 5, 7, 14.

Rev. Intern. de Philosophie (Brussels) VIII (1954).

Proceedings X Intern. Cong. Phil., Amsterdam, 1948.

Proceedings II Intern. Cong. Intern. Union Phil. Science, Zürich 1954 Neuchâtel, Éditions du Griffen, 1955.

E.W. Beth, *Fondements logiques de Mathématique,* Paris 1950

I.M. Bochenski, *Formale Logik.* Freiburg-München 1956, K. Alber.

A. Church, *Introduction to Mathematical Logic,* Princeton UP 1956.

R. Feys, *RPL,* 1946, 1947, 1950, 1953.

F. Gonseth, *Les mathématiques et la réalité,* Paris Alcan 1936.

G. Isaye, *Bijdragen,* 1952, 1953.

J. Ladrière, *RPL* 1950.

A.N. Prior, *Formal Logic.* Oxford Clarendon 1955.

W.V. Quine, *Mathematical Logic,* New York, Norton 1940.

B. Russell, *An Introduction to Mathematical Philosophy,* London 1919.
   Logic and Mysticism, London 1932 (7th ed.).
   An Inquiry into Meaning and Truth, London 1951 (4th ed.).

P.F. Strawson, *Introduction to Logical Theory.*

WORKS: M. Heidegger, K. Jaspers, G. Marcel, J. Sartre

L. Binswanger, *Le rêve et l'existence,* Desclée 1954.
   Grundformen u. Erkenntnis menschlichen Daseins, 1942.

Blackham, *Six Existentialist Thinkers,* Kegan Paul 1952.

I. Bochenski, *Europäische Philosophie der Gegenwart,* Bern 1947 Francke.

F.J.J. Buytendijk, *Phénoménologie de la rencontre,* Desclée 1952.

F.C. Copleston, *Contemporary Philosophy,* London 1956.

Dempsey, *Psychology of J.P. Sartre,* Blackwell.

W. Desan, *The Tragic Finale,* Harvard & Oxford 1954.

M. Dufrenne & P. Ricoeur, *Karl Jaspers et la philosophie de l'existence,* 1947.

J. de Finance, *Existence et liberté.* Paris Vitte 1955.

R. Grimsley, *Existentialist Thought*.

Th. Haecker, *Métaphysique du sentiment*, Desclée 1953.

Anteile. M. Heidegger zum 60 Geburtstag, Frankfurt Klostermann 1950.

F.H. Heinemann, *Existentialism and the Modern Predicament*, London A & C Black 1953.

E. Husserl, *Die Krisis der europäischen Wissenschaften und die transzendentale Phänomenologie*. Haag, M. Nijhoff, 1954.

H. Leisegang, *Denkformen*, Berlin W. de Gruyter, 1951 (2nd ed.).

H. Kipps, *Untersuchungen zu einer hermeneutischen Logik*, 1938.

M. Merleau-Ponty, *La structure du comportement*, 1942.
Phénoménologie de la perception, 1945.
de Waelhens on MP

E. Mounier, *Introduction aux existentialismes*, Paris 1947.
La personalisme, Paris 1950 PUF.

Max Müller, *Crise de la métaphysique, Situation de la philosophie au XXe siècle*, Paris Desclée 1953.

P. Ricoeur, *G. Marcel et K. Jaspers, deux maîtres de l'existentialisme* (Paris 1948).

G. Ryle, *The Concept of Mind*, London Hutchinson 1951 (4th ed.).

H. Stoffer, 'Die moderne Ansätze zu einer Logik der Denkformen,' *Z. f. phil. Forschung* X (1956) 442–66; 601–21.

J.O. Urmson, *Philosophical Analysis*, Oxford Clarendon 1956.

R. Troisfontaines, *De l'existence à l'être*, Paris Vrin 1953. 2 vols.

H.L. Van Breda, et al., *Problèmes actuels de la phénoménologie, Colloque intern. de phénoménologie*, Bruxelles 1951. Desclée 1952.

A. de Waelhens, *La philosophie de M. Heidegger*, Louvain, 1955 (4th ed.). With W. Biemel, Introd. et Trans. of *Vom Wesen der Wahrheit*, and *Kant und das Problem der Metaphysik*.

J. Wahl, *Études kierkegaardiennes*, Paris, 1930.
Esquisse pour une histoire de l'existentialisme, Paris Arche 1949.

L. Wittgenstein, *Tractatus Logico-philosophicus* Kegan Paul 1922.
Philosophical Investigations, Blackwell 1953.

Contemporary British Philosophy, London, Allen & Unwin, First Series, 1924, Second 1925, Third 1956.

A.G.N. Flew, ed., *Logic and Language*, Blackwell 1952, and *Philosophy and Psycho-analysis*, Blackwell, 1953.

A.J. Ayer, et al., *The Revolution in Philosophy*, London, Macmillan 1956.

Lexicon of Latin and Greek Words and Phrases

Latin Words and Phrases

actus: act

actus intelligendi: act of understanding

actus primus: first act

ad infinitum: endlessly

ad libitum: freely

adaequatio intellectus ad rem: the truth-equatability of the intellect with the real

anathema: cast out

ancilla: handmaid

an? whether

an sit? quid sit? is it? what is it?

ascensio mentis per intelligibile et verum ad ens: ascent of the mind through the intelligible and the true to being

assensus intellectus in verum: intellectual assent to truth

beatitudo imperfecta: imperfect beatitude, imperfect happiness

beatitudo naturalis: natural beatitude, natural happiness

beatitudo perfecta: perfect beatitude, perfect happiness

beatitudo perfecta est soli Deo naturalis: perfect beatitude is natural to God alone

beatitudo supernaturalis: supernatural beatitude, supernatural happiness

bonum: good

bonum et malum sunt in rebus: good and bad are in real things

bonum ordinis: the good of order

bonum particulare: particular good

bonum per essentiam: good essentially

causa cognoscendi: the cause of knowing
causa essendi: the cause of being
cogito: I think
constitutio Christi ontologica et psychologica: ontological and psychological constitution of Christ
cor ad cor loquitur: heart speaks to heart

de: regarding
de iure: by right
decem genera entis: ten categories of being
Deus existit: God exists
Deus scientiarum Dominus: God, Lord of Sciences
Deus se habet ad naturalia sicut artifex ad artificiata: God is to nature as the maker is to the artifact
divinarum personarum conceptio analogica: analogical conception of the divine persons
docta ignorantia: learned ignorance

ens: being
ens iuridicum: a being of law
ens per essentiam: being through essence, essential being
entia: beings
eo ipso: for that very reason
ethica more geometrico: ethics in a geometric style
exercite: as exercised, performatively

fides fiducialis: a faith of confidence
finis: end, goal
finis est propter naturam quae habet exigentiam finis: the end is for the sake of the nature that has an exigence for the end [a position that Lonergan calls 'nonsensical']
forma artificialis: an artificial form

homo viator: man the pilgrim
humani generis: of the human race

id quod est: that which is
id quod est ad finem est propter finem: what is headed to the goal is for the sake of the goal
in potentia: in potency
in sensu aienti: in an affirming sense
in sensu exclusivo: in an exclusive sense
in sensu negante: in a negating sense

in tota sua latitudine: in its full breadth
indocta ignorantia: unenlightened ignorance
intellectus: intellect, the habit of understanding, the faculty of understanding [noun], understood [participle]
intellectus agens: the agent intellect
intellectus possibilis: the possible intellect
intelligibile in sensibilibus: the intelligible in the sensibles

lumen naturale: natural light (of intelligence)

medium in quo ens cognoscitur: medium in which being is known

necessitas ex praesuppositione finis: necessity if the end is presupposed
nota theologica: theological note or qualification

odium: hate
odium philosophicum: philosophical opprobrium
odium theologicum: theological opprobrium
omnia: all, everything
omnia Deum appetunt: all things desire God
operabile: a do-able (plan)

pars prima: first part
participatio creata lucis increatae: created participation of uncreated light
patiens divina: suffering divine things
per accidens: accidentally
per contra: by contrast
perspicientia evidentiae sufficientis qua sufficientis: the grasp of sufficiency of evidence as sufficient
primum in aliquo ordine: first in some order
principia mathematica: mathematical principles
principia mathematica philosophiae naturalis: mathematical principles of natural philosophy
priora quoad nos: what are first for us
priora quoad se: what are first in themselves
prius quoad nos: what is first for us
prius quoad se: what is first in itself
propter quid A sit B? for what reason is A B?

quaestiones disputatae: disputed questions
quid: what
quid sit: what is it? what it is
quid sit Deus: what God is

re et ratione: in reality and in reason

reductio ad absurdum: reduction to the absurd

regina scientiarum: queen of the sciences

relatio alicuius ad seipsum est ens rationis: the relation of anything to itself is a being of reason

sapientia: wisdom

sapientis est ordinare: it is proper to the wise person to put (things) in order

secundum quid: 'according to some aspect,' secondarily

sensibile per accidens: accidentally capable of being sensed

si quis dixerit, anathema sit: if anyone speak (thus), let him be an outcast

signate et exercite: thematically and performatively

species intelligibilis: intelligible species

symbolicum technicum: symbolic-technical

tabula rasa: a clean slate

tamquam per medium in quo: as through a medium in which

unum per accidens: an accidental unity

unum per se: an intelligible unity

ut quo: as (that) by which

ut quod: as what

utrum A sit B? whether A is B?

verbum complexum: complete (inner) word

verbum incomplexum: incomplete (inner) word / concept

verbum interius complexum: complete inner word

verbum interius incomplexum: incomplete inner word

veritas formaliter est in solo iudicio: only in the judgment is there formally truth

verum: truth, true

verum est medium in quo cognoscitur ens, id quod est: truth is the medium in which being, that which is, is known

verum et falsum sunt in mente: true and false are in the mind

virtualiter et formaliter: virtually and formally

Greek Words and Phrases

aisthêta: sensibles

aisthêta [and] noêta: sensibles and intelligibles

aition tou einai: the essence, the cause of being

alêtheia: truth

dia: through

dialectikê: dialectic (art)
dia ti ti estin: through what something is
doxa: opinion
eidos: species (singular), form
epistêmê: science
epochê: restraint (Husserl: withdrawal, bracketting)
eristikê: debating (art)
legein: to read, let appear, unveil
mythos: myth
noêton, noêta: known, knowns (in the sense of intelligible)
ontôs on, ontôs onta: what truly is, are
ousia: the essence
phainomena legein: reading the appearances
physis: nature
pistis: belief
proton aition to einai: first cause of being
sophia: wisdom
syllogismos epistêmonikos: scientific syllogism
ti: what

Index

Some oddities and strategies of the index should be noted before its use. A major difficulty was the dense surveying character of both sets of lectures, with some sections over-rich in reference and bibliography. The problem was heightened by the inclusion, in Part Two, of the lecture notes, themselves obviously a form of indexing. A further density of reference occurs in the bibliographies of the appendices.

A first strategy for meeting these difficulties was to exclude index reference to Part Two except in cases where little or no treatment of a topic occurred in Parts One and Three. This is clearly the case regarding the lost lectures of the the second week, identifiable here as the notes running from page 187, section 7, to page 201, section 3. Those notes correspond to a gap between chapter 12 and chapter 13 in Part Three: the missing topics are Heidegger – the fragment in appendix D is a brief bibliographic introduction – and the beginning of the presentation on Horizon. So the notes in this case are fully indexed. This is true also of topics skimped or skipped in the lectures. However, where the notes stand to the lectures as topic mention to topic treatment, the clutter of reference to the lecture notes is avoided. Where there are divergences of order, or finer points in the notes, cross-referencing in the text was judged to be sufficient indexing.

Some further strategies are to be noted. The bibliographies of appendix E are not indexed, nor, in general, are the phrases and words that occur in the final lexicon. I also considered it unhelpful to index separately technical journals cited, or works casually referred to that are of fringe significance. Further, I have kept references to purely logical symbols to a minimum. Entries under Symbolism give sufficient indication of types and uses of such symbols: for instance, a reference there to the Kneale and Kneale classic points the reader towards a sufficient comparison of symbolisms, and noted also is Lonergan's preference for Polish symbolization. The mention of Lonergan brings me to my penultimate remark regarding index strategy. Footnote references to Lonergan are abundant throughout this book. I do not index all such references: under Lonergan, rather, I focus on significant aspects of his lecturing and on meanings within his other written works that enrich his presentation in this volume, referenced under their titles. Finally, I have omitted direct reference to page notes: a page reference, then, can be to either the text or a page note or, more usually, to both.

These simplifications opened up the possibility of a more focused, comprehensive, and integral index. Functional specialist indexing is clearly a thing of the future, but the selectivity of the present index is attributable to a foundational reaching. Foundational indexing is dominated by a normative pointing towards the actual and the potentially progressive structures of the elusive field (see the index under Field).

P.McS.

Abel, H.N., 61

Abschattung, 257–58, 262, 270

Absolute: certitude, 107; God as a., 262; a. ground of philosophy, 314–17; of judgment, 95, 343; in metaphysics, 79; necessity, 107; principles, 76–91; of truth, 337

Abstract: deductivism, 129; a. and existentialism, 227–28; man in a., 227; speculation, 227

Abstraction: of causes, 102–103, 136–37; of form from matter, 26–27, 32, 62, 71, 104, 267, 322–23, 328; from irrelevant, 333–34; of LF, 71; as selecting essential, 333; of universal, 71, 336

Acceleration: as free fall, 29–30; metaphysics of, 13

Accidental: events, 249; unity, 72

Accurate definition, 74

Achsenzeit, 224

Act: first, second, 352, 261; potency and, 111, 160, 334, 344, 352–53; of understanding, *see* Understanding

Action: diagrammed elements of, 323; effective, *see* Effective; possibilities of, 320

Advantage(s): of Christian context, 111, 122–23, 224; of ML, 4–5; of subjective focus, 313–17; of symbols, 4, 13–16, 21, 24–25, 30

Advertising, 189

Aesthetic, 235, 245

Aim: of existentialism, 219–20, 235, 240, 244, 297; of lectures, xv, 3, 193, 329, 353, 359; of ML, 5; of phenomenology, 266–69; of philosophy, 281–84, 289–95, 313–17; of science, 250; of Thomas, 344; at trivial, 337

Alchemy, 135

Alêtheia, 178, 250–51

Algebra(ic): equations, 29, 32–33, 60; formula, 10, 24; geometry, 29–30, 32–33; and isomorphism, 33; as model, 29–30

Ambiguity: of knowing, 137; of 'substance', 132, 137; of symbols, 29, 52; of technique, 96–98

Amsterdam, 51

Analogy: of canonicity, 65; of construction, 66; of 'enumerable,' 65–66; and existence, 31; of implication, 18–19; and knowledge of angels, 326; in ML, 63–66; with mechanics,

De ente supernaturali: on the natural desire to know God, 66

Defective: situation, 303; thinking, 303, 306, 309

Definable, 54

Definition(s): of act and potency, 111, 160, 334, 344, 352–53; of circle, 74; in Euclid, 10–13; implicit, 32, 328; operational, 320; of real numbers, 42; of straight line, 327–28; of truth, 72–75; and understanding, *see* Understanding

de Jongh, J.J., 36

de Lubac, H., 350

de Moivre's theorem, 57–58

Denkformen, 93–94

Deontic logic, 320

Depth psychology, and explanation, 270, 348

Derived: and nonderived, 39; and primitive (data), 266

Descartes, R.: and *Cogito*, 258–59, 316; and field, 316; and geometry, 32–33; and immutability of God, 119; and isomorphism, 31–33; and move to psychology, 330; and philosophy, 119

Description: and explanation, *see* Transition; and substance, 135–36; and technique, 328

Desire: for end, 352; to know God, 348; pure d., 192, 365

Detachment, 236, 242

Determinants of freedom, 292

Deus scientiarum Dominus, 252

Deuteronomy, 228

Development: dialectic d. of intelligence, 95; and distinctions, 119; leaps of, 258–59; linked d. of ML and mathematics, 6, 40–45, 48–49; of mathematics, 61; of ML systems, 40–45, 48–49; of mind, 330; of scientific method, 250; and soul, 258; of

subject, 280–88; of subject as subject, 132, 137, 314–17; systematic d., 122

de Waelhens, A., 248, 279

Diagonal: incommensurate, 59; squared, 26; theorem, 52–56

Diagram(s): Lonergan's d. of levels, xiv, 319–23; as model, 20–29; and real, 132; in text, 7, 8, 11, 12, 27, 28; and understanding, *see* Understanding: and image

Dialectic(s): in Aristotle, 299; and decline, 305; essence of, 308–10; in Hegel, 289; and horizons, 282–84, 300–10; of insufficient concept, 301; and levels, 245; of man, 301–302; in Middle Ages, 289; objective function of, 302–305; in Plato, 299–300; and policy, 302–304

Dialectica, 47–48

Dieudonné, J., 8, 48

Difference, fundamental axiom of, 62

Different: lines of logical development, 40–49; d. meanings, 64; d. philosophies, 88, 109–11, 345, 362–63; points as merely d., 62; d. schools of mathematics, 80

Differential equation, 347

Dilemma of subject, 195–98

Dilthey, W., 192, 249

Ding-an-sich, 226

Dionysius the pseudo-Areopagite, 118

Discovery: of fire, 224; and transfiguration, 244

Discrediting false position, 313

Discriminants, in algebra, 29

Discursive process, 364

Disputed questions, 245–46, 358

Distinction(s): of *an*, *quid*, 46; of content, 18; and deductive expansions, 17–18, 22, 121, 127–28; and development, 119; of inquiries, 123–33; and intuitionism, 45–46; and ML,